Springer Series on
Atoms+Plasmas 19

Editor: I. I. Sobel'man

Springer-Verlag Berlin Heidelberg GmbH

Springer Series on

Atoms+Plasmas

Editors: G. Ecker P. Lambropoulos I. I. Sobel'man H. Walther

Managing Editor: H. K. V. Lotsch

1 **Polarized Electrons** 2nd Edition
By J. Kessler

2 **Multiphoton Processes**
Editors: P. Lambropoulos and S. J. Smith

3 **Atomic Many-Body Theory**
2nd Edition
By I. Lindgren and J. Morrison

4 **Elementary Processes in Hydrogen-Helium Plasmas**
Cross Sections and Reaction Rate Coefficients
By R. K. Janev, W. D. Langer, K. Evans, Jr., and D. E. Post, Jr.

5 **Pulsed Electrical Discharge in Vacuum**
By G. A. Mesyats and D. I. Proskurovsky

6 **Atomic and Molecular Spectroscopy**
2nd Edition
Basic Aspects and Practical Applications
By S. Svanberg

7 **Interference of Atomic States**
By E. B. Alexandrov, M. P. Chaika and G. I. Khvostenko

8 **Plasma Physics** 2nd Edition
Basic Theory with Fusion Applications
By K. Nishikawa and M. Wakatani

9 **Plasma Spectroscopy**
The Influence of Microwave and Laser Fields
By E. Oks

10 **Film Deposition by Plasma Techniques**
By M. Konuma

11 **Resonance Phenomena in Electron-Atom Collisions**
By V. I. Lengyel, V. T. Navrotsky and E. P. Sabad

12 **Atomic Spectra and Radiative Transitions** 2nd Edition
By I. I. Sobel'man

13 **Multiphoton Processes in Atoms**
2nd Edition
By N. B. Delone and V. P. Krainov

14 **Atoms in Plasmas**
By V. S. Lisitsa

15 **Excitation of Atoms and Broadening of Spectral Lines**
2nd Edition
By I. I. Sobel'man, L. Vainshtein, and E. Yukov

16 **Reference Data on Multicharged Ions**
By V. G. Pal'chikov and V. P. Shevelko

17 **Lectures on Non-linear Plasma Kinetics**
By V. N. Tsytovich

18 **Atoms and Their Spectroscopic Properties**
By V. P. Shevelko

19 **X-Ray Radiation of Highly Charged Ions**
By H. F. Beyer, H.-J. Kluge, and V. P. Shevelko

20 **Electron Emission in Heavy-Ion-Atom Collision**
By N. Stolterfoht, R. D. DuBois, and R. D. Rivarola

21 **Molecules and Their Spectroscopic Properties**
By S. V. Kristenko, A. I. Maslov, and V. P. Shevelko

H. F. Beyer H.-J. Kluge V. P. Shevelko

X-Ray Radiation of Highly Charged Ions

With 79 Figures and 52 Tables

Springer

Dr. Heinrich F. Beyer
Professor H.-Jürgen Kluge
Gesellschaft für Schwerionenforschung,
Planckstrasse 1,
D-64291 Darmstadt, Germany

Dr. Viatcheslav P. Shevelko
Lebedev Physics Institute,
Russian Academy of Sciences,
Leninsky Prospekt 53,
117924 Moscow, Russia

ISSN 0177-6495

DOI 10.1007/978-3-662-03495-8

Library of Congress Cataloging-in-Publication Data.

Beyer, H. F. (Heinrich F.), 1950– . X-ray radiation of highly charged ions / H. F. Beyer, H.-J. Kluge, V. P. Shevelko p. cm. – (Springer series on atoms + plasmas; 19) Includes bibliographical references and index.
1. Ion sources. 2. X-ray spectroscopy. 3. Collisions (Nuclear physics)
I. Kluge, H.-Jürgen. II. Shevel'ko, V. P. (Vi͡acheslav Petrovich) III. Series. QC702.3.B49 1997 539.7'222–dc21
97-25461

Originally published by Springer-Verlag Berlin Heidelberg New York in 1997.

MyCopy version of the original edition 1997

Typesetting: Camera ready copy from the authors
SPIN 10519386 54/3144 - 5 4 3 2 1 0 - Printed on acid-free paper
www.springer.com/mycopy

Preface

The physics of highly charged ions continues to be one of the most active and interesting fields of atomic physics. A large fraction of the characteristic radiation of such ions lies in the x-ray region and its spectroscopy represents an important experimental tool. The field of x-ray spectroscopy grew directly from the discovery of x radiation by Wilhelm Conrad Röntgen in 1895. The early contributions to atomic physics that arose out of x-ray spectroscopy are well documented and are the subject of many centennial events. In the past, the gross features of most x-ray spectra in the hard x-ray region have been accounted for on a hydrogenic model. In many instances the gross spectral features recorded in the early days of x-ray physics match those observed with state-of-the-art techniques today and many of the early qualitative interpretations have remained unchanged.

It is in the details of the spectra that today's results are superior to those obtained many years ago, and it is in the quantitative and accurate descriptions that today's predictions are better. A rejuvenation of the field has occurred after the great achievements in the development of new ion sources for production of heavy ions with only one or few electrons. The new tools available to the experimenter allow the exploration of new states of matter and allow us to challenge new frontiers in our theoretical understanding of atoms and their interactions with other particles. The new fundamental atomic physics is related to atomic structure, interaction between particles in heavy-atom collisions, problems in quantum electrodynamics and cross-disciplinary interplay between atomic and nuclear physics. One of the most attractive aspects may be envisioned in the combination of the mature field of x-ray physics with the ability to employ the most advanced technology for the production of ions with the highest charge available in nature, the 92-fold ionized uranium. The power of the new achievements is beyond doubt. No atomic physicist can afford to ignore them.

This book has been written to support this new development. An up-to-date description is presented that should enable the reader to learn what is already known and to discover where many interesting problems still wait to be solved. For the first comprehensive monograph on the subject, the aim has been to produce a book that would take the reader up to the research frontiers without making severe demands on the reader's erudition. It should be equally useful to the advanced student as well as to the research scientist already specialized in one of the subfields of x-ray, atomic or plasma physics. At the same time, the book may serve as a nearly complete guide to the relevant research literature. We intend to give profound information on atomic techniques and ion sources used for investigation of electron-ion-atom collision processes as well as a broad overview of atomic structure and atomic characteristics required for many physics applications involving energy levels, Lamb shift, oscillator strengths and transition probabilities, photoionization and electron-ion recombination cross sections.

During the course of writing this book, we benefitted from a fruitful exchange of ideas with our co-workers from the Atomic-Physics Department of the GSI in Darmstadt and from the Optical Division of the P.N. Lebedev Physics Institute in Moscow. It is our pleasure to thank T. Kühl, L.N. Labzowsky, A. Müller, V.G. Pal'chikov, V.M. Shabaev, Th. Stöhlker, A.M. Urnov and W. Quint for valuable remarks. We are particularly grateful to I.I. Sobelman for useful comments and permanent interest in our work. Finally, we wish to record our thanks to H. Lotsch of Springer-Verlag for his patient cooperation.

Darmstadt - Moscow
May 1997

H.F. Beyer
H.-J. Kluge
V.P. Shevelko

Contents

List of Symbols

Fundamental Constants

$c_o = 299\,792\,458$ m s^{-1}	Velocity of light in vacuum
$\hbar = 6.582\,122\,0(20) \times 10^{-16}$ eV s	Planck constant divided by 2π
$hc_o = 1.239\,842\,44(37) \times 10^{-6}$ eV m	Conversion constant
$e = 1.602\,177\,33(49) \times 10^{-19}$ C	Elementary charge magnitude
$m_e = 0.510\,999\,06(15)$ MeV$/c_o^2$	Electron mass
$m_p = 938.272\,31(28)$ MeV$/c_o^2$	Proton mass
$u = 931.494\,32(28)$ MeV$/c_o^2$	Unified atomic mass unit (mass of ^{12}C atom)/12
$\alpha = e^2/\hbar c_o$	
$= 1/137.035\,989\,5(61)$	Fine-structure constant
$r_e = e^2/m_e c_o^2$	
$= 2.817\,940\,92(38) \times 10^{-15}$ m	Classical electron radius
$a_o = \hbar^2/m_e e^2$	
$= 0.529\,177\,249(24) \times 10^{-10}$ m	Bohr radius
$Ry = m_e e^4/2\hbar^2$	
$= 13.605\,698\,1(40)$ eV	Rydberg energy
$k = 8.617\,385(73) \times 10^{-5}$ eV K^{-1}	Boltzmann constant

The values given above are extracted from the set of constants recommended for international use by the Committee on Data for Science and Technology (CODATA) based on the "1986 adjustment of the Fundamental Physical Constants" by E.R. Cohen and B.N. Taylor, Rev. Mod. Phys. **59,** 1121 (1987). See also E.R. Cohen and B.N. Taylor, "The Fundamental Physical Constants", Phys. Today **48,** Pt. 2 (August 1995).

Basic Notation

α_r	Rate coefficient for recombination
$\boldsymbol{\alpha}$	Dirac matrix
β	$= v/c_o$
$\beta(s)$	Beta function

γ_{t}	Transition gamma
Γ	Level width
Γ_{a}	Auger width
Γ_{r}	Radiative width
ϵ	Electron energy, rest mass subtracted
η	Relative cooler length
λ_{D}	Debye length
$\boldsymbol{\mu}$	Magnetic moment
ξ	Chromaticity
$\rho_{\mathrm{min,max}}$	Minimum and maximum impact parameter
σ	Cross section
σ_{ν}	Photoionization cross section
σ_{r}	Recombination cross section
τ_{c}	Confinement time
$\Psi(s)$	Betatron phase
ω	Angular frequency
ω_{c}	Cyclotron angular frequency
A	Atomic number
A_{a}	Autoionization probability
A_{r}	Radiative decay probability
A_{ik}	Transition probability for transition $i \to k$
$\boldsymbol{B}$	Magnetic inductance
d	Characteristic trap dimension
E_{e}	Electron energy
$E_{i,k}$	Electronic binding energy of levels i and k
E_{rel}	Relativistic energy of bound electron
ℓ	Single-electron orbital angular-momentum quantum number
f_{ik}	Oscillator strength for transition $i \to k$
g	Statistical weight
	Lande factor
	Gaunt factor
$g(\eta_0, \eta_1)$	Gaunt factor for Bremsstrahlung
I	Nuclear-spin quantum number
$I_{n\ell}$	Absolute value of binding energy
I_q	Ionization potential
j	Single-electron total angular-momentum quantum number
J	Total angular-momentum quantum number
$\boldsymbol{k}$	Photon momentum
$K(s)$	Focusing strength
$\boldsymbol{\ell}_{\mathrm{e}}$	Electron angular momentum
L	Total orbital angular-momentum quantum number
L_{c}	Coulomb logarithm
m_{i}	Ion mass

M	Nuclear mass
n	Principal quantum number
n_e	Electron density
n_i	Ion density
n_q	Density of q-times ionized ions
N	Number of atomic electrons
$\boldsymbol{p}$	Electron momentum
P	Parity
P_I	Perveance
$P_{n\ell}(r)$	Radial wavefunction
P_s	Schottky power
$P(\theta)$	Degree of polarization
q	Ionic charge state
Q	Ionic charge qe
	Quadrupole moment
$Q_{h,v}$	Horizontal and vertical tune
R	Nuclear radius
s	Longitudinal coordinate
S	Total spin angular-momentum quantum number
S_d	Resonance strength for doubly excited state d
T	Thermal energy
T_r	Equivalent relative kinetic energy
U	Voltage
v	Velocity
V_a	Dielectronic capture probability
X^{q+}	q-times ionized atom
Z	Atomic number

1. Introduction

View. We start this book by developing a general view of the subject to be treated. The present book may retain its significance from the synthesis of x-ray science on one hand, and the physics and technology necessary for the production of highly charged ions on the other hand. More generally, *spectroscopy* refers to a process in which measurements of spectral lines and continua are used for mapping the energy-level structure of physical systems. As sketched in Fig. 1.1 the process of gaining a deeper understanding requires several ingredients (see, also, the introduction given by *Svanberg* [1.1]). We start from a suitable *ion* and *light source* which must provide the ionic species in their relevant excitation states. Control over the light source also relating to the thermodynamic state of the ions is essential for the adaption of an appropriate *analyzer*. The latter often is a spectrometer with a photon detector coupled to it. The reduction of the raw measurement data to interpretable results requires a *theory* or *model* plus, usually, control over the measurement process in such a way that the measured quantities can be related to previously obtained experimental and theoretical data. For the latter aspect the

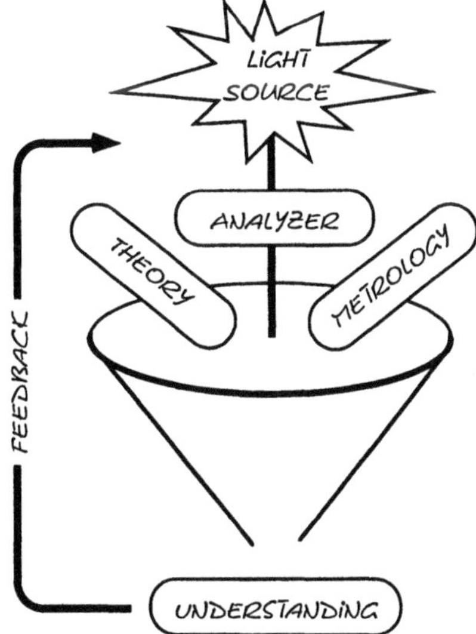

Fig. 1.1. Spectroscopy as an example for a feedback process in natural sciences

terms *metrology* or *calibration* are commonly used. From the *knowledge* of the interpretable and relatable results a new *understanding* of the physical system can result. As a consequence, a new measurement cycle may be enforced through a more or less tight *feedback* loop. The feedback loops are instrumental in achieving progress or in simply avoiding regression. Ideally, they lead to improved models and new technologies. The changes may proceed in evolutionary small or, occasionally, in revolutionary big steps. The process sketched for the example of spectroscopy may be visioned as embedded in still bigger feedback loops characteristic of developments in science. Next, we will return to the individual stations of the spectroscopy process.

Ion and Light Sources. The photon intensity per spectral bandwidth has often been emphasized as the most important parameter of a light source. High photon intensities allow the use of photon-detection devices of high resolving power which usually is inversely related to the detection efficiency. The most widely used light source for the x-ray region is the x-ray tube which since the discovery of x rays has been employed in many applications and still is in use today. Much effort has been spent to increase its light output. But new methods of generating electromagnetic radiation of both high spectral density and high photon frequency have been developed as manifested in the construction of several generations of synchrotron light sources world wide [1.2]. Using hard-photon or electron bombardment for excitation or ionization of matter, a wide field of x-ray research has been established including atomic spectroscopy, solid-state physics and material science as well as chemical analysis and radiation biology.

However, only single-hole states of atoms were accessible. With increasing photon intensity the experimental accuracy was increasing accordingly. The fine details like shake-up satellites or many-electron correlations could be experimentally investigated. For theory, however, it was very difficult to keep up in accuracy despite of the complexity of the problems involved. The situation changed considerably when multiply ionized atoms came into play by the development of powerful ion sources and accelerators. Initially, only light ions could be stripped to one and few-electron states leaving heavy atoms in intermediate charge states which again where difficult to be treated theoretically. With the recent development of novel powerful techniques involving advanced ion-source and accelerator technology it has become possible to experimentally investigate virtually any atom of the periodic system in any desired charge state up to completely stripped uranium. Among the successful devices are electron-cyclotron-resonance ion sources and electron-beam ion traps in which multiple electron collisions lead to a successive ionization. Successive ionization can also be achieved by using fast ion beams which get stripped in solid or gaseous targets.

There is a class of ion environments in which highly stripped ions may prevail, these are astrophysical and laboratory plasmas. The latter have received considerable attention in relation to thermonuclear fusion research.

Because of *a-priori* unknown distributions of temperature and density the process of investigation is reversed. For diagnostics of these plasmas a large demand for atomic data is created [1.3,4] which might be gained from studies of ions in more defined environments.

Analyzers. In the sense discussed above, analyzers can be regarded as filters which may be taken literally as frequency filters or more generally as photon spectrometers. In knowing the location in frequency of a spectral line, the difference between the binding energies of the electronic states involved may be deduced. In most practical cases this requires the measurement of further parameters in addition to the metrology to be discussed below. In the case of the presently exemplified photon spectroscopy, the additional parameters could involve the state of motion of the photon emitters. The additionally required quantities would then be the velocity vectors of the emitting ions or some measure of their distribution. In a hot plasma, for instance, the ion temperature and the respective ion velocities cause Doppler shifts which lead to a broadening and to a small frequency shift of the spectral line. Much more severe are the systematic shifts observed in the spectroscopy of fast ion beams. Only with the successful phase-space compression known as *cooling* of the ions in combination with a long accumulation in storage rings, has it become possible to control the ions state of motion sufficiently well for accurate measurements.

Metrology. For a very long time in the history of x-ray science, wavelength measurements were performed on their own wavelength scales which were difficult to compare to the length scale used in other regions of the electromagnetic spectrum. Hence, even measurements of high precision could not be directly related to theoretically calculated wavelengths based on the Rydberg energy, thus they were severely lacking accuracy. Only with the achievements of the combined x-ray-optical interferometry [1.5] has it become possible to record even hard x rays on a scale which is consistent with the adopted definition of the meter and of the other fundamental constants including the Rydberg energy. For wavelength calibration, a number of gamma-ray lines are now available that are linked to the *de-facto* secondary standard the 412-keV ^{198}Au line which in turn has been linked to optical wavelengths.

These achievements are a prerequisite for the accurate assessment of the level structure of highly charged one- and few-electron systems, where small energy shifts caused by quantum-electrodynamic effects are of interest. In this context, it is also very fortunate to be able to put different energy intervals of the same ion, like x-ray, fine, hyperfine and Lamb-shift intervals, on a common scale.

Theory. A theoretical model has to describe, in a general way, the types of states which have been shown to participate in transitions giving rise to x-ray spectra, and the manner in which the transitions take place. An understanding of the radiative processes requires knowledge of all types of states that can occur. Atomic-structure theories can be tested most stringently by studying simple atomic systems, i.e., ions and atoms with few electrons. Among the fundamental transitions are those in atomic hydrogen as for instance the $2p_{3/2} \rightarrow 1s$ Lyman-α transition. For atomic hydrogen, the transition energy of this line is, by the simple Bohr model, given as 3/4 times the Rydberg energy of 13.6 eV, i.e., approximately 10 eV. The same transition will be boosted to about 102 keV in a one-electron ion with a nucleus bearing Z=92 positive elementary charges. For an accurate description of these systems quantum-electrodynamic and nuclear-size effects have to be added to the relativistically correct solution of the Dirac equation. The strong Coulomb field of the nucleus causes these effects to be strongly enhanced in heavy atoms and ions. Normal x-ray lines like the $K\alpha_1$ line in heavy atoms with a large number of electrons are strongly influenced by the strong Coulomb field of the highly charged nucleus. Simultaneously many-electron correlations become very important. The magnitude of the charge-screening and correlation in uranium can be imagined when comparing the unscreened Lyman-α_1 transition in U^{91+} to the screened $K\alpha_1$ transition in U^{1+}. The corresponding x-ray energy is shifted from 102 keV to 98 keV. In the last decade, theory has made enormous progress in describing even the many-electron system accurately. This development has been triggered by repeated recourse to one and few-electron systems.

Understanding. It is too early to evaluate and to judge on the progress made in the field discussed above and it is even more inappropriate to do so in the introduction of a book, which can only be excused by the omission of a corresponding concluding section at the end. However, it is felt that the various activities devoted by a large number of scientists to the physics of highly charged ions will result in a more consistent description of the behavior of atomic systems in the domain of strong fields. In this book, many examples will be presented on current research frontiers where new and exciting, sometimes unexpected results are produced which forces one to thorough thinking and reformulation of questions to ask.

This introductory discussion, merely based on a spectroscopist's point of view, has been chosen to make matters simple. The scope given, however, will be widened to the methods adjacent to pure photon spectroscopy. This will include interactions between atomic particles and radiative and collisional properties of individual atoms and ions as well as useful properties for coming applications. Thereby we will cross borders and spread out to neighboring disciplines as necessary. In general, short-wavelength radiation of highly charged ions is successfully used in x-ray diagnostics, thermonuclear-fusion research and in astrophysics, x-ray spectroscopy, development of soft x-ray lasers plus in various fundamental areas of modern physics such as atomic structure and quantum electrodynamics, interrelation of atomic, nuclear and laser physics, as well as in different applications in natural sciences and medicine, x-ray microscopy, lithography and studies of biological objects.

What will Follow. The organization of the coming chapters is the following. In Chap. 2 the physical basis of ion sources is laid and technical solutions are introduced summarizing their performances. Furthermore, an introduction to accelerators and storage rings for heavy ions is included plus a short account on ion-trap projects. Chapter 3 is devoted to the physics and classification of the level-structure and spectra of highly ionized ions. Radiative and radiationless decay of the respective excited states is discussed in Chap. 4. Radiative processes are noted in Chap. 5 starting from the fundamental photoionization and Bremsstrahlung processes and concluding with polarization and x-ray lasers. Chapter 6 will look beyond collisional processes including dielectronic recombination and radiative electron capture. In order not to break up chapters, numerical data, helpful for many applications and quantitative estimates, have been shifted to the appendices.

2. Techniques

This chapter starts with a review of some of the techniques that are required in order to generate highly charged ions in the laboratory. It includes a basic introduction into the physical phenomena governing ionization processes in ion sources as well as a summary of the most important high-performance sources for production of ions with high charge states. The Electron-Beam Ion Source (EBIS) and the Electron-Beam Ion Trap (EBIT) are prominent examples of this category. Stripping of accelerated heavy ions to high charge states and their accumulation in storage rings have recently become available to the experimenter investigating the structure and interactions of highly charged ions with photons and atomic particles. Confining the ions in ion traps at very low laboratory velocities for an extended amount of time is a very challenging task. The present status of this technique is briefly summarized and possible future directions are indicated. X rays play a key role in the experimental investigation of the structure and radiative processes of the highly charged ions. The decay of excited ionic states or radiative recombination processes involve the emission of photons ranging from the ultra-soft to the hard x-ray region. The spectroscopic techniques for their measurement can be similar to those used with more conventional x-ray sources making use of only singly or few times ionized atoms. There are excellent monographs [2.1–5] covering x-ray techniques to which the interested reader is referred.

2.1 Ion Sources

The use of ion beams has much increased over the last several decades and is common to a large number of research fields and applications including material research, ion implantation, mass separation, thermo-nuclear fusion research and the wide field of atomic, nuclear and particle physics carried out with accelerators. Correspondingly, the development of ion sources has dramatically advanced. The wish to understand the underlying processes in ion sources has evolved into an important field of research spanning a wide range of different techniques. This is also manifested by a series of conferences, e.g., [2.6] where the impressive progress of the field can be examined. The term *ion source* refers to ion-beam formation devices but it is also used for plasma sources that do not have any provision for ion extraction. Quite exhaustive reviews [2.7–9] have been given on existing ion sources mainly used for ion beam injection into accelerators.

2.1.1 Elementary Processes in Plasmas

The constituents of a plasma are electrons, ions and neutral atoms or molecules. In order to create (ignite) and eventually sustain the plasma state for a certain amount of time it is necessary to provide some external energy as for instance through the electrical power in an arc discharge. In this section, elementary processes important for the creation of ions from the neutral state and their maintenance are listed and some of the quantities describing the plasma state are introduced at a basic level. For a more detailed introduction into plasma physics see [2.10–15].

Direct and Inverse Processes. Usually electrons play the most important role in excitation and ionization of heavy particles. The processes of interest are the following:
Ionization and three-body recombination:

$$X^{q+} + e \leftrightarrow X^{(q+1)+} + e + e \,, \tag{2.1}$$

where X^{q+} refers to a q-times ionized atom and the double arrow indicates that the process may proceed either from left to right or from right to left.
Collisional excitation and de-excitation:

$$X^{q+} + e \leftrightarrow X^{q+^*} + e \,, \tag{2.2}$$

where the asterisk denotes the excited state.
Radiative ionization and recombination:

$$X^{q+} + \hbar\omega \leftrightarrow X^{(q+1)+} + e \,, \tag{2.3}$$

where $\hbar\omega$ is the photon energy.
Dielectronic recombination and autoionization:

$$X^{q+} + e \leftrightarrow X^{(q-1)+^{**}} \rightarrow X^{(q-1)+^*} + \hbar\omega \,, \tag{2.4}$$

where the double asterisk denotes a doubly excited state.
Emission and absorption:

$$X^{q+^*} \leftrightarrow X^{q+} + \hbar\omega \,. \tag{2.5}$$

Each of the reactions defined in (2.1–5) is a pair of direct and inverse processes. The probability as expressed by the respective cross section for the individual process can be linked to its reverse by a simple formula. In the case of the dielectronic recombination, one electron is resonantly transferred to an excited ionic state in a first step by a simultaneous excitation of a bound electron. In a second step the doubly excited state stabilizes by a radiative decay below the autoionization limit. These processes are treated in more detail in Chap. 6.

Plasma Parameters. The description of a plasma with its large number of particles requires the concepts of statistical physics. Sufficient knowledge of the complete system would be achieved by knowing the distribution functions [2.10,16,17] of velocity and space for each plasma constituent (electrons, ions and neutrals). In practice, only a few moments of these distribution functions may be enough like the densities (zero order), the particle currents or velocities (first-order moments), the pressure or temperature (second order) and may be also the heat flux (third order). Of course, the moments are only defined on the basis of a known or approximated distribution function. If thermodynamic equilibrium is achieved in some volume of the plasma the following distributions are valid:
Maxwell distribution of electron or ion velocities:

$$\begin{aligned} \mathrm{d}n &= f(\boldsymbol{v})\mathrm{d}\boldsymbol{v} \\ &= f(v_\mathrm{x}, v_\mathrm{y}, v_\mathrm{z})\, \mathrm{d}v_\mathrm{x}\mathrm{d}v_\mathrm{y}\mathrm{d}v_\mathrm{z} \\ f(v_\mathrm{x}, v_\mathrm{y}, v_\mathrm{z}) &= m^{3/2}\,(2\pi T)^{-3/2} \exp\left[-\frac{m}{2T}\left(v_\mathrm{x}^2 + v_\mathrm{y}^2 + v_\mathrm{z}^2\right)\right], \end{aligned} \tag{2.6}$$

where m denotes either the electron or ion mass and $\boldsymbol{v}$ the electron or ion velocity. As is common practice in plasma physics, T denotes the thermal energy loosely speaking the temperature measured in electron Volt rather than the temperature measured in Kelvin. An energy of 1 eV corresponds to a temperature of 11 604 K. The velocity distribution is not necessarily isotropic. Especially if an external magnetic field is applied there will be a longitudinal and a transverse temperature $T_\parallel$ and $T_\perp$, respectively, with a two-temperature Maxwell distribution function.
Boltzmann distribution of excited atoms over energy levels:

$$\frac{n_j}{n_k} = \frac{\mathrm{g}_j}{\mathrm{g}_k} \exp\left(-\frac{E_j - E_k}{T}\right), \tag{2.7}$$

where E_j and E_k are electron binding energies in discrete levels j and k and g_j and g_k denote the statistical weights of those levels.
Saha distribution of atoms over degrees of ionization:

$$\frac{n_{q+1}}{n_q} = 2\left(\frac{mT}{2\pi\hbar^2}\right)^{3/2} n_\mathrm{e}^{-1}\, \frac{\mathrm{g}_{q+1}}{\mathrm{g}_q} \exp\left(-\frac{I_q}{T}\right), \tag{2.8}$$

where I_q denotes the ionization energy of the q-times ionized atom and n_e the electron density; g_q is the partition function for charge state q and may be obtained by summing over the statistical weights of individual states in this charge state. It is important not to confuse the degree of ionization of the individual ion in the plasma with the relative degree of ionization of a certain plasma volume. There may be individual ions carrying a high charge or even being completely stripped. However, the plasma as a whole is rarely ionized to a high degree (more than about 5–10%) because usually the neutral component is still the largest fraction and highly charged ions make up only a very small fraction.

Plasma Frequency. Despite local deviations the plasma volume as a whole is not electrically charged. The charge neutrality may be expressed as

$$\sum_q q n_q = n_\mathrm{e} \,, \tag{2.9}$$

where the sum runs over all charge states q present in the plasma. Small deviations from charge neutrality give rise to restoring forces which lead to oscillations in modes that are fundamental to the plasma. Most important are the electron plasma oscillations with the characteristic electron plasma frequency corresponding to

$$\omega_\mathrm{pe}^2 = \frac{e^2 n_\mathrm{e}}{m_\mathrm{e}} \,, \tag{2.10}$$

where m_e denotes the electron mass. The ions may oscillate too. Their ion plasma frequency is given by

$$\omega_\mathrm{pi}^2 = \frac{q^2 e^2 n_\mathrm{q}}{m_\mathrm{i}} \,, \tag{2.11}$$

where now the electron mass and density have been replaced by the ionic mass m_i and ionic density n_q for charge state q.

Debye Screening. The charged particles of the plasma interact via their Coulomb forces. However, the corresponding electric fields only extend a certain distance through the plasma. This shielding distance or screening distance is called *Debye length* [2.18] and is given by

$$\lambda_\mathrm{D}^2 = \frac{T_\mathrm{e}}{e^2 n_\mathrm{e}} \,. \tag{2.12}$$

For distances larger than the Debye length the electric field of a charged particle is shielded by surrounding particles of opposite charge. An analogue behavior is observed with external fields being shielded from the interior of the plasma volume. The phenomena in the boundary layer also known as plasma sheath are of the same nature. Inside the plasma sheath the charge neutrality is violated. The charged particles, in essence the electrons because of their higher mobility, assume a distribution that cancels the external field.

Magnetic Fields. In ion sources magnetic fields are frequently used in order to make efficient use of the ionizing particles, notably of the electrons, by confining them to the plasma volume. A particle of electric charge Q experiences the Lorentz force $\boldsymbol{F} = Q\boldsymbol{v} \times \boldsymbol{B}$ where $\boldsymbol{B}$ denotes the magnetic flux density. This leads to a circular motion of the particle around the magnetic field lines. Equating the Lorentz force to the centripetal force yields the radius of the gyromotion

$$\rho = \frac{m v_\perp}{QB} \,, \tag{2.13}$$

where $v_\perp$ denotes the velocity component perpendicular to the direction of the magnetic field. Equation (2.13) can be expressed in terms of the ion or electron temperature through the relation $T = 1/2\, m v_\perp^2$. Because of the velocity distribution (2.6) the gyration radii also have some distribution. Their mean value can be obtained by averaging over the Maxwell distribution:

$$\rho_\mathrm{e} = 14\,\mu\mathrm{m}\, T_{\mathrm{e}\perp}[\mathrm{eV}]^{1/2} B[\mathrm{T}]^{-1}\,, \tag{2.14}$$

$$\rho_\mathrm{i} = 3.3\,\mu\mathrm{m}\, (\mathrm{A} T_{\mathrm{e}\perp}[\mathrm{eV}])^{1/2} (qB[\mathrm{T}])^{-1}\,, \tag{2.15}$$

where A is the atomic mass number of the ion and the temperatures are expressed in eV and the magnetic flux density in Tesla. The corresponding gyration or cyclotron frequencies are obtained by equating the Lorentz force to the centripetal force in the form $Q\rho\omega B = mr\omega^2$ which results in

$$\omega_\mathrm{c} = \frac{QB}{m}\,. \tag{2.16}$$

Here again the charge for electrons is $Q = e$ and for ions $Q = qe$. For practical purposes (2.16) may be expressed as

$$f_\mathrm{ce} = 28\,\mathrm{GHz}\, B[\mathrm{T}]\,, \tag{2.17}$$

$$f_\mathrm{ci} = 15\,\mathrm{MHz}\, qB[\mathrm{T}]/\mathrm{A}\,. \tag{2.18}$$

The electrons of a plasma can efficiently be heated by coupling microwave power into the plasma just at the electron cyclotron frequency. Plasmas heated resonantly this way are called ECR *plasmas* and ion sources making use of this principle are called ECR *ion sources.* The electrons may reach energies exceeding 10 keV. For the efficient heating to high temperatures, however, one has to make sure that the collision frequency, i.e., the number of electron-atom collisions per time unit, is small compared to the electron cyclotron frequency. Therefore a low gas pressure is necessary. In an ECR source highly charged ions are produced this way. In contrast, low charge states and low electron temperatures prevail in high-pressure microwave discharges operated with or without magnetic fields.

2.1.2 Classification of Ion Sources

The characteristics of existing ion sources vary over a very large range (cf. [2.17]) as shown in the map of electron density and temperature displayed in Fig. 2.1. The temperature and density range is over many orders of magnitude. Within this map the sources for the production of highly charged ions are at high electron temperature and at moderate density. Ion sources for high current and low charge states use a high density but low temperatures.

Collisions. The number of collisional events per unit of time for the processes discussed in Sect. 2.1.1 may be expressed as

$$W = n_\mathrm{e}\, \langle v\sigma \rangle$$

with

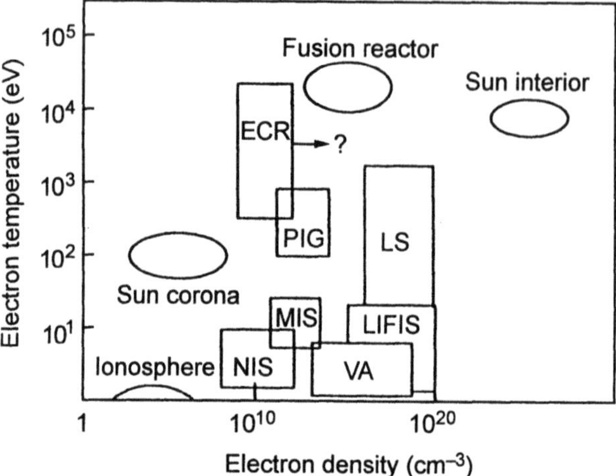

Fig. 2.1. Temperature-density map locating existing ion sources. ECR: electron-cyclotron resonance ion source. PIG: Penning ion gage. MIS: metal-ion source. NIS: Negative-ion source. LS Laser-ion source. LIFIS: Light-ion fusion ion source. VA: vacuum arcs. [2.17] © 1992 AIP

$$\langle v\sigma \rangle = \int_{E_{\min}}^{\infty} v\sigma(E)f(E)\mathrm{d}E\ , \tag{2.19}$$

with σ denoting the cross section for the respective process and $E = mv^2/2$ being the collision energy. For the production of highly charged ions successive ionization by electron impact very often plays a dominant role. There is a threshold energy $E_{\min}$ in (2.19) which is equal to the absolute value of the ionization potential I_q for a certain charge state q. A complete table of ionization potentials of the elements in all their ionization stages has been compiled by *Carlson* et al. [2.19]. Examples for the ionization potential as a function of charge state are shown in Fig. 2.2 for some selected ions. Whereas the first ionization potential amounts to a few eV only a minimum energy of 132 keV is necessary in order to remove the last electron of uranium.

The cross section for electron-impact ionization is zero at threshold and increases with increasing collision energy reaching its maximum at about 2–3 times the binding energy of the electron to be ionized. Figure 2.3 shows examples of the cross section for successive electron removal. A survey of available cross-section data may be found in [2.20,21]. Over a large range, estimates within a factor of two are possible with the use of the semi-empirical *Lotz* [2.22] formula. Simple semi-empirical cross-section formulae for multiple ionization of atoms and ions by electron impact have been reported [2.23,24]. According to (2.19) the average time necessary for building up the charge state q from the neutral state, by successive stripping, can be expressed as

$$\tau_{\mathrm{i}}(q) = \sum_{q'=0}^{q-1} \langle n_{\mathrm{e}} v_{\mathrm{e}}\, \sigma^{\mathrm{I}}_{q',q'+1} \rangle^{-1}\ , \tag{2.20}$$

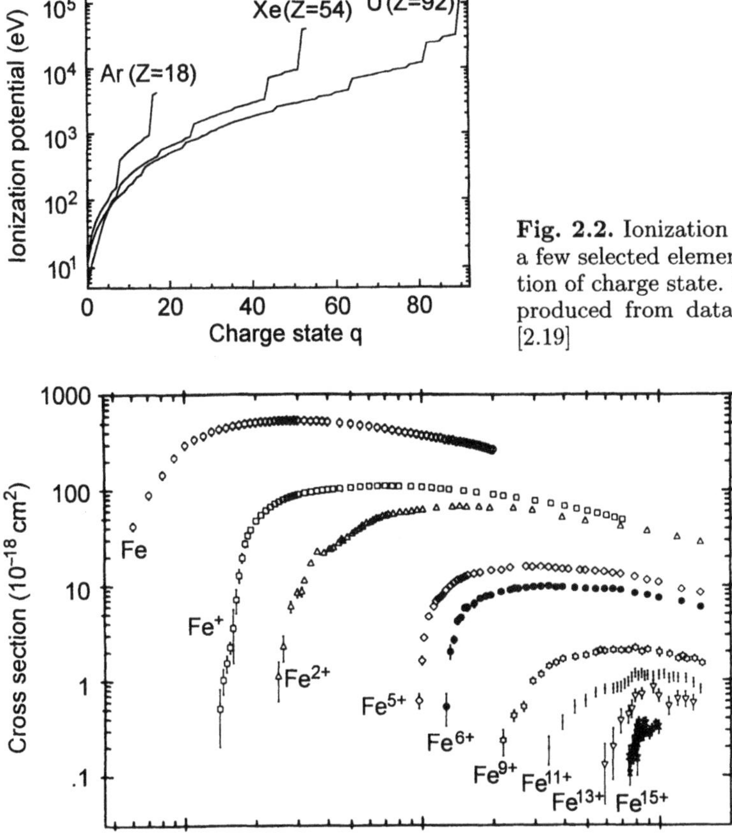

Fig. 2.2. Ionization potentials of a few selected elements as a function of charge state. The plot was produced from data reported in [2.19]

Fig. 2.3. Cross section for ionization of iron Fe^{q+} ($q = 0, \ldots, 15$) by electron impact [2.20] © 1991 Springer

where $\sigma^{\mathrm{I}}_{q',q'+1}$ is the cross section for single ionization. The averaging is over the random velocity distribution. For an electron beam the product $n_{\mathrm{e}}v_{\mathrm{e}}$ equals the current density divided by the elementary charge $j_{\mathrm{e}}/\mathrm{e}$. The build-up time $\tau_{\mathrm{i}}(q)$ is only significant as long as the ion confinement time τ_{c} for the particular ion source is large enough. A balance equation, for a single atomic species with atomic number Z, takes the general form

$$\begin{aligned} \frac{\mathrm{d}n_q}{\mathrm{d}t} &= \sum_{q'=0}^{q-1} n_{q'} \langle n_{\mathrm{e}} v_{\mathrm{e}}\, \sigma^{\mathrm{I}}_{q',q} \rangle - \sum_{q'=q+1}^{Z} n_q \langle n_{\mathrm{e}} v_{\mathrm{e}}\, \sigma^{\mathrm{I}}_{q,q'} \rangle \\ &\quad + \sum_{q'=q+1}^{\mathrm{Z}} n_{q'} \langle n_{\mathrm{o}} v_{q'}\, \sigma^{\mathrm{C}}_{q',q} \rangle - \sum_{q'=0}^{q-1} n_q \langle n_{\mathrm{o}} v_q\, \sigma^{\mathrm{C}}_{q,q'} \rangle \end{aligned} \tag{2.21}$$

$$+ n_{q+1} \langle n_e v_e \sigma^{RR}_{q+1,q} \rangle - n_q \langle n_e v_e \sigma^{RR}_{q,q-1} \rangle - \frac{n_q}{\tau_c} ,$$

where gain and loss terms are taken into account due to single and multiple ionization (I), single and multiple capture (C) and radiative recombination (RR). The last term accounts for a finite confinement time τ_c of the ions.

In practice, some simplifications are possible. For instance, ionization and capture of single electrons are more effective than the processes involving two and more electrons in a single collision. This would restrict the summations in (2.21) to a few terms only with small values of $|q - q\prime|$. The balance of charge states is obtained by setting $dn_q/dt = 0$ for all charge states q and solving the set of resulting simultaneous equations. Such a calculation was carried out [2.25] for the electron beam ion trap EBIT for which an example is shown in Fig. 2.4. There the equilibrium population fraction of the high charge states of uranium ions is displayed as a function of electron-beam energy. At the highest energy of 300 keV the helium-like fraction amounts to more than 50% followed by the lithium-like and hydrogen-like ions.

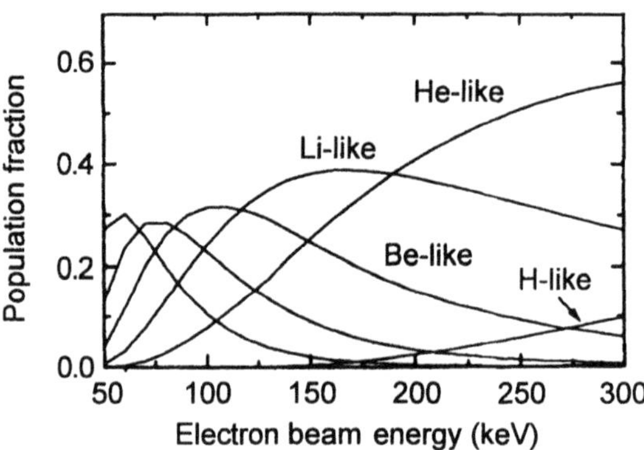

Fig. 2.4. Calculated equilibrium ionization balance of uranium ions in an EBIT as a function of the electron energy [2.25] © 1995 Plenum

Performance of High-Charge-State Ion Sources. Several attempts [2.8,17,26,27] have been made to classify the large number of different ion sources according to the physical processes responsible for their operation and according to their performance figures – mainly ion currents and charge states. Table 2.1 is merely intended to give an impression of the large variety of different operation principles used and the sort of ions that can be extracted. The ion sources are grouped corresponding to whether the ionization process occurs inside a plasma or at a surface of a solid. The energy carrier can be microwaves, photons, electrons, high electric fields and energetic ions for sputtering a metal surface. The ion species indicated in Table 2.1 should be taken as typical examples. In particular for the low charge states, virtually any ion of the periodic table has been produced often with a high ion current as for instance in the *Freeman* [2.28] arc-discharge sources.

Table 2.1. Some frequently used ion sources and their typical outputs

Production mechanism	Energy carrier	Ion-source type	Maximum charge state	Current
Volume	Microwave	ECR (hexapole)	Pb^{32+}	6 μA
		ECR (multicusp)	Ar^{4+}	0.3 μA
		High-pressure Microwave source	Ar^{+}	100 mA
	RF	RF source	Ar^{+}	50 mA
	e-beam	EBIS	Kr^{30+}	6 μA
		EBIT	U^{91+}	-
		e-beam plasma discharge	Ar^{+}	20 mA
	Laser	Photoionization	Ar^{+}	0.1 μA
	Arc discharge	Multicusp	Bi^{+}	30 mA
		PIG	U^{10+}	50 μA
		Freeman	Cr^{+}	2 mA
		Duoplasmatron	U^{3+}	30 μA
Surface	Ion	Multicusp		
	Current	Sputter	Pt^{-}	6 mA
	Arc discharge	MEVVA	Ti^{+}	1 A
	E-field	GFIS	H^{+}	0.2 μA
	Laser	LPIS	Ta^{10+}	10 mA

Moderate to high charge states with appreciable intensity can be generated in the PIG (Penning Ion Gauge), LPIS (Laser Produced Ion Source), ECR (Electron-Cyclotron Resonance) ion source and in the EBIS and EBIT [Electron-Beam Ion Source (Trap)]. The charge states available from these

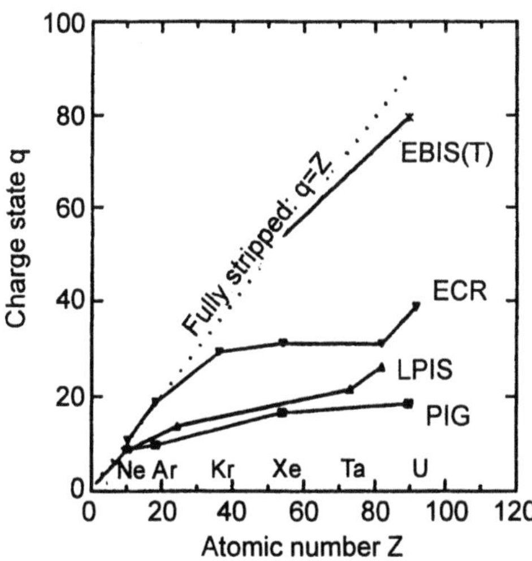

Fig. 2.5. Available charge states in various ion sources and stripping of fast accelerator beams [2.27] © 1994 AIP

sources [2.27] are compared in Fig. 2.5. For the PIG, LPIS and ECR the charge states range from about 10 to 20. With those devices it is possible to produce one- to three-electron ions from atoms between neon and argon. The highest charges are presently available in the EBIS and EBIT where also heavy atoms can be stripped to few-electron states. They approach the limit $q = Z$ of fully stripped ions otherwise only accessible by stripping of accelerated heavy-ion beams.

Figure 2.6 [2.27] shows the available particle intensities in these sources as a function of charge state for Xe, Ta and Pb. The curves clearly show the intensity decreasing with the charge state. The most intense beams can be produced with the PIG source for $q < 10$ whereas the high charge states can be produced in ECR and EBIS.

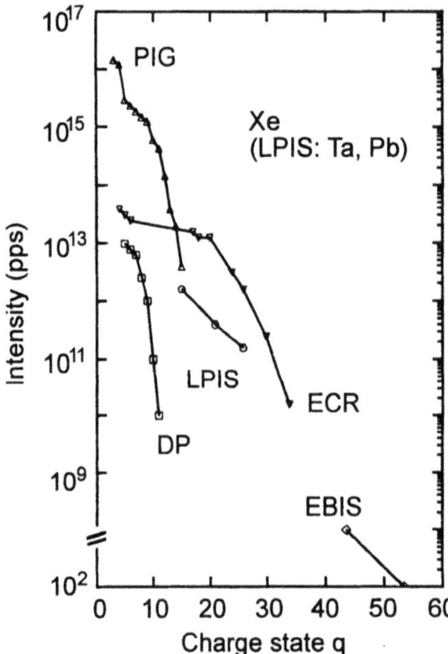

Fig. 2.6. Particle currents of some ion sources versus charge state for Xe, Ta, and Pb [2.27] © 1994 AIP

In the following sections we will introduce the operation principles of the PIG, ECR, EBIS, EBIT and LPIS ion sources and discuss their specific features offered to the experimenter using these devices.

2.1.3 Penning Ionization Gauge

The Penning Ionization Gauge (PIG) is based on the ionization vacuum gauge invented in 1937 by *Penning* [2.29]. It has a long tradition in accelerator technology and can be traced back to the early days of cyclotrons where it was used as an internal ion source [2.30–34]. Later it has become widely used mainly because of its simplicity and the high ion current available. A number of excellent reviews [2.35–41] on the Penning ion sources have been published.

The principal layout of a PIG ion source is illustrated in Fig. 2.7. It consists of a cathode and an anticathode, both made of metallic blocks, plus a cylindrically shaped anode. A magnetic field is applied parallel to the cylinder axis. Through a gas inlet the working gas is introduced. The cathode may be cold or it may be indirectly heated by a stream of electrons from a nearby filament in order to increase the electron emission out of the cathode surface. The anticathode can be held on the same potential as the cathode or it is floating. For high-power versions the anticathode must be cooled.

Electrons emitted from the cathode are radially confined by the magnetic field. They are spiraling around the magnetic field lines as described in Sect. 2.1.1. In the longitudinal direction they are confined by the electrical field of the cathode and anticathode. The radial and longitudinal potentials are indicated in Fig. 2.7. Most of the potential drops within the narrow plasma sheath. Electrons may oscillate several times between cathode and anticathode ionizing the heavy particles before they eventually get lost from the plasma volume. On the average, each primary electron can produce 5–10 secondary charges. The ion current can be extracted radially, or axially as shown in Fig. 2.7 through a small hole in the cathode or anticathode. Radial extraction is more often used because in the case of axial extraction the size of the extraction aperture and consequently the total ion current is very limited.

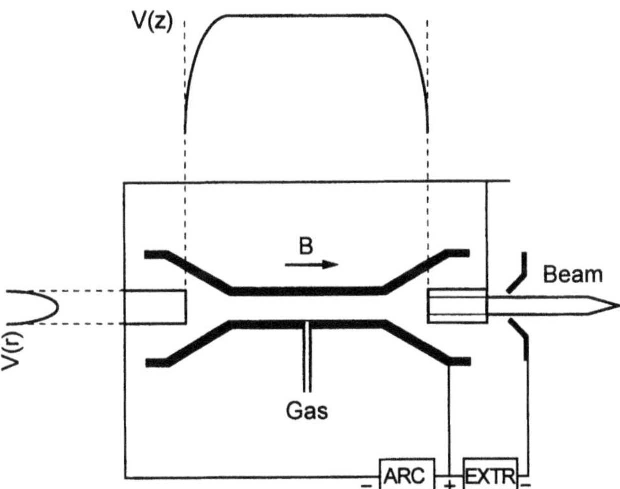

Fig. 2.7. Schematic layout of a Penning ion source with axial beam extraction

In order to optimize the ion current it is necessary to regulate the arc impedance [2.41]. The discharge can be operated in a regime where the voltage decreases with increasing arc current allowing very high currents at still moderate electrical power. Over a wide range the ion current that can be extracted is proportional to the arc current. A typical set of operating parameters of a heated-cathode Penning ion source are given in Table 2.2. For optimizing the yield of multiply charged ions, the neutral density or the gas pressure is an important parameter. If the gas pressure is reduced and a constant current is ascertained by the external circuitry, the arc responds by increasing the arc voltage thereby producing multiply charged ions to compensate for the lower number of ions.

Table 2.2. Typical operation parameters for a heated-cathode Penning ion source

Arc voltage	2 kV
Arc current	10 A
Magnetic field	0.2–1 T
Gas pressure	1–10 Pa
Electron density	10^{12}–10^{13} cm^{-3}
Ion confinement time	10–100 μs

A rare gas is usually used as the carrier gas from which a high ion current can be drawn. However, sputtering of the cathode and anticathode material – usually tantalum or tungsten – by the gas ions occurs during normal operation limiting the lifetime of the source. Sputtering is the most universal method to introduce metal atoms into an ion-source discharge. A special sputter electrode kept on a negative potential is frequently used to increase the number of metal atoms in the discharge. Metals with a high melting point like Mo, Ta, W, Pt, etc. can withstand the heatload when used as a sputter material. Other, low-melting-point materials have to be sufficiently cooled. The fraction of the metal component in the extracted ion beam can be as high as 10–20%. For a review on metal-ion production see [2.42].

For the formation of an ion beam, a single or a set of apertures are used which have to be carefully designed [2.43–46] to the actual needs. The existing magnetic field can be used to preselect a desired ion species corresponding to its charge-to-mass ratio. According to the *Child-Langmuir* law [2.47,48] the maximum current that can be extracted under space-charge limited conditions is proportional to $U_{\text{extr}}^{3/2}$ where U_{extr} denotes the extraction voltage. For an extraction system, the same proprtionality has been found empirically [2.45,49] and the extracted ion current can be expressed as

$$I_{\text{extr}} = P_I \, (q/\text{A})^{1/2} \, U_{\text{extr}}^{3/2} \,, \tag{2.22}$$

where q and A denote the charge state and the atomic number of the extracted ions, respectively. The proportionality factor P_I is called the *perveance* of the ion gun.

Although the Penning ion source is also a bright light source in the UV and ultra-soft x-ray region one of its main applications has been the use for heavy-ion injection into accelerators [2.50,51]. Despite their great success as ion injectors over several decades, arc-discharge ion sources have been gradually replaced by more powerful sources of highly charged ions.

2.1.4 Electron Cyclotron Resonance Ion Source

The Electron Cyclotron Resonance (ECR) ion source was first proposed by *Geller* in 1969 [2.52] and by *Potsma* in 1970 [2.53] and the first operational ECR source was presented by Geller and his co-workers in 1971 [2.54]. From that time on the ECR ion-source technology for producing substantial amounts of highly charged ions has rapidly advanced. The status of development has been regularly evaluated in a series of workshops, e.g., [2.55] started in 1978. Several splendid reviews [2.56–63] on the subject can be recommended.

Operation Principle. In an ECR ion source, the energy source for plasma generation and maintenance is resonant electron heating by microwaves mentioned in Sect. 2.1.1. The basic principles [2.60] upon which the ECR ion source is predicated are demonstrated in Fig. 2.8. A metallic vessel serves both as a multi-mode cavity and as a plasma chamber. The dimensions of the chamber must be larger than the microwave wavelength being used. For a frequency of $f = 10$ GHz, for instance, the wavelength is $\lambda = 3$ cm. The microwave power is coupled to the cavity through a waveguide. From a first stage or injector stage serving as a reservoir of electrons a cold plasma diffuses into the main second stage.

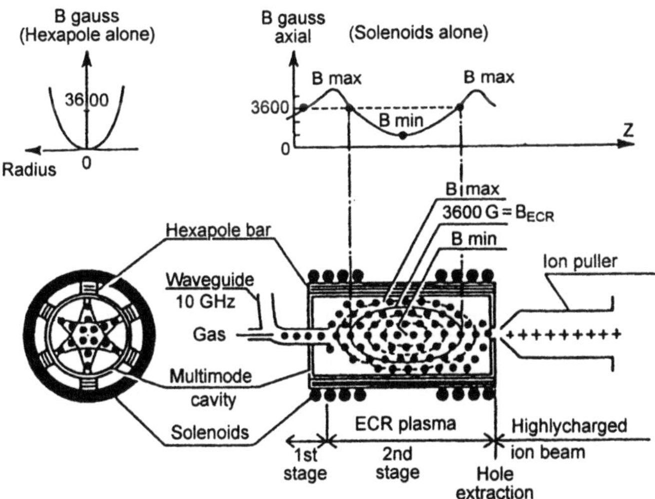

Fig. 2.8. Principle of a 10 GHz ECR ion source. The plasma electrons are trapped in a B_{min} structure and are energized at the magnetic surface where the magnetic field is B_{ECR} [2.60] © 1991 Springer

In order to produce multiply charged ions through step-by-step ionization it is essential to maximize the product of electron density, electron velocity and ion confinement time $n_e v_e \tau_c$. At the same time the neutral density has to stay low for avoiding excessive ion-charge reduction by electron-capture processes. Theoretical estimates [2.64] show that ion confinement times of 10 ms or more are required to effectively produce highly charged ions like Ar^{16+}. Such long ion lifetimes can be achieved with a magnetic trap or bottle realized by a so-called *minimum-B structure*. This is illustrated in Fig. 2.8 where the magnetic configuration is a superposition of an axial-mirror and a radial-sextupole field used for the axial and radial confinement of electrons. Indicated are the B_{min} and B_{max} values of the axial field and the ellipsoidal ECR surface fulfilling the resonance condition. In the numerical example the resonance frequency is chosen as $f = 10$ GHz corresponding to $B_{\text{ECR}}=0.36$ T according to (2.17). Upon passing the ECR surface, the electrons may be stochastically heated if a component of the electric-field vector of the microwave is perpendicular to the direction of the magnetic field. The ions are confined by the space-charge potential created by the electrons. Production of ions from metallic materials can be similar to that in a Penning source. Especially miniaturized ovens [2.65,66], which can be integrated in the coaxial guide for the microwave power, have proved very successful.

Magnetic Confinement. Various magnetic-field configurations have been developed [2.60,63,64,67,68] in order to accomplish the minimum-B structure for the magnetic confinement. The first ECR ion source producing substantial amounts of multiply charged ions, called superMAFIOS, is the basis of all ECR ion sources. Its power consumption, however, was 3 MW because conventional copper coils were used to produce the strong solenoidal and multipolar magnetic fields. Subsequently the power consumption was drastically reduced by employing superconducting magnets, permanent magnets and iron yokes. Table 2.3 gives an overview of the magnet technologies used for the prototype ECR ion sources developed at Grenoble.

Table 2.3. Magnetic-field configurations of the prototype ECR sources; warm: conventional copper coils, cold: superconducting coils, perm: permanent magnets, yoke: iron yoke

Ion source	Hexapole			Solenoid			
prototype	warm	cold	perm	warm	cold	perm	yoke
superMAFIOS	√			√			
cryoMAFIOS		√			√		
miniMAFIOS			√	√			
CAPRICE			√	√			√
neoMAFIOS			√			√	

Ion sources based on the miniMAFIOS or CAPRICE technology need less than 0.1 MW of power. The multipoles can be made of hard permanent-magnetic materials like $SmCo_5$ or NdFeB with high remanent fields. With these two materials it is possible to use resonant magnetic fields of B_c = 0.36 T and B_c = 0.51 T, respectively. The ECR frequencies correspond to 10 GHz and 14 GHz allowing one to use reasonably priced microwave generators developed for satellite communication. Figure 2.9 schematically shows the layout of a modern miniMAFIOS source [2.56] where the configuration of the solenoidal coils and the hexapole, made from permanent magnets, can be seen. The feeds for the microwaves and the gas are indicated on the left side of the figure. This version of an ECR source uses a biased probe for supplying electrons to the plasma. The mirror ratio of presently operating ECR sources is in the range of B_{max}/B_{min} = 2–7. A high mirror ratio is essential [2.69] for a long confinement time and consequently for the production of highly charged ions.

It is possible to construct very compact sources like the neoMAFIOS [2.70] by employing only permanent magnets for all the magnetic fields. The magnet configuration [2.71] of such an economic source is displayed in Fig. 2.10 where three slices of six permanent magnets are shown. The inner set of six magnets is used to generate the hexapolar radial field and the outer slices for production of the solenoidal field.

The magnetic multipoles used for radial confinement leave an area near the magnet axis where the magnetic field strength does not change appreciably. This area is important for establishing the ECR tuning zone and for

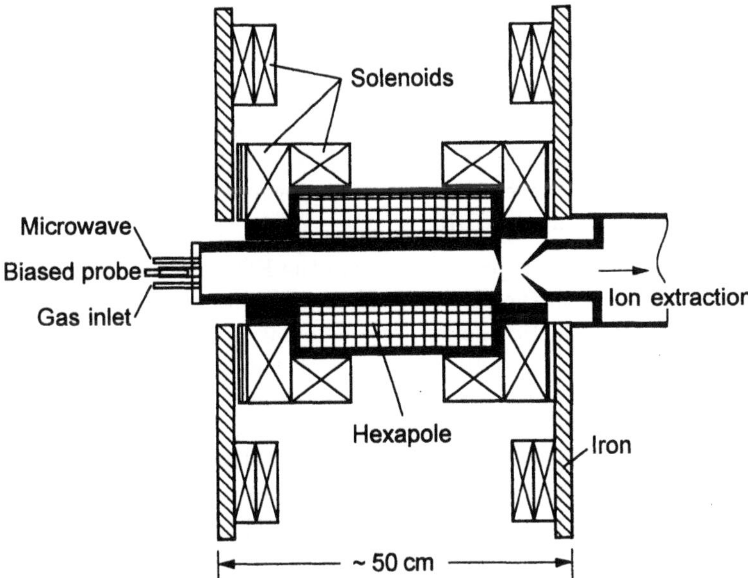

Fig. 2.9. MiniMAFIOS ECR ion source designed for 10–18 GHz. Microwave power is coupled radially off axis and a biased probe is used for supplementary electron production

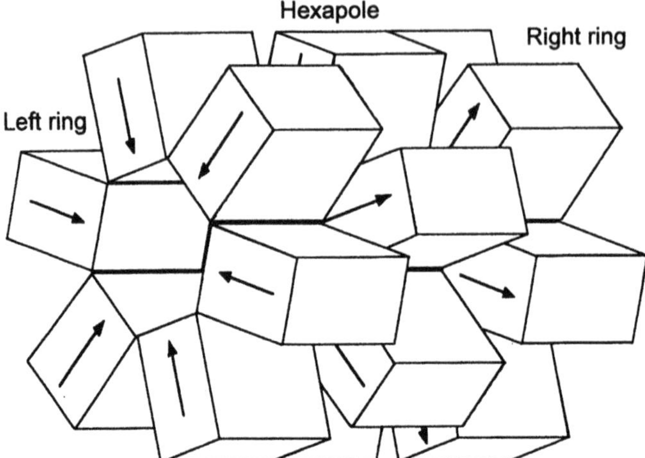

Fig. 2.10. Compact versions of the ECR ion source use a magnetic configuration accomplished totally by permanent magnets. The *arrows* indicate the polarity of the magnets. [2.71] © 1992 AIP

ion extraction. It has been pointed out [2.72,73] that the production of high charge states could be improved when this area increases. For a multipole with N_{mul} cusps the magnetic field strength varies with the radius r as $r^{N_{\mathrm{mul}}/2-1}$. Therefore, the area of practically constant magnetic field near $r = 0$ can be increased by choosing a higher number of cusps. A simulation of electron orbits demonstrating this effect is exposed in Fig. 2.11. The electrons were assumed to have energies of 1 keV in order to make their gyro orbits visible. The results of these simulations [2.72] clearly show that the number of electrons confined to orbits in the uniform-field region of the source increases with an increase in the number of cusps from $N = 6$ to $N = 12$. This is supported by experimental investigations on the form of the plasma glow of octopoles [2.74] and dodecapoles [2.75].

Electron Heating. Electrons passing through the ECR surface, and which happen to be in phase with the electric-field component, are accelerated by the transfer of electromagnetic energy perpendicular to the magnetic field. Electrons accidentally arriving out of phase with the electric field are decelerated. After many passes through the ECR surface, the electrons gain a net energy which is called stochastic heating [2.57,60,76–78]. Because the electrons preferentially acquire kinetic energy in the direction perpendicular to the axial magnetic field, $T_{\perp} \gg T_{\parallel}$, they are effectively trapped between the mirror fields at the ends of the plasma chamber. The enhanced trapping, also called *mirror plugging*, is a consequence of the conservation of angular momentum and of energy in the absence of collisions. It results in a small longitudinal thermal energy at the high-field end regions. The electrons oscillate between the mirror regions because they are reflected by the force

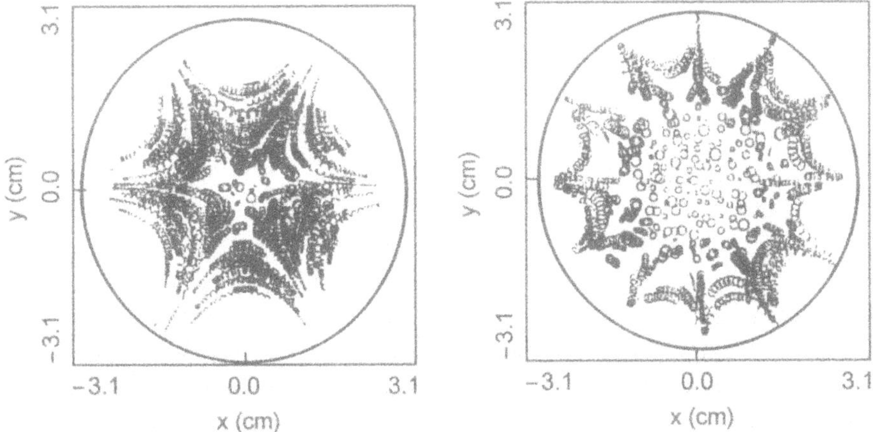

Fig. 2.11. Electron orbits in a multicusp field with six *(left)* and twelve *(right)* cusps and a superimposed uniform solenoidal magnetic field along the z axis. The area of constant magnetic field present in the central region of the plasma increases with the number of cusps as does the number of electrons that have stationary orbits within this region of field [2.72] © 1994 AIP

$\boldsymbol{F} = -\boldsymbol{\mu}\,\mathrm{grad}\boldsymbol{B}$, where $\boldsymbol{\mu}$ denotes the magnetic moment of the circular currents.

Although a transverse launching of microwaves has been frequently used, the most efficient way for heating the plasma electrons is by injecting right-hand circularly polarized waves along the direction of the magnetic field. The direction of rotation of the plane of polarization is the same as the gyration direction of the electrons. The left-hand circularly polarized waves do not have a resonance because of the opposite sense of electron gyration.

Power can only be coupled into the plasma as long as the plasma electron density n_e does not reach the critical value n_c at which the plasma, excitation and ECR frequencies are equal, $\omega_\mathrm{pe} = \omega_\mathrm{c}$. According to (2.10) the critical density n_c scales as $\propto \omega_\mathrm{rf}^2$, or through (2.16) as $\propto B^2$,

$$n_\mathrm{c} = 1.2 \times 10^{10}\,\mathrm{cm}^{-3} f[\mathrm{GHz}]^2 \,.$$

To increase the maximum electron density requires increases in both the microwave frequency and the magnetic-field strength.

Enhancement of Highly Charged Ions. A high frequency and a correspondingly strong magnetic field together with a high mirror ratio generally assure the production of very highly charged ions. The strong fields are easily achieved with the superconducting variants like the source at Michigan State University [2.69] which can be tuned to a mirror ratio of more than 7, or the new Superconducting Electron Cyclotron Resonance Source SERSE [2.79–81] at Catania with an axial field tunable up to 2.7 T. There are a number of specific upgrade technologies which have successfully been used to improve the performance for high charge states:

- electron-beam injection
- biased electrodes
- wall coating
- gas mixing
- two-frequency heating
- light-ion cyclotron resonance heating
- afterglow pulse
- pulsed magnetic fields.

The first three methods increase the plasma electron density, for details see [2.56,82–86]. When a second lighter gas is mixed into the plasma of an ECR ion source a substantial increase of the currents of highly charged ions is found. This effect has been widely used in many sources and attempts [2.87–90] have been made to explain its origin. The effect might be the combination of several factors of which ion cooling plays a key role. The lighter ions take over energy from the heavier ions on leaving the potential well. They must be constantly refueled in order to sustain the cooling.

An ECR source can be operated, very well, in a pulsed mode by pulsing the microwave generator. The source can be tuned to release a very intensive burst (Fig. 2.12) of highly charged ions during a small time of 0.5–2 ms after the microwave power is switched off [2.58,91,92] This 'afterglow effect' is a consequence of the electrostatic trapping of the ions in the space-charge well of the magnetically confined hot electrons. After the break down of the discharge, electrons leave the plasma and are no longer refueled. Therefore, the ions have to leave the plasma because of the charge neutrality condition. When tuned to this mode of operation, very high peak currents can be achieved which is very useful for injection into RF accelerators. The heavy-ion injector at CERN [2.93], for instance, can produce a current of more than 100 μA of Pb^{27+}.

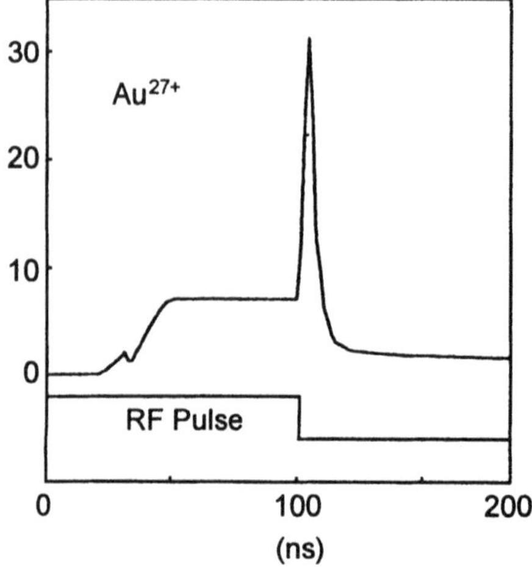

Fig. 2.12. Afterglow pulse of Au^{27+} ion current from a 10 GHz CAPRICE source [2.58] © 1994 AIP

A more defined release of the highly charged ions from their trapping potential has been investigated at the Gesellschaft für Schwerionenforschung (GSI) in Darmstadt [2.94] by pulsing the magnetic field of a miniMAFIOS ECR source. The pulse height and length of the extracted current can be controlled that way.

As an example for the high state of performance reached in modern ECR sources we show, in Fig. 2.13, a charge-state spectrum [2.65,66] of uranium obtained in an upgraded CAPRICE source at Grenoble when optimized for $q = 32$. The electrical current of U^{32+} has been 5 μA. In Table 2.4, the ion currents obtained with the aforementioned methods of electron-beam injection, wall coating and two-frequency heating are compared. The results have been obtained at the Lawrence Berkeley National Laboratory (LBNL) with its Advanced ECR ion source AECR [2.85,86].

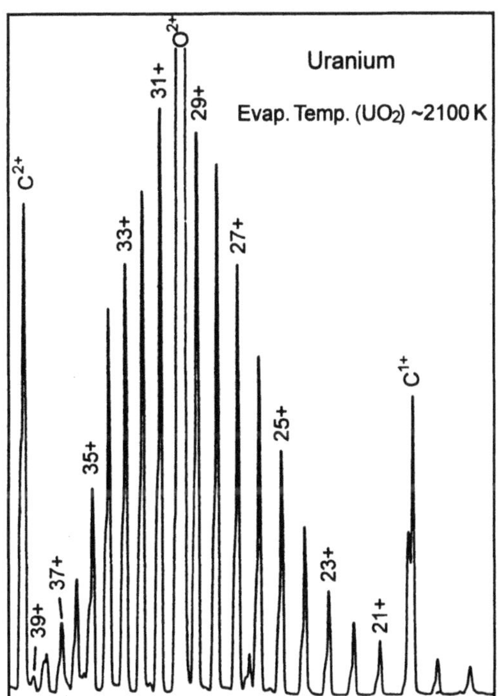

Fig. 2.13. Charge-state distribution of uranium ions measured at Grenoble with an upgraded CAPRICE ion source that has been optimized for high charge states [2.66] © 1996 AIP

Radiation. Through the ECR process, electrons can be heated to very high energies up to ~ 100 keV. These hot electrons cause x-ray emission due to electron-ion collisions which can be used to diagnose [2.17] the ECR plasma. The dominating x-ray emission processes are *first*, the electron-ion Bremsstrahlung generated if an electron is accelerated in the Coulomb field of an ion and *secondly*, the characteristic x-ray radiation emitted when an inner-shell hole is filled. The measurement of the x rays is one important way to

Table 2.4. Electrical ion currents, in μA, of high charge-state Bi^{q+} and U^{q+} ions extracted from the AECR source of the Lawrence Berkeley National Laboratory [2.85]. The influence of an additional electron gun, coating of the walls with aluminum and two-frequency (TF) versus single-frequency (SF) was investigated

	SF	SF		TF	
	e^- Gun	Al Coating		(10 + 14 GHz)	
q	Bi	Bi	U	Bi	U
25	3.8	8.7			
27	5.5	16.6			
28	6.0	19.5			
29	5.7	20.0			
31	4.5	15.7			
32	3.5	12.8	3.6		11.
33	2.6	8.4	3.8	12.1	12.
34	1.5	6.0		10.2	
36	0.7	2.5	2.4	6.5	8.7
37	0.4	2.2	1.6	5.0	7.1
38	0.2	1.3	1.2	3.2	5.4
39			0.7		3.7
40		0.5		1.8	
41		0.25		1.1	1.5
42					1.0
43					0.5
44				0.24	
46				0.08	

diagnose the ECR plasma. This requires, besides the measurement, an adequate plasma model.

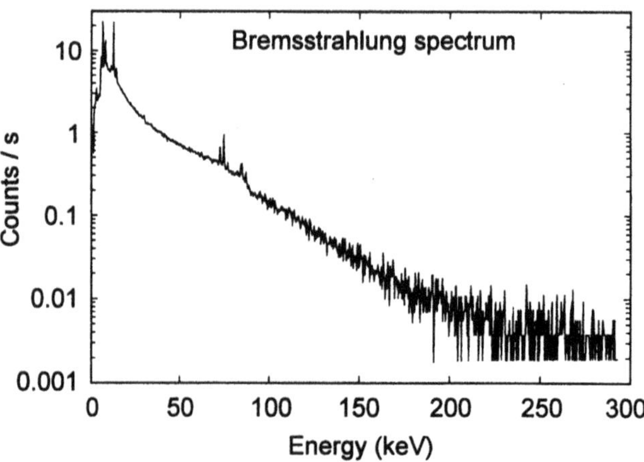

Fig. 2.14. Bremsstrahlung spectrum in the miniMAFIOS source showing the existence of high-energy electrons [2.95] © 1992 AIP

As an example, Fig. 2.14 shows the Bremsstrahlung continuum measured at a miniMAFIOS source running with oxygen [2.95]. The Bremsstrahlung extends up to an energy of 200 keV. From the slope, or more generally from the shape of the spectrum, information about the transverse electron energy $T_\perp$ is obtained. Because the fast electrons do not have a Maxwell distribution in energy one has to be careful [2.96] with the definition of a temperature parameter.

Other x-ray diagnostic tools are provided by the characteristic radiation [2.97]. If there are heavy ions in the source their K-shell radiation, for instance, is also an indicator of the electron temperature. There one makes use of the energy dependence for K-shell excitation by electron impact beyond threshold. With increasing degree of ionization the energy of the characteristic x-ray lines increases because of the removal of outer electrons shielding the nuclear charge. From the energy shifts the degree of ionization of the ions can be inferred.

2.1.5 Electron-Beam Ion Source and Trap

The principal of the electron-beam ion source was proposed [2.98] and demonstrated [2.99] for the first time by *Donets.* This device is now very efficient for the production of very highly charged ions. Ions trapped inside a dense electron beam are continuously bombarded by electrons and are sequentially ionized to high charge states. The physical parameters, like bombarding energy and confinement time, responsible for the production of very highly charged ions seem to be better under control in the EBIS than in other ion sources. That is why it has attracted many researchers and there is now an increasing number of these devices spread around the world. A very powerful extension of the EBIS concept was introduced with the electron-beam ion trap (EBIT) [2.100] where ions with the highest charge states can now be produced and stored for very long times. The history of the EBIS(T) developments and the present knowledge of their physical basis has been reviewed by *Donets* [2.101,102] and by *Becker* [2.103]. An overview of operating electron-beam ion sources has been given by *Stöckli* [2.104] and the electron-beam ion trap has been reviewed by *Knapp* [2.25] and by *Marrs* et al. [2.105].

Basic Principle. In an EBIS(T) a high-density electron beam is launched along the magnetic axis of of a strong magnetic field forming a radial space-charge well for ions. Cylindrical electrodes surrounding the beam are positively biased as to confine the ions in the longitudinal direction. The length of the trap is typically near one meter for an EBIS and a few centimeters for the EBIT. Figure 2.15 shows a typical EBIS configuration [2.106]. Atoms ionized by electron impact or singly charged ions delivered from an external ion source are stepwise ionized to consecutively higher charge states until equilibrium with electron-capture and other loss processes is reached or the ions are extracted by a suitable change of the electrode potentials. Ionization times vary over a large range depending on the current density and on the desired charge state. For the higher charge states several seconds can be regarded as typical in an EBIS whereas trapping times in an EBIT can be several hours. In an EBIS the highest charge states are obtained in a *batch* mode [2.101,107] where ions of lower charge state are initially trapped in a seed trap (Fig. 2.15). At the beginning of a production cycle the barrier between seed trap and main trap is lowered flooding the main trap with seed ions. After the confinement time necessary for the desired charge state the potential of the main trap is raised in order to extract the ions. A high magnetic field of up to ~5 T usually produced by a superconducting solenoid (EBIS) or by a pair of superconducting Helmholtz coils allows the compression of the electron beam to a diameter of less than 100 μm and current densities of a few thousand A/cm^2. The dynamics of space-charge compensation [2.103,108,109] of the electron beam by ions produced from the residual gas plays an important role in the ion confinement. With increasing degree of compensation the confinement of cold ions is improved whereas hot ions are lost. To make the time constant for compensation comparable to the ionization time requires an ultra-high vacuum ($\sim 10^{-7}$ Pa) inside the trap. There is a smooth transition [2.110] between EBIS and EBIT modes mediated by the compensation time in relation to ionization time.

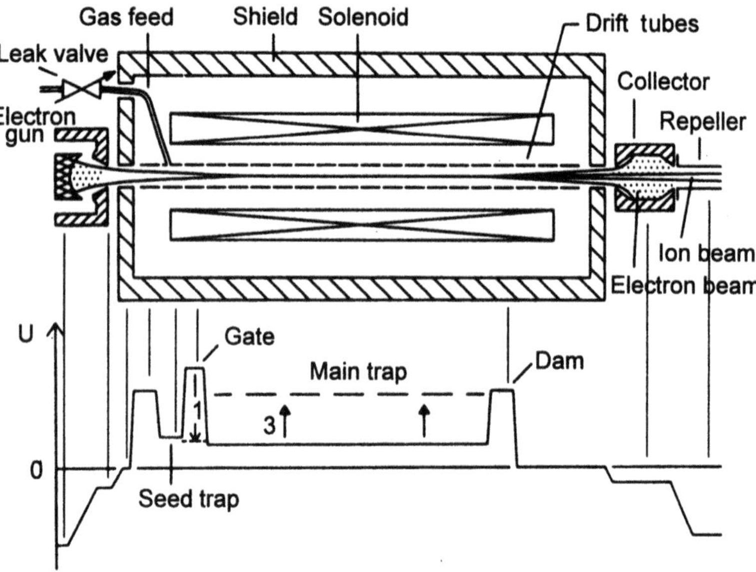

Fig. 2.15. Schematic of an electron-beam ion source and its electrical potentials applied to the drift tubes and other electrodes. During the injection period the gate potential is lowered to fill the main trap with seed ions, where they continue to be ionized until the desired charge state is reached. Then the ions are expelled over the dam by raising the potential of the main trap [2.106] © 1992 AIP

Ionization. As already discussed in Sect. 2.1.2 the evolution of the ionic charges is governed by the atomic collision processes. A simplification of the balance equations (2.21) is obtained when only singlestep ionization and radiative recombination is taken into account as for the results shown in Fig. 2.4. Here, we complete the discussion by displaying the corresponding time evolution in Fig. 2.16. The calculation [2.25] was performed for the highest charge

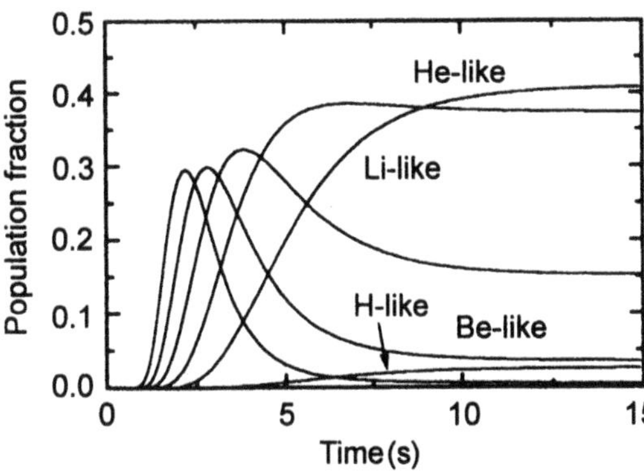

Fig. 2.16. Calculated evolution of uranium charge states determined only by ionization and radiative recombination at an electron energy of 200 keV [2.25] © 1995 Plenum

states of uranium in the EBIT. Ionization cross sections were approximated by the *Lotz* formula [2.22] and for radiative recombination the theory of *Kim* and *Pratt* [2.111] was used. After a period of about 10 s an equilibrium of ions with one to four electrons is established. Lithium- and helium-like ions constitute the most abundant charge states with a small fraction of hydrogen-like uranium. For the latter a K-shell electron has to be ionized which has the lowest cross section. Because of the large step of the ionization potential observed whenever a closed electron shell has to be opened an enhanced population of one single charge state can be obtained by using an electron energy just below the ionization potential of that shell.

Ion Heating. Long trapping times require a careful control of the thermal energy of the ions. Several mechanisms [2.112] can contribute to the heating of heavy ions in a trap, like direct heating by elastic electron-ion collisions, ionization heating by the sudden change of the ions' charge in the potential well or plasma instabilities propagating along the beam axis. At least for the EBIT, with a small trap volume, ion heating due to ion-electron elastic collisions is dominating. According to *Spitzer* [2.113] the rate of heating is

$$\frac{\mathrm{d}E_q}{\mathrm{d}t} = \frac{0.442\, q^2 j_\mathrm{e} \ln(\rho_\mathrm{max}/\rho_\mathrm{min})\, n_q}{\mathrm{A}E_\mathrm{e}}\ \mathrm{eVs^{-1}cm^{-3}}, \tag{2.23}$$

where n_q is the number density of ions and A the atomic mass number of the ions and ρ_max and ρ_min denote the maximum and minimum impact parameter, respectively. ρ_max is the Debye length or the electron-beam diameter, whichever is smaller. For heavy highly charged ions the heating rate can be several keV per second and per ion, whereas the trapping-potential depth is below one keV. Therefore, a strong cooling mechanism is required to keep highly charged ions in the trap.

Evaporative Cooling. Similar to the gas-mixing effect in ECR ion sources, cooling of highly charged heavy ions by light ions is very effective in EBIS(T) sources. This evaporative cooling [2.114,115] is illustrated in Fig. 2.17 indicating the distribution of ions with low and high ionic charges. The ions are trapped with a potential energy proportional to their charge. If we assume

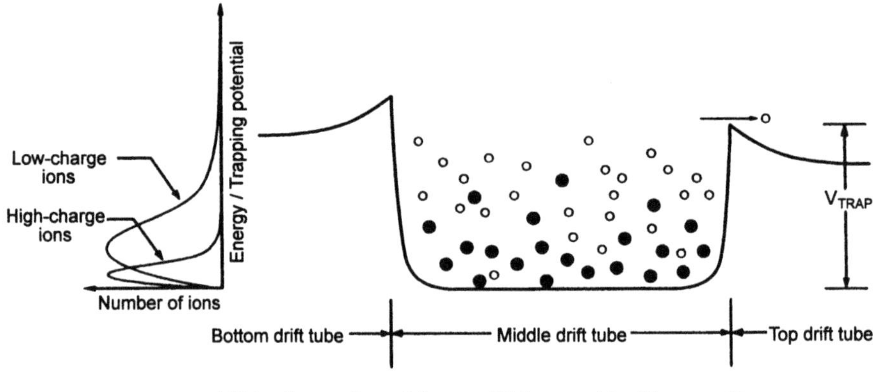

Fig. 2.17. Evaporative cooling in an electron-beam ion trap. Highly charged ions with charge Q are more strongly trapped than ions with a small charge q. Heat exchange by ion-ion collisions brings the energy distributions of the two ion species into equilibrium, so the low-charge ions preferentially escape from the trap, cooling the high-charge ions. [2.25] © 1995 Plenum

that all ions in the trap interact sufficiently long, then all ions will have the same temperature T_i. If the ion velocity, furthermore, can be approximated by a Boltzmann distribution their loss rate is

$$R_\mathrm{loss}(q) = \tau_\mathrm{comp}^{-1} \exp\left(-qeV_\mathrm{trap}T_\mathrm{i}^{-1}\right) , \qquad (2.24)$$

where V_trap denotes the effective potential depression in the compensated beam and τ_comp is the mean compensation time. Equation (2.24) shows that, at the same temperature T_i, ions with low charge are preferentially lost. Therefore, light ions are an ideal coolant because they are rapidly ionized to their highest charge state while being still cold enough to cool down the heavy ions. The light ions may either be provided by the residual gas or by a continuous flow of light ions or neutral atoms from an external source.

As implied by (2.24) the mean temperature of ions trapped is limited by the trapping potential V_trap because hot ions boil off from the trap. It has been impressively demonstrated by *Beiersdorfer* et al. [2.116] how low the ion temperature of the ions can be. They measured the thermal Doppler broadening (Fig. 2.18a,b) of K-shell emission lines of highly charged titanium ions.

Storing helium-like Ti^{20+} ions the following dielectronic recombination process

$$1s^2 + e \rightarrow 1s2p^2 \rightarrow 1s^22p + \hbar\omega$$

is evoked by choosing an appropriate electron energy near 3.3 keV. The two lines labeled j and k in Fig. 2.18a,b are due to the doublet $1s2p^2, J = 5/2, 3/2 \rightarrow 1s^22p, J = 1/2$. They were measured with a high-resolution crystal spectrometer for two different experimental conditions. Going from the high-temperature to the low-temperature case the applied external trapping potential and the beam-current were drastically lowered. The result is the measured reduction of the ion temperature from 685 eV down to 70 eV.

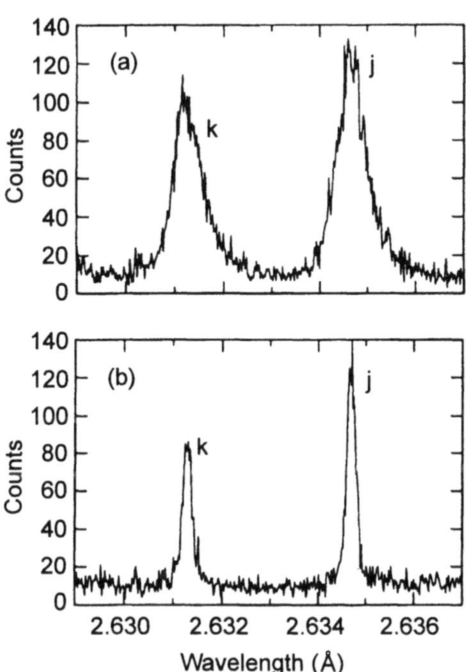

Fig. 2.18a,b. X-ray line profiles (k and j satellites) produced by dielectronic recombination of beam electrons with trapped helium-like ions. The width of each line is determined by the thermal motion of the emitting ions. In the high-temperature case (a) the line width corresponds to an ion temperature of 685 eV and in the low-temperature case (b) the linewidth corresponds to 70 eV ion temperature [2.116] © 1995 Elsevier

EBIT for Internal Atomic Spectroscopy. The electron-beam ion trap developed at the Lawrence Livermore National Laboratory has been used for various atomic-physics experiments including x-ray spectroscopy of few-electron ions and electron-ion cross-section measurements. Because of its magnetic-field configuration employing a pair of Helmholtz coils it is relatively easy to have access through several view ports for x-ray spectroscopy. The inner section of the EBIT device is schematically shown in Fig. 2.19.

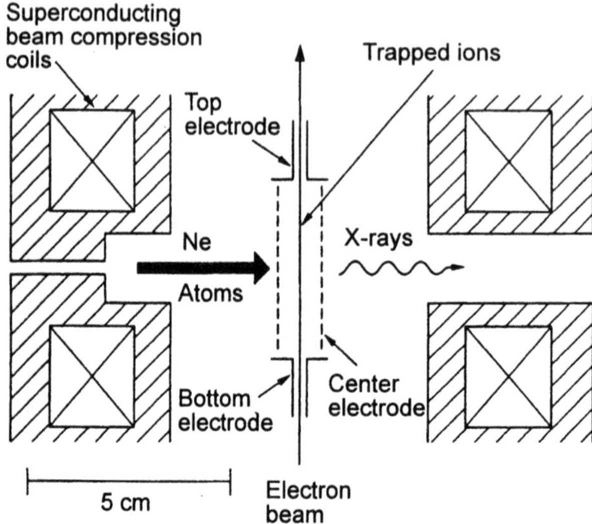

Fig. 2.19. Outline of the high-energy EBIT. The high-current electron beam is compressed to a diameter of less than 80 μm by the strong magnetic field which is produced by a pair of superconducting coils. [2.117] © 1994 APS

Table 2.5. Typical EBIT and SuperEBIT operating parameters [2.25,117]

Parameter	EBIT	SuperEBIT
Electron energy	5–25 keV	10–200 keV
Electron beam current	0–160 mA	0–200 mA
Electron Beam diameter in trap	80 μm	70 μm
Effective electron current density	0–3200 $\mathrm{A\,cm^{-2}}$	0–5000 $\mathrm{A\,cm^{-2}}$
Trap length	2 cm	2 cm
Magnetic field in trap	3 T	3 T
Axial trap depth	10–400 V	10–400 V
Highest uranium charge state	U^{82+}	U^{92+}
Total number of trapped ions	10^5–10^6	10^5–10^6
Ion density	10^9–10^{10} $\mathrm{cm^{-3}}$	10^9–10^{10} $\mathrm{cm^{-3}}$
Trapping time	< 5 h	< 2 h

The original EBIT has produced ionization states up to neon-like U^{82+} as limited by the available beam energy of less than about 30 keV. Because of this limitation a second, high-energy trap, called SuperEBIT, was built that can run at an energy of more than 200 keV. Typical EBIT and SuperEBIT operating parameters are given in Table 2.5. The trapping times indicated result from intermittent high-voltage breakdowns.

The evaporative ion-ion cooling technique has been successfully used to produce and store hydrogen-like and bare uranium [2.117] ions in the SuperEBIT. A collimated beam of neutral neon atoms, launched perpendicular to the electron beam, was used as a cooling medium. The neon atoms reach a high average charge before escaping the trap with the removal of an estimated energy of about 300 eV per neon ion. The density of neutral neon had to be carefully controlled to avoid excessive charge-exchange losses. The total compensation of the electron-beam space charge by neon and uranium charges was less than 10%.

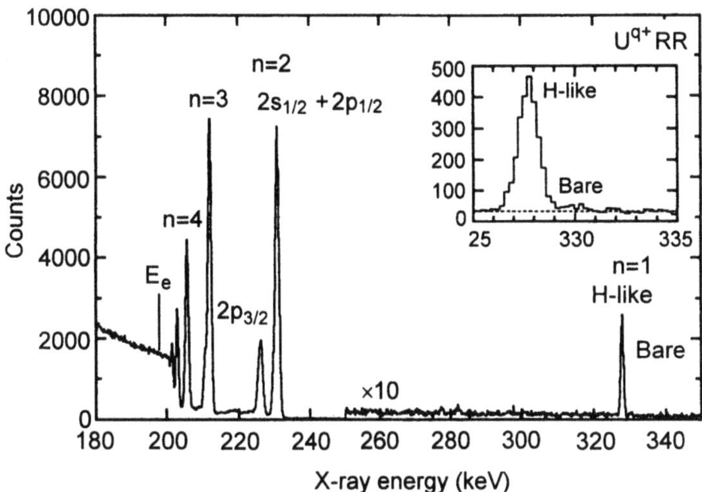

Fig. 2.20. X-ray spectrum of uranium ions in SuperEBIT recorded at an electron beam energy of 198 keV. The high-energy lines are produced by radiative recombination of hydrogen-like and bare uranium with electrons. [2.117,25] © 1994 APS

Figure 2.20 shows the x-ray spectrum of uranium at an electron energy of 198 keV. The lines are produced by radiative capture into different electronic states of highly ionized uranium. The high-energy peaks are produced by radiative recombination of beam electrons with hydrogen-like and bare uranium ions. At equilibrium, the abundance ratio $N_{\text{bare}}/N_{\text{H}}$ can be inferred from the corresponding intensity ratio in the radiative-recombination spectrum. Knowing the radiative-recombination cross sections theoretically [2.118], electron-impact ionization cross sections can be determined from the ionization balance. The ionization cross section can be written as

$$\sigma^{\text{ion}}_{\text{H}\to\text{bare}} = N_{\text{bare}}/N_{\text{H}} \left(\sigma^{\text{RR}}_{\text{bare}\to\text{H}} + \langle\sigma^{\text{CX}}\rangle\right), \qquad (2.25)$$

where $\langle\sigma^{\text{CX}}\rangle$ is an effective charge-exchange cross section which is small as compared to the recombination cross section $\sigma^{\text{RR}}_{\text{bare}\to\text{H}}$. The total number of uranium ions in the trap for that experiment have been estimated to be $N_{\text{total}} \approx 5 \times 10^4$ and the number of hydrogen-like and bare uranium ions are estimated as $N_{\text{H}} \approx 500$ and $N_{\text{bare}} \approx 10$, respectively.

2.1.6 Laser Ion Source

In this section we refer to the ion source based on the plasma which is created when a solid surface is irradiated by intense laser light. The term *laser ion source* is also used in another context, where the laser light is used for direct photoionization rather than for plasma heating. In resonance ionization spectroscopy (RIS) [2.119,120] a specific atomic species is selectively ionized through resonance excitation. This kind of ion source can produce isotopically pure beams of low-charge ions for mass measurements [2.121] in ion traps and accelerators. We will not further discuss this interesting issue but concentrate on the highly charged ions of laser-produced plasmas.

Plasma Generation by Laser Light. When a pulsed laser beam is focused on a solid surface with a small spot of typically less than 100 μm diameter resulting in a high power density in excess of 10^8 W cm^{-2} a hot plasma [2.122–125] is created from which multiply charged ions can be extracted. The incoming light can penetrate the surface and the originated plasma only a small distance before reaching the critical electron density [2.123] at which the plasma frequency equals the frequency of the laser light. For the two most commonly used high-power lasers, the CO_2 and Nd:YAG lasers, the wavelengths and hence the critical densities are

$$\begin{aligned} \mathrm{Nd:YAG} \quad & \lambda = 1.05\ \mu\mathrm{m} \quad n_\mathrm{e} = 10^{21}\ \mathrm{cm}^{-3}, \\ \mathrm{CO_2} \quad & \lambda = 10.6\ \mu\mathrm{m} \quad n_\mathrm{e} = 10^{19}\ \mathrm{cm}^{-3}. \end{aligned}$$

Beyond the critical density no wave propagation can occur. Within this layer the electrons are strongly heated mainly via inverse Bremsstrahlung. At high power density, resonance absorption [2.126], coupling the electromagnetic and electron plasma waves, might play an important role. The hot electrons produce a very intensive ionization forming a hot and dense plasma. Target material is explosively ablated perpendicular to the surface and an extremely dense plasma plume is formed. In the early phase of the expanding and cooling plasma three-body, radiative and dielectronic recombination limit the ionic charge states. Later, a freezing of the charge-state distribution is observed which is attributed to the presence of fast electrons emerging from three-body collisions. The mean charge state produced, increases with the power density of the laser. At a power density in the range of 10^{11}–10^{12} W cm^{-2}, the electrons may acquire kinetic energies near 1 keV which is sufficient to multiply ionize the L shell of third-row elements. With further increased power density, charge separation [2.127] between ions and electrons provides an accelerating field for the ions which helps to transport the ions from the surface. The expanding plasma has been found [2.128] to be confined to a narrow cone. There are two favorable directions [2.129] relative to the target surface and incoming laser light where one can find highly charged ions: perpendicular to the surface and in the direction of the mirror reflection of the laser light.

Extracted Ions. The directional plasma plume is a useful feature for constructing an ion source from a laser-produced plasma device. Other advantages are the versatility in producing various ion species by selecting an appropriate target material, high states of ionization, high number of ions per pulse, relatively simple design making use of the widely spread high-power laser technology. Obvious problems with this ion source are related to the limited repetition rate and pulse width of the laser. Because target material is evaporated at the focal spot there will be a hole burned into the target after several shots. That is why the target has to be rotated for a long-term operation. Furthermore one has to prevent degradation of optical lenses and mirrors used for focusing the laser light. Figure 2.21 shows an ion-source scheme which was developed at the *Joint Institute for Nuclear Research* (JINR) in Dubna [2.130,131] for the injection of ions into their Synchrophasotron accelerator. In this device, the expanding plasma is compressed along a magnetic field perpendicular to the target. Ion extraction occurs through a slit in the anode of the extractor electrode assembly. The slitwidth is somewhat larger than the plasma diameter. By optimizing the delay and shape of the extraction pulse relative to the laser pulse the output of highly charged ions can be optimized. Table 2.6 summarizes laser ion sources giving their typical ions together with the kind of application.

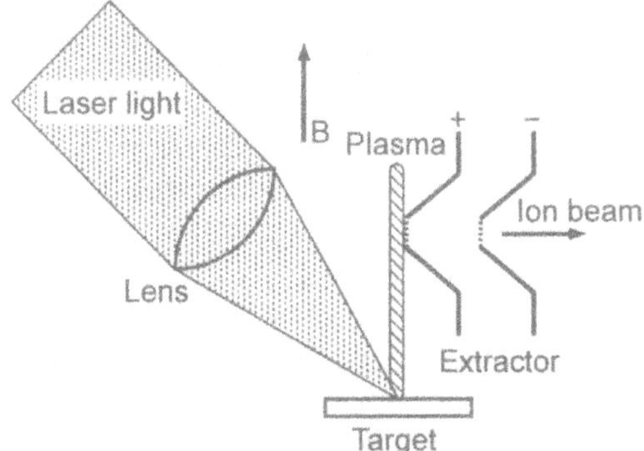

Fig. 2.21. Schematic of a laser ion source with magnetic field perpendicular to the target

Although the use of the laser ion source for ion injection into accelerators was proposed [2.132,133] a long time ago it is not yet routinely used in many laboratories for that application. Renewed interest is coming from the needs [2.134] of high-energy heavy-ion accelerators which require a certain minimum charge-to-mass ratio for the injected ions. Besides the systematic studies on external magnetic fields [2.130], other methods have been explored to increase the rate of highly charged ions. Compensation of the ion space charges has been investigated [2.135] by active electron injection and by applying a radial image-charge field with the help of an axial wire. A two-laser scheme [2.136] has been used to additionally heat the expanding plasma thereby increasing the maximum charge state.

X-Ray Generation in Laser-Produced Plasmas. The laser-produced plasma is a bright x-ray source which has been extensively studied. Conversion of laser light into x rays can be very efficient for both continuum and

Table 2.6. Some laser ion sources in different laboratories. Listed are laser power densities P_{laser}, repetition frequencies f_{laser}, pulse currents I_{max}, ion species and applications for which the sources are developed

Laboratory	P_{laser} (Wcm^{-2})	f_{laser} (Hz)	I_{max} (μA)	τ_{pulse} (μs)	Ion	Application
ITEP Moscow [2.137]	1×10^{10}	0.3	0.006[a]		Si^{4+}	Synchrotron
JINR Dubna	1×10^{9}	1	100	1	Cr^{13+}	
[2.130]	2×10^{9}	25	0.1[a]	1	Li^{2+}	Cyclotron
CERN Geneva [2.138]	3×10^{11}	0.25	5	6	Pb^{26+}	Test bench
TU München	6×10^{8}		1	5	Fe^{4+}	Van de Graaff
[2.139,135]	4×10^{13}		2	5	Ta^{21+}	Test bench
Univ. Arkansas [2.140]	1×10^{11}	50	75	6	C^{3+}	Test bench
GSI Darmstadt [2.141]	1×10^{11}	1	15		Ta^{15+}	Test bench

[a] average current

line radiation. Total conversion efficiencies may exceed 50% whereas transfer into single spectral lines may reach $(1–5)\times10^{-4}$ of the incident laser energy. The investigation of the laser-produced plasma is motivated by:

- inertial confinement fusion
- intense ion beams
- x-ray flash lights
- x-ray lasers.

In the fusion research [2.142] high-power lasers, often with a very short pulse duration, are used as drivers. The indirect drive method for inertial confinement fusion converts incident laser energy to x rays which compress and heat spherical capsules containing fusion fuel. Due to the complexity of the laser-plasma interaction, the interpretation of many diagnostic measurements relies on sophisticated plasma models. There are hydrodynamic simulation codes [2.143,144] including models for laser absorption, energy transport and hydrodynamic expansion and compression. X-ray line radiation, as in many other plasma-diagnostic applications, is used to get information about the rapidly changing plasma environment from which the x rays are emitted. The x-ray diagnostics require a time and spatially resolved measurement with high spectral resolution. Many studies are made in the soft x-ray region as for instance on hydrogen-like aluminum [2.145,146] but also MeV x-ray generation was found [2.147] with a femtosecond laser. Escape of (line) radiation from the plasma can represent an unwanted cooling mechanism for the heating and compression of a fusion target. For soft x-ray lasers [2.148] pumped by recombination, it helps to sustain a population inversion. Resonance-line photons are no longer reabsorbed in the plasma when the velocity gradients [2.149] in the expanding plasma are high enough to produce large Doppler shifts.

2.2 Heavy-Ion Accelerators

From the very beginning, nuclear and high-energy physics have been the driving forces for the development of accelerators. Accelerator development, on the other hand, is closely linked to technological progress in other fields like high-vacuum, radio-frequency or low-temperature techniques. The small laboratory devices used in the early experiments have been turned into large-scale facilities serving large communities. The x-ray radiation discovered in 1895 by Wilhelm Conrad Röntgen is a prominent example for the production of secondary radiation. Today, the use of synchrotron light is a rapidly expanding field. Atomic physics with particle beams can be traced back to the famous experiments in 1913 by James Franck and Gustav Hertz, who investigated atomic excitation by electron impact. More recently, atomic spectroscopy at particle accelerators has become an important technique since it is now possible to produce any desired ionization stage of virtually any element of the periodic table. Accelerator physics and technology, e.g., [2.150–153] in themselves constitute a mature field of research in their own right.

2.2.1 Acceleration of Charged Particles

A particle *beam* is a flow of a continuous stream or a bunch of particles that move along a straight or curved path defined as the *longitudinal* direction. The transverse velocity components and the spread in longitudinal velocities are generally small compared to the mean longitudinal velocity of the beam. Examples are the *straight* beams in linear accelerators and the *curved* beams in circular accelerators like betatrons, cyclotrons and synchrotrons.

Acceleration Mechanisms. Direct-voltage accelerators employ a *DC acceleration* in a static electric field as maintained in an accelerating tube. Connected to the accelerating tube are high-voltage devices of which the famous *Cockcroft Walton* [2.154] generator (up to 800 kV) or the *Van-de-Graaff* [2.155] generator (up to 1.5 MV) have been in use for a long time. However, a quite restrictive limitation was imposed by the maximum voltage of these devices. An alternative had been proposed already in 1924 by

Ising [2.156] who planned to repeatedly apply the same voltage to the particle using alternating fields. The beam in this case consists of short bunches with a pulse length that is small compared to the wavelength of the radio frequency (rf). This *resonant acceleration* was to become the principle of all modern high-energy accelerators. Figure 2.22 shows the principle of the rf

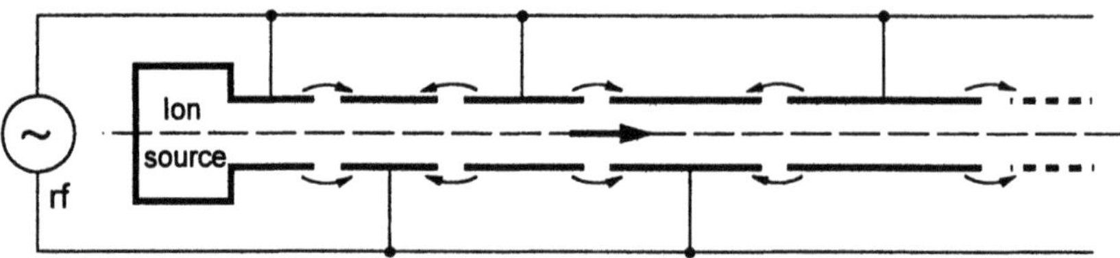

Fig. 2.22. Schematic of a Wideroe-type RF linac. Whenever the bunch of particles enters the gap between the drift tubes a positive field accelerates the particles

linear accelerator in which the beam bunches are in phase with the positive electric field between the drift tubes. When the same rf frequency is used for all accelerating gaps the length of the drift tubes have to increase with the particle velocity. The accelerating structures may be in separate tanks as first used by *Wideroe* [2.157], or in a common resonant vacuum tank such that the field has the same phase in all gaps. The latter scheme was invented by *Alvarez* in 1946 [2.158] and is still in use for protons and heavy ions. The operating frequency is in the order of 100–200 MHz and energies of up to 200 MeV are reached. For relativistic particles ($\beta \approx 1$) the velocity is nearly constant as is the length of drift tubes. The high velocities require rf power at high frequency, typically at 3 GHz, and accelerating cavities in which the rf phase velocity matches the particle velocity.

In circular accelerators, repeated use of the same rf source is an elemental feature. The fixed-frequency cyclotron conceived by *Lawrence* in 1929 [2.159] makes use of a constant guiding magnetic field that keeps the particles on a spiral path during acceleration. Acceleration is accomplished by an rf electric field applied across the small gap between two *D*-shaped electrodes. Cyclotrons are easy to operate but are limited in energy because of unacceptably large magnet poles and because of relativistic effects at very high velocities.

In the *betatron acceleration* due to R. Wideroe the beam effectively forms the secondary winding of a transformer. By the changing magnetic field a circumferential electric field is induced which accelerates the particles. The circulating particles are held on a constant orbit by increasing the magnetic field in the vicinity of the beam according to the increase in particle energy. It can be easily shown that the inductive field has to increase at twice the rate of the guiding field. The betatron has the advantage of being robust and simple. The one active element is the power converter driving the large inductive load of the main magnet.

In the *synchrotron* [2.160,161] the guiding magnetic field increases with particle velocity in order to maintain a stationary orbit as in the betatron. Acceleration, however, is accomplished by an rf voltage fed to one or more cavities operated at the revolution frequency or at a higher harmonic. Many high-energy machines around the world are of this type including the 500 GeV Super Proton Synchrotron (SPS) at the Conseil Européen pour la Recherche Nucléaire (CERN).

Focusing Mechanisms. In the early circular accelerators transverse stability was achieved by slightly decreasing the guiding magnetic field with increasing radius. The field gradient was constant around the circumference of the machine and the scheme is known as *weak* or *constant-gradient focusing.* Severe limitations, especially at large machines, arise from the small tolerances of the magnetic field and the large apertures required.

Practically all modern synchrotrons and storage rings use magnetic elements that bend the particles and that focus them by alternating between strong positive and negative gradients. This is known as *strong* or *alternating-gradient focusing* and has an analogy in geometrical optics where it is also possible to achieve an overall focusing by combining a sequence of concave and convex lenses. In accelerator design, magnetic quadrupole lenses provide horizontal focusing and vertical defocusing or vice versa. An overall focusing is achieved by combining them to doublets or triplets. The magnetic elements are periodically arranged around the circumference forming a *lattice.* If the focusing magnets are not superimposed on the bending magnets the lattice is known as a *separated-function lattice.* The periodic arrangement facilitates the modular construction of large accelerators and also their theoretical description. The equation for the particle motion through the segments of an accelerator may be approximated [2.152] by

$$\frac{\mathrm{d}^2 y}{\mathrm{d}s^2} + K(s)y = \frac{1}{\rho}\frac{\Delta p}{p_\mathrm{o}} . \tag{2.26}$$

Here, y and s denote the transverse and longitudinal coordinates as defined in Fig. 2.23 and y can be replaced by the horizontal or vertical coordinate, y_h or y_v, respectively. The focusing strength $K(s)$ depends on whether a particular segment is a drift space, a quadrupole lens, a bending magnet or a combined-function element. The quantity ρ denotes the bending radius, which for simplicity is assumed constant and Δp denotes the momentum deviation from the mean particle momentum p_o.

Emittance. A general solution of (2.26) may be written as

$$y(s) = C(s)y_\mathrm{o} + S(s)y'_\mathrm{o} + D(s)\frac{\Delta p}{p_\mathrm{o}} . \tag{2.27}$$

The functions C, S and D (Cosine-like, Sine-like, Dispersion) are called the principal trajectories. For $\Delta p = 0$ a real solution of (2.26) is

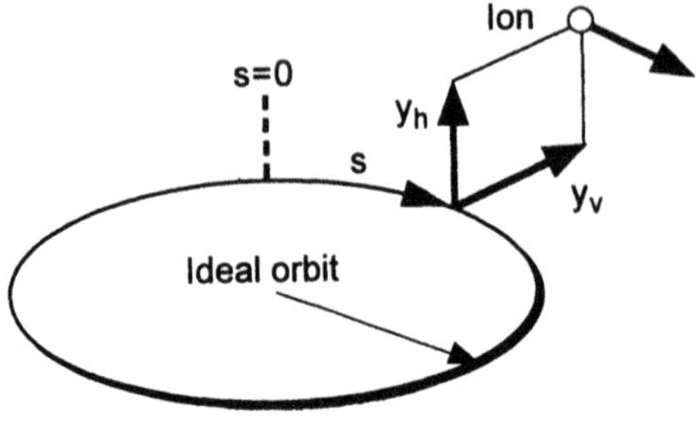

Fig. 2.23. Transverse (y_h and y_v) and longitudinal (s) coordinates in a circular accelerator referenced against the ideal orbit of an equilibrium particle

$$
\begin{aligned}
y(s) &= a_y\sqrt{\beta(s)}\,\cos(\psi(s)+\delta)\,,\\
y'(s) &= -\frac{a_y}{\sqrt{\beta(s)}}\left[\sin(\psi(s)+\delta)-\frac{1}{2}\beta'(s)\cos(\psi(s)+\delta)\right],
\end{aligned}
\tag{2.28}
$$

where the prime denotes the derivative with respect to the longitudinal coordinate s. Equations (2.28) describe *betatron oscillations* with a betatron amplitude $a_y\sqrt{\beta(s)}$ and with a betatron phase $\psi(s)=\int \mathrm{d}s/\beta(s)$. The quantity $\beta(s)$ is called the *beta function* and is an intrinsic property of the lattice. Equations (2.28) are a parametric representation of an ellipse in the y-y' plane which is illustrated in Fig. 2.24 giving also the maximum values of y and of y' in terms of the *emittance* defined as $\varepsilon=a_y^2$. The phase-space ellipse may be viewed either for a single-particle motion along the circumference or, at a fixed position, for many particles with different phases. Because the particle density has some distribution in y and y' it is convenient to take only particles with an amplitude corresponding to one standard deviation of this distribution. The quantity $\beta\gamma\varepsilon$, often referred to as the *normalized emittance*, is an invariant.

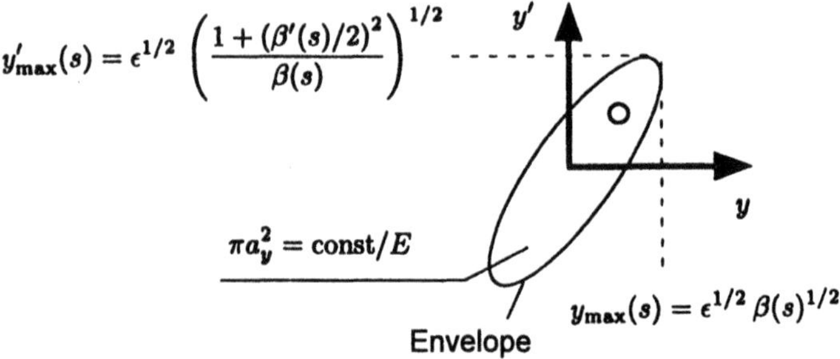

Fig. 2.24. Phase-space ellipse defined in (2.28)

The betatron wavelength or tune defined as

$$
Q_{\mathrm{h,v}}=\frac{1}{2\pi}\oint\frac{\mathrm{d}s}{\beta(s)}
\tag{2.29}
$$

is an important operating parameter. Because of unavoidable imperfections in the manufacturing of the lattice, disturbances occurring at a position along the ring could periodically build up if the ions encounter one of the betatron resonances

$$k\,Q_h + \ell\,Q_v = m \quad \text{with } k, \ell, m = 0, \pm 1, \pm 2, \ldots \tag{2.30}$$

For stable operation, especially the lower resonances have to be avoided. Tune shifts of the betatron oscillation in circular accelerators caused by the space charge of the beam particles is of fundamental importance for the achievable intensities in such machines.

Momentum Compaction. Particles with a deviation Δp from the mean momentum follow a dispersed orbit with a displacement $\Delta y = D(s)\Delta p/p_o$ given by the dispersion function $D(s)$. Changes in the momentum lead to different paths around the ring and to different revolution frequencies. The fractional frequency change is related to the fractional momentum change through

$$\frac{\Delta f}{f} = \left(\frac{1}{\gamma^2} - \frac{1}{\gamma_t^2}\right)\frac{\Delta p}{p_o}\,. \tag{2.31}$$

The first term within the brackets accounts for the relativistic velocity change with change of momentum. The second term, known as *momentum-compaction*, is the fractional change of the orbit length divided by the fractional change of the momentum. The quantity γ_t is known as the *transition gamma* and $\gamma_t m_i c_o^2$ as the *transition energy.* From (2.31) one finds that the frequency increases with momentum if the particle energy remains below the transition energy ($\gamma < \gamma_t$). The transition point $\gamma = \gamma_t$ is very critical for momentum focusing, especially in synchrotrons.

Phase Stability. Since the bending and focusing forces depend on the particle momentum there is also a tune shift related to a momentum shift

$$\Delta Q = \xi\,\frac{\Delta p}{p_o}\,. \tag{2.32}$$

The factor ξ is called the *chromaticity* and is, like the dispersion, an intrinsic lattice parameter.

The principle of momentum focusing or bunching by an rf cavity operated at a harmonic frequency of the revolution frequency is shown in Fig. 2.25. Below the transition point, an increase in energy leads to an increase in revolution frequency. A particle lagging in energy with respect to the equilibrium particle will be traveling more slowly than the accelerating wave $F(\varphi)$. If the rf phase is tuned such (as in Fig. 2.25) that by lagging the particle rides up the wave, then the particle can catch up the equilibrium particle which has the phase φ_o. Above the transition energy, the situation is reversed and the lagging particle has too much energy. Therefore, one has to jump the rf phase by an amount of $2\varphi_o$ when crossing the transition energy.

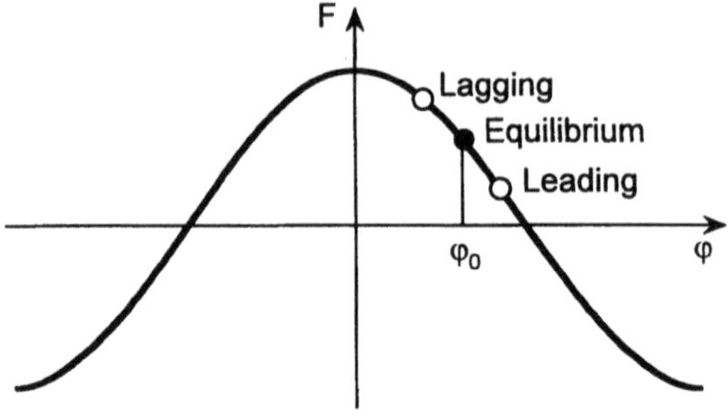

Fig. 2.25. Phase stability below the transition energy is secured by accelerating lagging and decelerating leading particles relative to the equilibrium particle which is in phase with the accelerating field $F(\varphi)$

2.2.2 Accelerator Laboratories for Heavy Ions

There is a considerable number of large accelerators [2.162] around the world operated mainly for high-energy physics. The accelerated particles in most cases are electrons, protons and their antiparticles. Some laboratories provide heavy ions either for their main field of research or by an upgrade of existing accelerators initially designed for light ions. Table 2.7 gives an overview of several such laboratories with emphasis on high atomic numbers and energies that are high enough to strip the ions to high ionic charges. The information given is the name and location of the facility, the type of accelerator and a typical ion beam characterized by the ionic species, by its specific energy in MeV/u and by its beam intensity. The latter is given as the total number of stored particles (p), or in particles per second (pps). Because of the continued progress in accelerator technologies and various upgrading programs going on in the laboratories, ion numbers and energies tend to increase with time. Some of the high-energy laboratories such as BNL [2.163] at Brookhaven and CERN [2.164] are upgrading accelerators from protons to heavy ions. Online documentation is available for most of the facilities and can be retrieved by starting for instance at the World-Wide-Web site maintained at the University of Bonn [2.165].

As an example we show, in Fig. 2.26, the layout of the heavy-ion accelerator facilities [2.166,167] at GSI Darmstadt. The linear accelerator UNILAC has been in operation since 1975 and provides ion beams of practically all elements between Ne and U with specific energies up to 20 MeV/u. The UNILAC can run with a Penning or MEVVA ion source and a Wideröe structure or with a high-charge state injector consisting of an ECR source and an RFQ/IH injector linac. The UNILAC serves as an injector for the heavy-ion synchrotron SIS having a circumference of 216 m. In the SIS the heavy ions are accelerated from 11.7 MeV/u up to a maximum of 2 GeV/u. Space-charge limits of this machine are near 2×10^{11} neon and 4×10^{10} uranium ions per cycle. Connected to the SIS is the Fragment Separator (FRS), for production and separation of radioactive isotopes, and the Experimental Storage Ring ESR. The ESR accepts ions of up to 830 MeV/u Ne^{10+} or 560 MeV/u U^{92+}. In the ESR the ions can be accumulated, cooled and decelerated. Experi-

Table 2.7. A Selection of Accelerator Laboratories providing highly stripped heavy ions

Facility	Accelerator Type	Typical Ion Beam		
		Species	Mev/u	p (pps)
ANL Argonne, USA				
ATLAS	S.C. Tandem	Kr^{34+}	10	1×10^{10}
BNL Brookhaven, USA				
Booster	Synchrotron	Au^{33+}	192	6×10^{8}
AGS	Synchrotron	Au^{77+}	10 200	3×10^{8}
RHIC[a]	Collider	Au^{79+}	100 000	3×10^{9}
CEN Saclay, F				
SATURNE	Synchrotron	Kr^{36+}	1 150	1×10^{6}
CERN Genève, CH				
PS[b]	Synchrotron	Pb^{53+}	4 250	
SPS[b]	Synchrotron	Pb^{82+}	160 000	1×10^{8}
GANIL Caen, F				
CSS1	Cyclotron	U^{24+}	3.7	2×10^{11}
CSS1	Cyclotron	U^{59+}	24	6×10^{10}
GSI Darmstadt, D				
UNILAC	Linac	U^{70+}	18	1×10^{11}
SIS	Synchrotron	U^{92+}	1 000	1×10^{7}
ITEP Moscow, Ru				
Synchrotron	Synchrotron	Si^{4+}	3 500	3×10^{9}
JINR Dubna, Ru				
U-400M	Cyclotron	Ar^{5+}	10	1×10^{13}

[a] under construction, to be completed by 1999
[b] design parameters for heavy-ion upgrade

ments are conducted with internal beams in the ESR and with ion beams extracted directly from SIS or from ESR. The GSI accelerator facility serves a broad range of experiments including nuclear physics, nuclear chemistry, atomic physics, radiation biology and material science.

2.2.3 Ion Stripping and Charge States

The distribution of charge states for projectiles traversing solid or gaseous targets is of fundamental interest for designing accelerators and for planning experiments. From measured charge-state distributions one may also gain information on various charge-exchange processes in the stripper target. Technologically important is the economy of ion acceleration and, related to that, the attainable beam energies and intensities. The evolution of charge states inside the target is characterized by a large number of collisions in which the projectile may loose or gain electrons. This process can be modeled by an ionization balance similar to the successive ionization in ion sources as

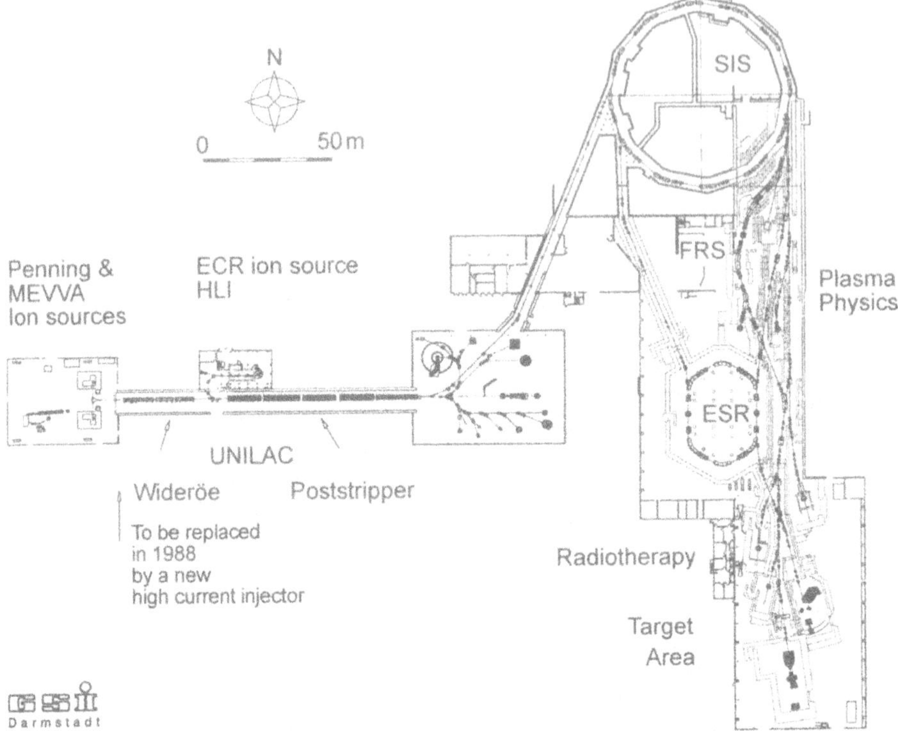

Fig. 2.26. Planview of the heavy-ion accelerator facilities at GSI Darmstadt. The main components are the linear accelerator UNILAC, the synchrotron SIS and the storage ring ESR. Ion beams can be used for experiments in the low-energy experimental area located at the end of the UNILAC or in the target area supplied with high-energy beams from SIS/ESR [2.166]

analyzed in Sect. 2.1.2. The physical processes to consider can be classified as:

- ionization
- excitation
- radiative decay
- Auger decay
- electron capture
- radiative electron capture.

For ions with *many* electrons, all processes mentioned are important and a large number of different cross sections have to be considered. In addition, the number of ground and excited states which contribute can be excessively large precluding a precise theoretical prediction of ion populations. Because of many small uncertainties in the cross sections it is also difficult to explain the greater effectiveness of solid stripper foils over gas strippers [2.168–170]. For practical purposes, empirical formulae have been developed, for instance those [2.171] used for accelerator tuning at GANIL at energies between 3.8 and 10.6 MeV/u.

The situation changes when very heavy ions are considered and the beam velocities are high enough to produce substantial amounts of bare ions. For uranium at energies above about 0.5 GeV/u it may be sufficient to include only ions with zero, one and two electrons. Because of the high decay rates at high nuclear charge, excited ions can decay to their ground state in a time which is short when compared with the mean time between collisions. In this situation, the number of compulsory states is small and only a small number of different processes have to be considered. Therefore, the precision in theoretical models is increased and also the interpretation of measured charge-state distributions becomes more reliable. Such investigations were carried out at the BEVALAC at the Lawrence Berkeley National Laboratory (LBNL) [2.172] and at GSI [2.173–175]. Figure 2.27 shows the equilibrium charge-state distribution of uranium ions emerging from a solid copper foil as a function of the projectile energy where the incident ions were completely stripped. The target was thick enough (equilibrium thickness) so that a change in the target thickness no longer changes the emerging charge state distribution. The measured data are compared to theoretical calculations based on the ground-state model [2.172] which assumes that all excited states formed, instantaneously decay to their ground state. The shape of the curves is determined by the energy dependence of the radiative and of the non-radiative electron capture. The measurement of the x rays emitted from the excited projectiles and of the photons originating from radiative electron capture provide a means to determine the corresponding cross sections under rather clean experimental conditions [2.175].

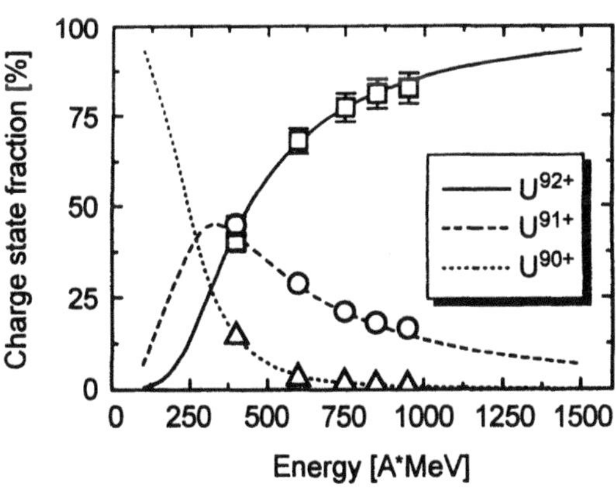

Fig. 2.27. The equilibrium charge-state distributions of a uranium beam after penetrating a copper target as a function of the incident energy. The measured data (symbols) are compared to theoretical predictions (lines) [2.173] © 1994 Elsevier

The charge state inside the target also determines the slowing down of the projectile governed by the energy transfer to the target electrons. With Xe ions that are fully ionized during their penetration of the target, it has been possible to investigate [2.176] the details of the energy loss of relativistic ions in solids.

2.2.4 Accelerator-Based X-Ray Sources

Fast-Beam Light Source. When a fast beam of ions traverses a solid target, fluorescence is excited in a wide spectral range. The measurement of the corresponding spectra, referred to as *Beam-Foil Spectroscopy* [2.177], has contributed to our present knowledge of basic atomic physics. Three important properties of the beam-foil source have been exploited. First, numerous excited states of various charge states are populated. Secondly, the time of excitation is well defined relative to the lifetimes of many excited states. This enables lifetimes to be measured directly by time-of-flight techniques. The third property refers to the source geometry which has only cylinder symmetry when the beam axis coincides with the surface normal of the foil. Therefore, the ions are generally produced in an anisotropic state. The light emitted through the subsequent decay exhibits the anisotropy through the angular distribution and the polarization of the light [2.178].

The light emitted by the beam reveals a rich spectrum of lines ranging from the optical to the x-ray region. The identification of numerous lines and the study of the corresponding level structure represents one of the main applications of beam-foil spectroscopy. Extended tabulations of level energies and transition wavelengths [2.179] of highly charged ions resulted from this work.

The methods of beam-foil spectroscopy have been widely accepted initially at small accelerators with severe limitations in the charge states accessible. With the development of powerful heavy-ion accelerators, it is now possible to completely strip even the heaviest elements and to study the high-Z few-electron systems. For the heavy ions a large fraction of the radiative decays proceed via x-ray emission.

Two experimental difficulties have to be faced when aiming at a precise spectroscopy. First, spectator electrons with high principal and angular-momentum quantum numbers may be present in the ion under study. The spectator electrons lead to small line broadenings and shifts that normally cannot be resolved spectroscopically. Secondly, the radiation is emitted in flight from ions having velocities that are not small compared to the velocity of light. This leads to severe Doppler shifts and Doppler uncertainties.

Population of spectator electrons can be avoided [2.180] when the excited state is produced by single-electron capture from a thin gas target of light atoms. In addition, the charge state of the ion can be magnetically analyzed after the photon emission as schematically illustrated in Fig. 2.28. By a delayed coincidence between the x ray and the ion the spectrum is cleaned up from unwanted ionic states.

The Doppler effect can be brought under control [2.181] by replication of the wavelength measurement at several observation angles and at several

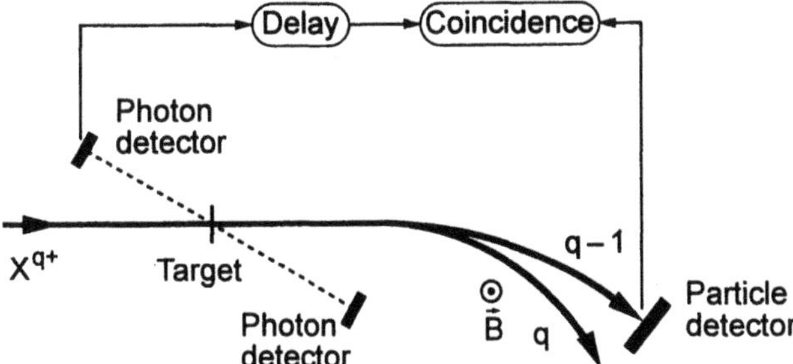

Fig. 2.28. Typical geometry for the measurement of x rays emitted from fast ions

beam velocities. Deceleration of the stripped ion beam in linear or circular accelerators reduces the size of the Doppler corrections considerably. Imaging properties of x-ray spectrometers may also be turned to advantage if carefully adopted to the emission characteristic of the fast-beam source.

Recoil-Ion Light Source. A different approach [2.182,183] to the study of highly charged ions is the bombardment of a lighter target gas with very highly charged heavy projectiles. It is possible to ionize the target atoms already at large impact parameters where the heavy projectiles do not impart a high momentum to the target ions produced. In this recoil-ion light source the Doppler effects are very small allowing, in principle, a very high spectral resolution. Detection efficiencies of high-resolution spectrometers can be limiting. Unless the recoil ions are extracted from the target zone, there is a mixture of all excitation and ionization states produced in a common small target volume. Again, population of high-lying spectator-electron states can be a limitation. In the past, the spectral resolution of x-ray spectrometers used where not sufficient to completely resolve these influences.

2.3 Storage Rings

2.3.1 Overview

Storage rings can provide high collision energies and high luminosities for the study of rare events in colliding beams including proton-antiproton ($p\bar{p}$) and electron-positron (e^-e^+) colliders. Antiparticle beams, produced by bombarding special targets with primary beams of high energy, have inherently large emittances and momentum spreads. It was a precondition to reduce the phase space volumes of antiproton beams before their successful use in high-energy collision experiments. This led to the invention of *electron cooling* by *Budker* [2.184] at Novosibirsk and of *stochastic cooling* by *van der Meer* and coworkers [2.185] at CERN. Stochastic cooling was an essential tool in the discovery of the vector bosons W and Z [2.186].

The pioneering developments, in particular those at the proton storage ring NAPM at Novosibirsk [2.187] and at the Low-Energy Antiproton Ring LEAR at CERN [2.188], inspired many laboratories to build their own storage rings for various demands. Several small rings [2.189,190] have become operational for nuclear and atomic physics [2.191,192] with heavy ions. Table 2.8 lists a selection of small storage rings which provide heavy ions and gives their main operation parameters.

Table 2.8. Operating heavy-ion storage rings and their main parameters

Name	**CRYRING**	**TSR**	**ASTRID**	**LEAR**	**ESR**
Institute	MSI	MPI	ISA	CERN	GSI
Location	Stockholm S	Heidelberg D	Århus DK	Genève CH	Darmstadt D
C[m]	48.6	55	40	76.6	108.4
$B\rho$[Tm]	1.4	1.5	1.93	6.6	10
B_{max}[T]	1.1	1.3	1.6	1.6	1.6
Injector	CRYEBIS + RFQ	Tandem or RFQ + Linac	Isotope Separator	Linac	UNILAC + SIS
Injection Energy	0.3 MeV	0.5–15 MeV/u	200 keV	4.2–180 MeV/c	50–830 MeV/u
Ion Mass A	20–208	12–130	4–238	16–208	20–238
q/A	0.5–0.3	0.5–0.37	0.5–0.1	0.5–0.4	0.5–0.39
U_{cool}[kV]	0.2–13	0.5–16	0.5–3	1–30	2–300
Stochastic Cooling at	-	-	-	2–1270 MeV/c	500 MeV/c
Reference	[2.193–195]	[2.196–199]	[2.200–203]	[2.204]	[2.205,167]

These storage rings are versatile instruments for atomic and nuclear physics. The main purpose of these machines is the accumulation of a high ion current and the phase-space cooling providing high-quality ion beams for experiments. Therefore, they include all capabilities of synchrotrons plus the devices needed for in-ring experiments. As an example, the lattice of the experimental storage ring ESR is shown in Fig. 2.29. Besides the dipole magnets and focusing elements there are indicated rf cavities for beam acceleration or deceleration and, at the two long straight sections comprising the electron cooler and the internal gas jet, respectively. The range of ions available for experiments depends on the acceptable charge-to-mass ratio according to the magnetic rigidity of the bending magnets and on the kind of ions delivered by the injector coupled to the storage ring.

Efficient accumulation of ions is assisted by the cooling techniques. The horizontal phase space of the ring can be filled by multiturn injection sweeping the injected beam. To reach high phase-space densities the rf stacking method can be combined [2.206] with electron cooling. By switching the cooler voltage

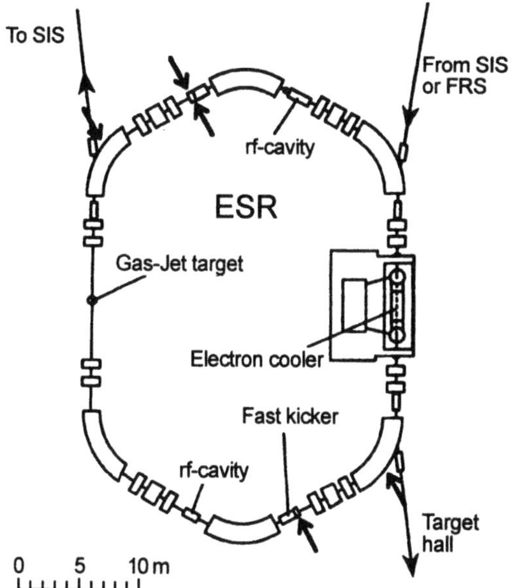

Fig. 2.29. Layout of the Experimental Storage Ring (ESR) at GSI

the ions can be cooled both on their injection orbit and on their stack orbit. Another technique called ECOOL *stacking* [2.207] makes use of the drag force of the slightly detuned electron beam which pulls the ions into the stack.

There are several effects limiting the storage time of heavy ions and the total number of ions that can be accumulated. Beam losses are due to collisions with the residual gas with the gas jet and with the electrons in the electron cooler, to intra-beam scattering as well as space-charge and microwave instabilities. Without gas target and with an ultra-high vacuum with pressures typically near 10^{-9} Pa, storage times can reach several hours. Because the probability for electron capture drastically increases at low velocities, decelerated highly charged ions can be lost within a minute or less in experiments using a gas jet. When the stacking rate is balanced by the total loss rate the limit in the number of stored ions is approached. Actual ion numbers vary between 4×10^{10} O^{8+} ions at 6 MeV/u accumulated in the TSR and 1×10^{8} U^{92+} ions at 240 MeV/u in the ESR.

2.3.2 Beam Cooling

Electron Cooling. When a low-temperature electron beam is superimposed on a hot ion beam and both particle streams travel with the same average velocity, $\langle \boldsymbol{v}_{\mathrm{i,lab}} \rangle = \langle \boldsymbol{v}_{\mathrm{e,lab}} \rangle$, Coulomb collisions between ions and electrons will lead to temperature relaxation. The ion beam is cooled down by heating up the electron beam until thermal equilibrium, $T_{\mathrm{i}} = T_{\mathrm{e}}$, is reached. There are several reviews [2.208–212] covering theory and techniques of electron cooling. In the storage ring, the electron beam interacts with the ion beam only over a small fraction of the ring circumference typically in the order of $\eta \approx 0.01$. During each traversal through the cooling section, the ion beam imparts only a small fraction of its thermal energy to the electron beam which is constantly refreshed. Because of the high revolution frequency of about 1 MHz, efficient relaxation is obtained with time constants of the order of 0.1 s. The electron beam is produced in an electron gun with a thermionic cathode. It therefore has a transverse temperature in the order of $T_{e\perp} \approx 0.1$ eV whereas the longitudinal temperature $T_{e\parallel}$ is much smaller due to the acceleration

$$T_{e\parallel} = \frac{T_{e\perp}^2}{2\beta^2\gamma^2 m_e c^2} \ll T_{e\perp} \,. \tag{2.33}$$

The electron beam is guided by a longitudinal magnetic field usually in the order of 0.1 T. As an example we show in Fig. 2.30 the design of the ESR electron cooler.

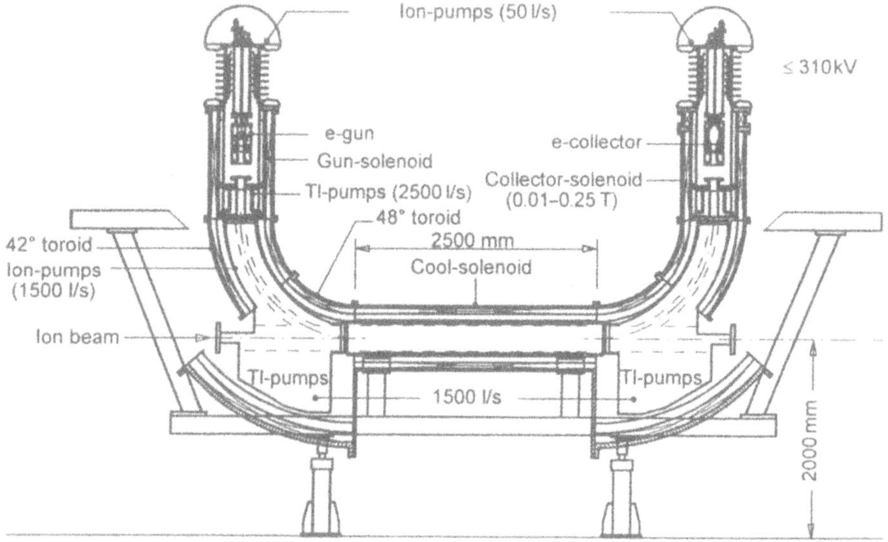

Fig. 2.30. The electron cooler of the ESR designed for voltages up to 300 kV. Electrons extracted from the gun are guided by the magnetic field produced in solenoidal and toroidal coils. The length of the straight cooling section is 2.5 m

In the coordinate system moving with the average particle velocity the electron velocity has an anisotropic Maxwell distribution

$$f(\boldsymbol{v}_e) = \frac{m_e}{2\pi T_{e\perp}} \left(\frac{m_e}{2\pi T_{e\parallel}} \right)^{1/2} \exp\left(-\frac{m_e v_{e\perp}^2}{2T_{e\perp}} - \frac{m_e v_{e\parallel}^2}{2T_{e\parallel}} \right) \,. \tag{2.34}$$

For cooling and beam manipulation the cooling or drag force is an important quantity. It can be estimated by considering the stopping of a charged particle in a cold electron gas expressed as a *Bethe-Bloch* [2.213] equation

$$F(v_i) = -4\pi \frac{q^2 e^4 n_e}{m_e} L_c \int f(v_e) \frac{v_i - v_e}{|v_i - v_e|^3} \, d^3 v_e \,, \tag{2.35}$$

where L_c is the Coulomb logarithm defined as

$$L_c = \log \left(\frac{b_{max}}{b_{min}} \right) \tag{2.36}$$

in terms of the minimum and maximum impact parameter, b_{min} and b_{max}, respectively. It originates from an integration of the momentum transfer within the limits of b_{min} and b_{max} where the lower limit usually is taken as $b_{min} = qe^2/m|v_i - v_e|^2$ and the upper limit is identified with the Debye shielding length $b_{max} = \lambda_D$ as defined in (2.12).

The strong guiding magnetic field confines the electrons on helical orbits with a small Larmor radius ρ given by (2.13). For slow collisions with impact parameters $b > \rho$ the cooling effect is much enhanced because the transverse electron motion is frozen. This is known as the *magnetized* electron cooling effect. From expression (2.35) a quadratic increase of the cooling force with the ion charge q is noted and a suitable integration over the velocity distribution yields [2.214] a longitudinal cooling force scaling like

$$F_{\|} \propto q^2 \begin{cases} v_{i\|}, & v_{i\|} < v_{e\|} \\ v_i^{-2}, & v_i > v_{e\perp} \end{cases} \,. \tag{2.37}$$

In the theoretical estimate, it is tacitly assumed that the ion introduces only a small perturbation to the local electron velocity distribution which might no longer be the case if highly charged ions are to be cooled. Noting that the Larmor radius ρ can be much smaller than the Debye shielding length λ_D suggests a reduction of the magnetized cooling effect in slow collisions. Theoretical considerations [2.215,216] and experimental results [2.217] support a charge dependence close to $q^{3/2}$ instead of q^2. In Fig. 2.31 the longitudinal drag force measured [2.217,218] for heavy ions as a function of the relative velocity is shown. At low velocities the measurements were performed by observing the response of the ion beam after a stochastical heating with a set of electrodes. At higher velocities the drag force was measured after the cooler voltage has been detuned by a preset amount. In practice the cooling process is in equilibrium with a number of heating mechanisms most important intra-beam scattering [2.219] and scattering with residual and eventually with target-gas atoms. Figure 2.32 shows the measured [2.217] momentum spread of highly charged heavy ions which have been stored and cooled in the ESR. The observed dependence on the number of stored ions is approximately $\propto N_i^{1/3}$ characteristic for the balance of intra-beam scattering and electron cooling.

Limitations of the electron cooling efficiency and of the associated resolution in experiments due to the transverse electron temperature have been much reduced by an adiabatic acceleration and transverse expansion of the electron beam. Because $T_{e\perp}/B$ is an invariant if B changes adiabatically, a reduction of $T_{e\perp}$ has been achieved by slowly decreasing the magnetic field

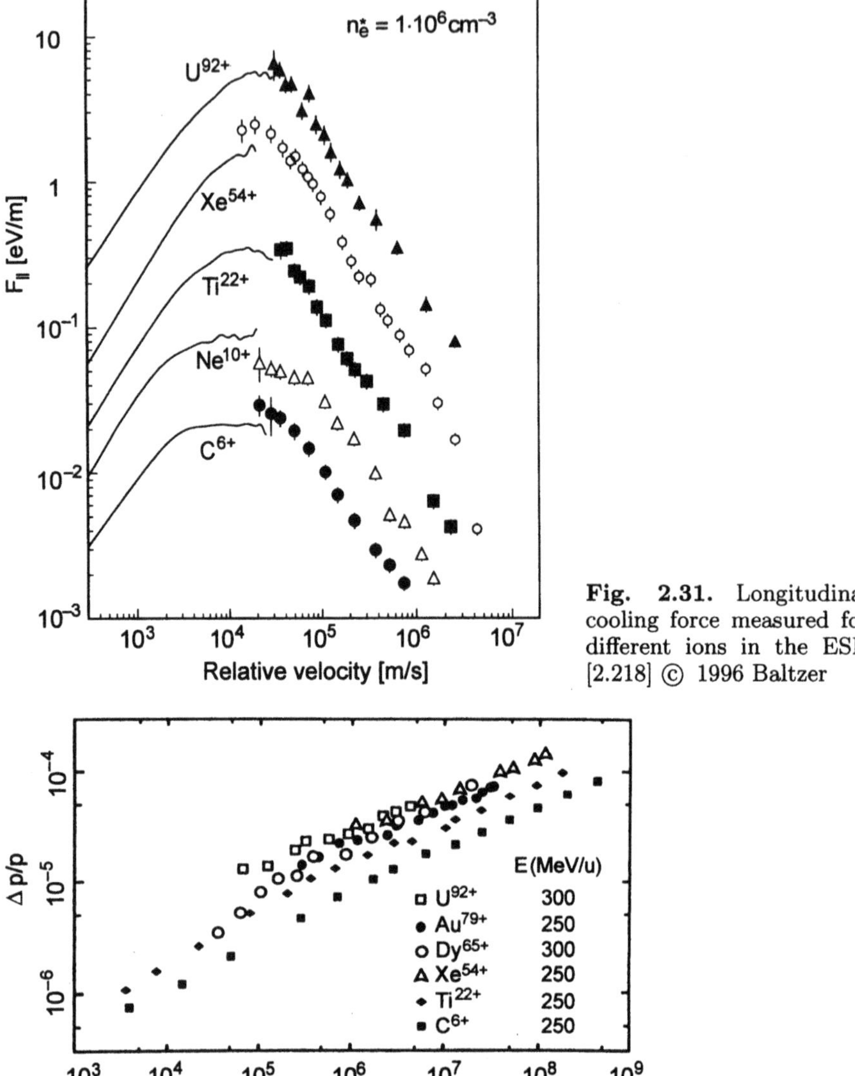

Fig. 2.31. Longitudinal cooling force measured for different ions in the ESR [2.218] © 1996 Baltzer

Fig. 2.32. Relative momentum spread of various ion species, stored and cooled in the ESR, as a function of the number of ions [2.217] © 1994 World Scientific

along the electrons path from the electron gun to the cooling section. The effect has been demonstrated at the CRYRING with D^+, D_2^+ and H_3^+ ions [2.195,220] and at the TSR with Se^{23+} ions [2.221] reducing the transverse electron temperature down to 10 meV and 19 meV, respectively. Another approach avoids a hot cathode altogether by making use of a cold photocathode [2.222]. A further concern is the drift velocity induced by the space charge of the dense electron beam. It can be reduced by a controlled trapping and neutralization [2.223] by residual ions.

Stochastic Cooling. For completeness we give a short account on stochastic cooling which is a complementary method to electron cooling as it is effective for rather hot ion beams where the electron-cooling force decreases. The principle of transverse stochastic cooling is illustrated in Fig. 2.33. It is based on the idea that it might be possible to detect the statistical fluctuation of the center of gravity of the beam and then to amplify this signal and apply it with the appropriate phase to a corrector some way downstream in the ring. After some time, the feedback loop between pickup and corrector should reduce the statistical fluctuations and increase the beam density.

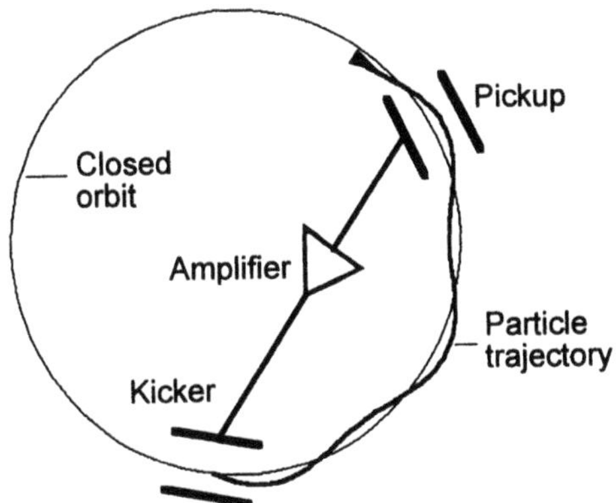

Fig. 2.33. Principle of stochastic cooling. The displacement of a sample of ions from the ideal orbit is detected by the pick-up electrodes. This signal is amplified and used as a correction signal in the kicker located an odd number of quarter betatron wavelengths downstream from the pick-up

A single particle in the storage ring with a revolution frequency f_o represents an electric current which is an infinite (Fourier) series of lines at multiples of the revolution frequency. The root-mean-square current of a beam of particles having some random phase is known as the *Schottky* current. Because there is also some distribution of the revolution frequency, Schottky signals arrange themselves into Schottky bands around the average revolution frequency. The total power per band for a beam of N_i particles is $P_\mathrm{s} = 2N_\mathrm{i}q^2e^2f_\mathrm{o}^2$ and increases with the square of the ion charge state q. Schottky signals can be observed with a pair of electrodes as indicated in Fig. 2.33. The beam current is measured by adding the two signals on the pickups. When the two pickup signals are subtracted the displacement times the current is obtained. Schottky measurements represent a very powerful and convenient diagnostic tool at a storage ring.

For the purpose of stochastic cooling, it is important to realize a small sampling time and a low thermal noise. Because of the statistical fluctuations in the beam, the sample of ions measured in the pickups is not conserved after many revolutions and a mixing with other beam particles occurs. There is an unwanted (but unavoidable) mixing on the path between pickup and kicker and a wanted mixing between kicker and pickup. Analysis of the coherent cooling and incoherent heating mechanisms [2.224] which occur, yields for the cooling rate, at an optimum setting of the amplifier gain,

$$\frac{1}{\tau_s} = \frac{2W_s}{N_i(M_s + U_s)}, \tag{2.38}$$

where M_s and U_s denote the mixing and thermal noise factor and W_s the bandwidth of the cooling system. The rate decreases with increasing number of ions. For highly charged ions the relatively high signal levels are helpful in making U_s small. Stochastic cooling can also be applied for the longitudinal phase space. Momentum deviations are detected by differences in the revolution frequencies and off-momentum particles will receive a longitudinal kick by an appropriate sensing and feedback system. For a more detailed discussion we refer to several reviews [2.224–228] on the subject. At the ESR, stochastic cooling will be used [2.167] to precool radioactive beams from the fragment separator on the injection orbit followed by electron cooling on the stack orbit.

Laser Cooling. The interaction of laser light with ions and atoms provides a unique possibility for phase-space cooling and the technique has been applied for a long time to ions in electromagnetic traps [2.229,230] and to thermal atoms [2.231,232] which even could be captured into laser traps. In a storage ring, one can make use of the spontaneous force [2.197,233–235] schematically illustrated in Fig. 2.34. A laser beam with a narrow bandwidth is superimposed on the ion beam in a straight section of the storage ring and the Doppler-shifted laser frequency is tuned near a resonance of a two-level scheme which does not have any branchings to other levels. The ion will absorb a photon which spontaneously will be reemitted. Since the laser photons all have the same direction but the emission process is isotropic the ion will gain the momentum $\hbar\boldsymbol{k}$ of one photon in each absorption/emission cycle. This leads to a net repulsive force away from the resonance and does not yet represent a cooling mechanism. The cooling requires a stable velocity point v^* with a restoring force for particles slightly deviating from this point. As illustrated in Fig. 2.34 this can be achieved with an auxiliary force counteracting the laser force.

The average force resulting from repeated scattering resembles the absorption probability having a Lorentz shape. It can be expressed [2.197] as

$$\langle \boldsymbol{F} \rangle = \hbar\boldsymbol{k}\,\Gamma\,\frac{S/2}{1 + S + (2\Delta\omega/\Gamma)^2} - \boldsymbol{F}_{\mathrm{aux}}\,, \tag{2.39}$$

where Γ and S denote the decay rate and the optical saturation parameter, respectively, and $\Delta\omega = \gamma\,(1-\beta)\,\omega - \omega_o$ is the frequency detuning from the resonance ω_o in the ion rest frame. A constant auxiliary force can be provided by an rf cavity or by a simultaneous electron cooling. The laser-cooling rate is simply given as

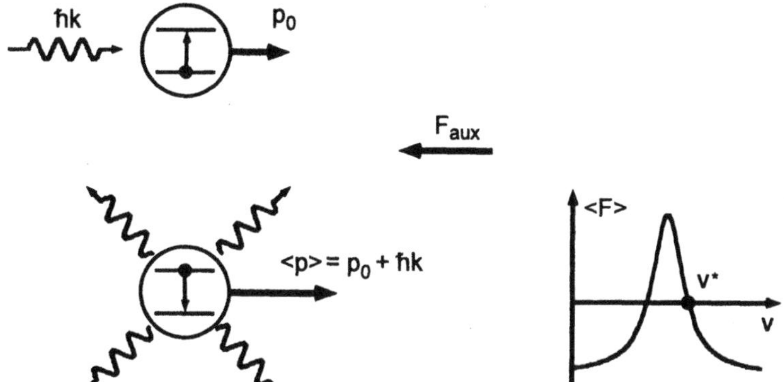

Fig. 2.34. Principle of laser cooling

$$\frac{1}{\tau_{\mathrm{c}}} = -\frac{2}{m_{\mathrm{i}}} \left(\frac{\partial F}{\partial v} \right)_{v=v^*} \tag{2.40}$$

in terms of the laser force gradient. Another way of creating a stable point is to use two counterpropagating laser beams. The principal limit of the ion temperature, in the absence of heating processes, is set by the recoil from a single photon.

The scheme described, obviously works only in the longitudinal direction. Applying one or more lasers perpendicular to the ion beam severely lacks geometrical overlap. For bunched beams, a scheme has been proposed [2.236] to couple the longitudinal and transverse degrees of freedom of the stored ions.

Laser-cooling experiments have been conducted at the TSR with Li^+ [2.233] and with Be^+ ions [2.197] and at the ASTRID with Li^+ [2.201] and with Mg^+ [2.203] ions. For beams of low intensity, momentum spreads below 10^{-6} have been obtained.

To access also highly charged ions one can pursue the magnetic dipole transitions in the ground-state finestructure of the boron and carbon isoelectronic sequence or between hyperfine levels of very heavy one-electron ions. Because of the lower spontaneous decay rate for these schemes it will be necessary to use stimulated emission [2.237–239].

2.4 Ion Traps

The purpose of ion traps is to confine charged particles to a small volume where they can be cooled to a very low temperature and be used for experiments during a considerable amount of time. Since the original application of magnetic and electric fields for increasing the electron-ion interaction time in electrical discharges [2.29] the technique has been enormously refined. Owing to the detailed understanding of the operational principle and known systematic limitations [2.240] electromagnetic traps have been turned into instruments for high-accuracy measurements [2.241,242]. They have been applied for tests of QED, of fundamental symmetries and nuclear models and for metrology. Impressive accuracies have been demonstrated in the measurements of the electron and positron g factors [2.243] and their mass ratio [2.244], and of atomic-mass ratios involving the proton and the antiproton [2.245,246], and singly charged radioactive heavy ions [2.247–249].

Trapping of *highly charged* heavy ions is motivated by a further increased accuracy of mass measurements and extensions to new physical domains [2.250]. Proposed measurements of the g factor [2.251] and of the mass of heavy one-electron ions such as U^{91+} would test fundamental theory. Other applications like electron-ion and ion-atom interactions [2.252,253] do not require the ultimate accuracies but explore the realm of high-charge-low-velocity collisions.

Ion sources for highly charged ions like EBIS or EBIT with an appropriate transfer line are well suited for feeding an ion trap. Starting with fast ions from accelerators and storage rings needs a careful design of deceleration and cooling mechanisms to keep ion losses at a tolerable level. An alternative is the trapping of slow recoil ions [2.252] produced by heavy-ion impact.

Major experimental programs for the trapping of highly charged ions have been initiated at two laboratories as will be explained below in more detail. At the *Manne Siegbahn Institute* (MSI) at Stockholm, ions from their cryogenic electron-beam ion source CRYSIS are used to feed a Penning trap and at the Livermore National Laboratory ions extracted from their EBIT are transferred to a Penning trap. There also is a proposal at the GSI at Darmstadt, named HITRAP [2.254], to decelerate bare very heavy ions and to capture them in an electromagnetic trap.

Penning Trap Operation. In a Penning trap a three-dimensional static trapping potential is created by the superposition of a homogeneous magnetic field on an electric quadrupole field provided by the electrode configuration sketched in Fig. 2.35. A constant voltage U_0 is applied between the

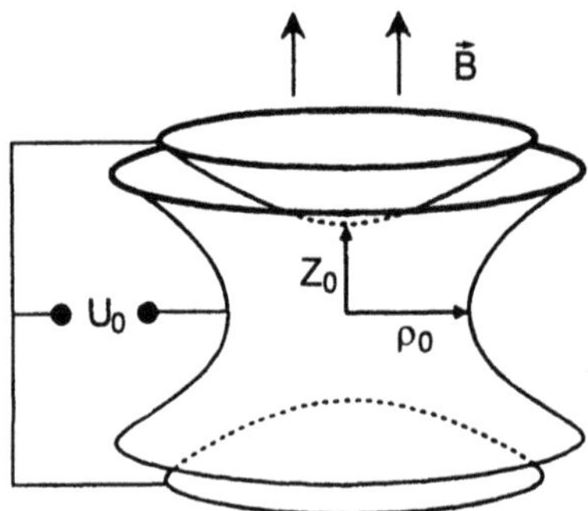

Fig. 2.35. Basic electrode configuration of a Penning trap. A homogeneous magnetic field is superimposed on a constant electric quadrupole field generated by a DC voltage applied between the ring electrode and the two end caps

center electrode ring and the two end caps. A nearly perfect quadrupole field is obtained by letting the (hyperbolic) electrode surfaces follow the desired quadrupole potential. In practice, several correction electrodes are used in addition, to compensate for the finite dimensions of the electrodes and for apertures needed for particle injection and for experiments.

In the ideal case, the motion of an ion in the trap can be described by a superposition of three independent harmonic eigenmotions [2.240] characterized by their angular frequencies

$$\omega_+ \gg \omega_z \gg \omega_- .$$

The eigenmotions are a fast cyclotron motion (ω_+), a longitudinal oscillation (ω_z) and a slow azimuthal drift (ω_-) known as magnetron motion. Their angular frequencies are given by

$$\omega_z = \left(\frac{qeU_o}{m_i d^2}\right)^{1/2} , \tag{2.41}$$

$$\omega_\pm = \frac{1}{2}\left[\omega_c \pm \left(\omega_c^2 - 2\omega_z^2\right)^{1/2}\right] , \tag{2.42}$$

where ω_c denotes the ion cyclotron-resonance frequency (2.16) and the characteristic trap dimension d is defined by

$$d^2 = z_o^2/2 + \rho_o^2/4$$

with the geometrical dimensions given in Fig. 2.35. The eigenfrequencies can be measured by exciting the corresponding motions with oscillating dipole fields or with an azimuthal quadrupole rf field. Because of the large differences between the three frequencies it is important to measure the highest of the frequencies, i.e., the reduced cyclotron frequency ω_+, with the highest accuracy. A high measurement precision can be achieved if a very high cyclotron resonance frequency is chosen through a high magnetic inductance employing superconducting solenoids. As the frequency is proportional to the ionic charge, a corresponding increase in frequency can be gained by using highly charged ions.

A number of different cooling techniques like laser cooling [2.255] (compare also Sect. 2.3.2), resistive cooling [2.256] and sympathetic cooling [2.257] (compare also Sects. 2.1.4, 1.5) have been developed which are central to the successful operation of ion traps as precision instruments.

2.4.1 SMILETRAP

A precision Penning trap has been set up as a mass spectrometer by researchers from Mainz and Stockholm [2.258–260]. The apparatus named SMILETRAP (Stockholm-Mainz-Ion-LEvitation TRAP) is designed for precision mass measurements of highly charged ions delivered by the Cryogenic Electron-Beam Ion Source CRYSIS. A schematic view of the experimental arrangement is depicted in Fig. 2.36 showing the transfer line between CRYSIS and the ion-trap facility which consists of the pre-trap and the precision Penning trap.

Filling of the trap occurs in two steps. For this purpose, the ion source is operated in a pulsed mode with a variable extraction time. The ions start from the operating potential of the ion source of +2.8 kV. After magnetic analysis part of an ion pulse is accepted by the pre-trap. Its potential can be varied between +2.8 kV and 0 V within 3 ms. The ions are ejected from the pre-trap at 0 V and transferred through a set of drift tubes at -1 kV to the precision

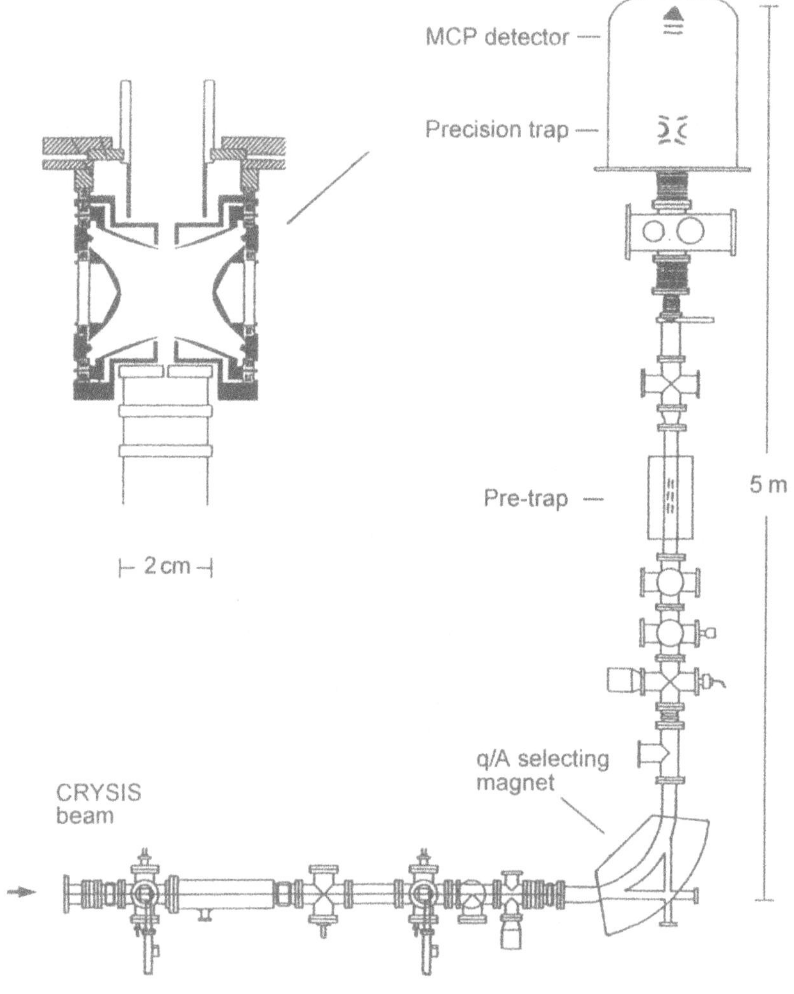

Fig. 2.36. Schematic view of the SMILETRAP apparatus showing the transfer line with a 90° bending magnet plus the pre-trap and the precision trap [2.259] ©1995 IEEE

trap. The pre-trap consists of seven cylindrically shaped electrodes positioned in the magnetic field of 0.25 T of a warm solenoid. The precision trap [2.261] is designed for minimum imperfections of the magnetic and electric field and is located in a 4.7 T magnetic field of a superconducting coil. After capture, hot ions are boiled off from the trap by lowering its potential from 5 V to 0.15 V.

Mass comparisons are made by exciting the sum frequency [2.262] $\omega_c = \omega_+ + \omega_-$ by an azimuthal quadrupole rf field applied to the center electrode ring which is split into eight segments. The resonance signal is recorded by a time-of-flight technique [2.263]. Ions in resonance have a higher kinetic energy and thus an increased magnetic moment due to the orbital motion. When the ions are extracted from the high magnetic field region they have to pass a high magnetic gradient. Because the acceleration depends on the magnetic moment, a resonance signal is observed in the flight time to an external detector. Measurements have been performed with $^{40}Ar^{18+}$ and with $^{86}Kr^{29+}$ ions with relative accuracies near 10^{-9}. The freedom to vary the charge state of the ion has the advantage to perform mass comparisons between different regions of the nuclear chart while keeping the charge-to-mass ratio nearly constant.

2.4.2 RETRAP

The electron-beam ion trap introduced in Sect. 2.1.5 functions both as an ion source and as an ion trap. For precision experiments, however, it is highly desirable to separate the highly charged ions from the intense electron beam. Therefore, a Retrapping apparatus, called RETRAP [2.264], has been built that can extract short ion pulses from the EBIT and capture them in a cryogenic Penning trap. Very highly charged ions such as Xe^{44+} or Th^{72+} have been retrapped.

The transport system consists of electrostatic einzel lenses, steerers and two 90° deflectors. The ions are electrostatically extracted and are decelerated again in a deceleration tube directly before the Penning trap. A pair of superconducting Helmholtz coils provide a magnetic flux density of up to 6 T. Both, the deceleration tube and the trap are in thermal contact with the liquid-helium reservoir used to cool the Helmholtz coils. Two different trap configurations [2.265,266] have been used which are schematically shown in Figs. 2.37,38. In the first case an open-ended cylindrical trap is used with a tuned circuit produced by an inductor connected to the compensation electrodes. The circuit is used for detection of the ions' axial motion and to damp the oscillations which initially have large amplitudes. Trapping and charge-state identification can be diagnosed through excitation of the parametric resonance at $2\omega_z$ using amplitudes sufficient to drive the ions out of the trap. The second configuration, sketched in Fig. 2.38, is a tandem trap consisting of two closely coupled hyperbolic Penning traps. With this design, a systematic cooling of the highly charged heavy ions becomes possible using a combination of a resistive precooling and a sympathetic cooling with cold $^{9}Be^{+}$ ions. The lower trap is for catching, electronic precooling and laser cooling of the $^{9}Be^{+}$ ions. The upper trap is used to catch and electronically precool the highly charged ions. After transferring the highly charged ions into the lower trap they will be further cooled via Coulomb collisions .

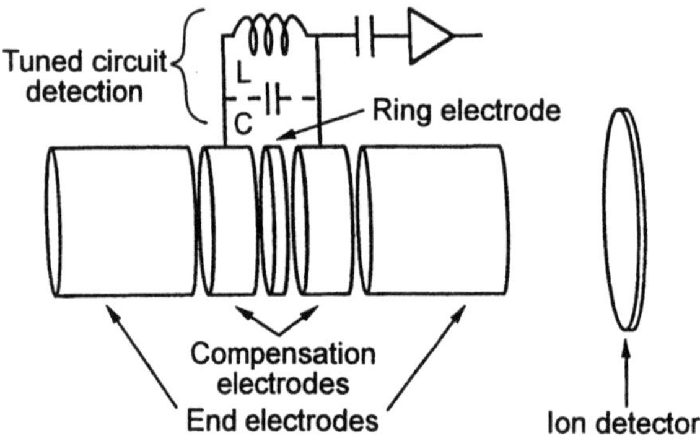

Fig. 2.37. Schematic of the open-ended cylinder trap used for the RETRAP experiments. The tuned circuit can be used for diagnostics and for resistive cooling [2.265], reprinted with permission

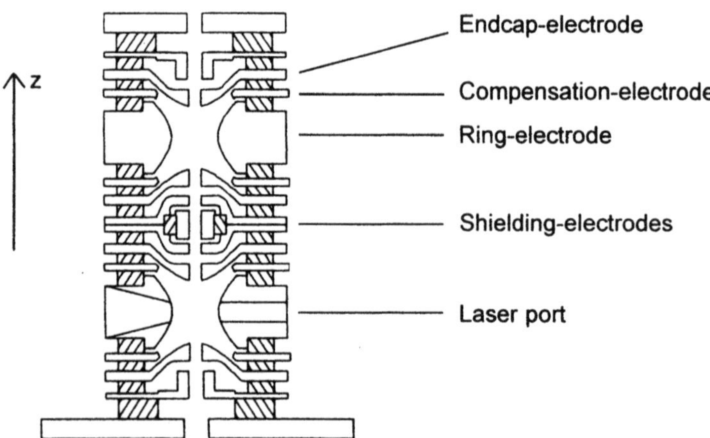

Fig. 2.38. Tandem trap used in the RETRAP experiments. Light ions are captured into the lower trap where they can be cooled by lasers. Highly charged heavy ions are captured into the upper trap and are cooled by collisions with the light ions [2.266], reprinted with permission

3. Atomic Structure and Spectra

The influence of electron correlation, relativistic and quantum electrodynamical effects on the electronic binding strongly increases with the ionic charge. In this chapter a review on the progress in measurements and calculations of the binding energies, fine and hyperfine structure intervals and the Lamb shift in highly charged ions is given. The schemes of coupling of the atomic orbital and spin angular momenta into the total angular momentum are considered as well.

3.1 Classification of Spectral Lines

The spectrum of an atom as determined by its electronic configuration dramatically varies with the degree of ionization of the atom considered. For neutral atoms or ions in *low* ionization states one is often faced with the problem of many interacting electrons. If only one active electron outside a closed shell is present, like in the alkali atoms, the electronic structure approaches that of a simple one-electron atom. For a one-electron system the gross properties are determined by the binding energies of the main shells given by the well-known Bohr model as

$$E_n = -\frac{Z^2}{n^2}\,\mathrm{Ry}, \tag{3.1}$$

where Ry denotes the *Rydberg* energy, 1 Ry $=$ 13.605 698 1(40) eV, n the principal quantum number and Z the nuclear charge number, respectively.

For an electron outside a core, consisting of the nucleus plus a number of core electrons, the Coulomb potential at large distances r can be approximated as

$$U(r) \approx -\frac{Z-N+1}{r} = -\frac{Z_{\mathrm{eff}}}{r}, \qquad r \to \infty, \tag{3.2}$$

if N denotes the total number of atomic electrons. Correspondingly, (3.1) may be used to approximate the energy levels of the core-excited states. For this purpose the nuclear charge Z has to be replaced by an effective charge that is approximated by the net charge of the core

$$Z_{\mathrm{eff}} = Z - N + 1 = q + 1\,, \tag{3.3}$$

where q is the ionic charge state. Such an approximation works best for high n states and for a high net core charge. Here, we will call ions with $q > 5$ Highly Charged Ions (HCI). The core charge $q + 1$ is also known as the *spectroscopic symbol* and is used to classify the atomic spectra according to the ionization stage of the atom from which the spectra are observed. As a common practice, $q + 1$ is attached, in the form of a Roman number, to the symbol for the chemical element. For example, the neutral Fe atom is written as Fe I, the Fe^{25+} ion as Fe XXVI, and so on. Ions with the same number of electrons N arranged in an increasing order of the nuclear charge Z belong to the *isoelectronic sequence* of the corresponding atom A and are termed A-like ions or [A] ions. For example, if $N = 1$ one has H-like ions, [H]; if $N = 2$ He-like ions, [He], etc..

In the hydrogenic approximation, scalings for many atomic characteristics can be found such as those for frequencies, transition probabilities, effective cross sections and rate coefficients. In Table 3.1 we summarize approximate scalings with the nuclear charge. In the case of multielectron ions an estimate is obtained by substituting the nuclear charge Z by the core charge $q + 1$ which is used as an effective nuclear charge $Z_{\text{eff}} = q + 1$ in an hydrogenic approximation. Of particular interest are one and few-electron ions of very high charge since their rigorous theoretical description is possible allowing a systematic study of fundamental questions. Because of the different dependencies upon the atomic number Z observed in the quantities listed in Table 3.1, their relative importance drastically changes with increasing Z. For instance, the leading term of the radiative corrections responsible for the Lamb shift increases in proportion to Z^4 whereas the electronic binding energies scale only as Z^2. Therefore relative Lamb shift contributions are much enhanced at high values of the nuclear charge.

The energy levels are often described in the $\boldsymbol{LS}$-coupling scheme (Sect. 3.2) and labeled in the form:

$$^{2S+1}L_J \ , \tag{3.4}$$

where L and S are the quantum numbers of the orbital and spin angular-momenta of an ion, respectively, and J is the total angular-momentum quantum number which according to the general rule for vector addition can take the values $\boldsymbol{J} = \boldsymbol{L} + \boldsymbol{S}$, i.e.,

$$|L - S| \leq J \leq L + S \ . \tag{3.5}$$

A spectral line is radiated if an ion undergoes a transition from an excited initial state into a lower final state. Transitions with a change of spin $|\Delta S| = 1$ are called *intercombination* transitions. Transitions which are not allowed by the selection rules (Sect. 4.1) are called *forbidden* transitions. Notations used for electric and magnetic multipole transitions are given in Table 3.2. The notations from x-ray spectroscopy are also used and are listed in Table 3.3. The notations of the basic spectral lines in H- and He-like ions and corresponding dielectronic satellites are given in Tables 3.4,5. The spectral

Table 3.1. Approximate scaling behavior of radiative and collisional characteristics of HCI with the nuclear charge Z. The quantities α, m_e and m_p denote the fine structure constant and the electron and proton mass, respectively

Atomic radius	Z^{-1}
Ionization potential	Z^2
Transition energy (frequency)	Z^2
Fine-structure (multiplet) splitting	$\alpha^2 Z^4$
Hyperfine splitting	$(m_e/m_p)\alpha^2 Z^3$
Lamb shift	$\alpha^5 Z^4$
Oscillator strength	Z^0
Radiative transition probability	Z^4
Radiative lifetime	Z^{-4}
Relative Doppler width $\Delta\omega_D/\omega$	Z^1
Autoionization probability	Z^0
Static dipole polarizability	Z^{-4}
External electric field strength for ionization of HCI	Z^3
Photoionization cross section	$\alpha^1 Z^{-2}$
Photorecombination rate coefficient	Z^1
Bremsstrahlung cross section	$\alpha^3 Z^2$
Electron-impact excitation and ionization cross sections	Z^{-4}
Electron-impact excitation and ionization rate coefficients	Z^{-3}
Electron capture cross section of HCI on neutral atoms	$Z^a, a = 1-5$
Excitation and ionization cross sections of neutral atoms by HCI	Z^2
Ion temperature	Z^2

Table 3.2. Notations for electric and magnetic 2^κ-pole transitions

Notation	Transition	Example	
E1	Electric dipole transition	$2\,^1P_1$ - $1\,^1S_0$	in [He]
M1	Magnetic dipole	$2s_{1/2}$ - $1s_{1/2}$	in [H]
		$2\,^3S_1$ - $1\,^1S_0$	in [He]
E1M1	Two-photon electro-magnetic	$2\,^3P_0$ - $1\,^1S_0$	in [He]
E1M2	Two-photon electro-magnetic	$2p_{3/2}$ - $1s_{1/2}$	in [H]
E2M1	Two-photon electro-magnetic	$2p_{3/2}$ - $2p_{1/2}$	in [H]
E2	Electric quadrupole	$2p_{3/2}$ - $2p_{1/2}$	in [H]
M2	Magnetic quadrupole	$2\,^3P_2$ - $1\,^1S_0$	in [He]
2E1	Two-photon electric dipole	$2s_{1/2}$ - $1s_{1/2}$	in [H]
		$2\,^1S_0$ - $1\,^1S_0$	in [He]
2E2	Two-photon electric quadrupole		
2M1	Two-photon magnetic dipole		
2M2	Two-photon magnetic quadrupole		

lines of He-like ions and their satellites are usually identified using *Gabriel*'s notation [3.1].

Table 3.3. X-ray spectroscopy notations for atomic shells

Shell	Atomic notation	Shell	Atomic notation
K	$1s_{1/2}$	N_6	$4f_{5/2}$
L_1	$2s_{1/2}$	N_7	$4f_{7/2}$
L_2	$2p_{1/2}$	O_1	$5s_{1/2}$
L_3	$2p_{3/2}$	O_2	$5p_{1/2}$
M_1	$3s_{1/2}$	O_3	$5p_{3/2}$
M_2	$3p_{1/2}$	O_4	$5d_{3/2}$
M_3	$3p_{3/2}$	O_5	$5d_{5/2}$
M_4	$3d_{3/2}$	O_6	$5f_{5/2}$
M_5	$3d_{5/2}$	O_7	$5f_{7/2}$
N_1	$4s_{1/2}$	P_1	$6s_{1/2}$
N_2	$4p_{1/2}$	P_2	$6p_{1/2}$
N_3	$4p_{3/2}$	P_3	$6p_{3/2}$
N_4	$4d_{3/2}$	P_4	$6d_{3/2}$
N_5	$4d_{5/2}$	Q_1	$7s_{1/2}$

Table 3.4. Notations of the basic spectral lines in H- and He-like ions

Sequence	Transition	Notation
H	$2p_{1/2}$ - $1s_{1/2}$	$Ly\alpha_2$
H	$2p_{3/2}$ - $1s_{1/2}$	$Ly\alpha_1$
He	$2s_{1/2}$ - $1s_{1/2}$	$M1$
He	$2p_{1/2}$ - $1s_{1/2}$	$K\alpha_2$
He	$2p_{3/2}$ - $1s_{1/2}$	$K\alpha_1$
He	$1s2p\,^1P_1$ - $1s^2\,^1S_0$	w, resonance line
He	$1s2p\,^3P_2$ - $1s^2\,^1S_0$	x, magnetic quadrupole M2 transition
He	$1s2p\,^3P_1$ - $1s^2\,^1S_0$	y, intercombination transition
He	$1s2p\,^3S_1$ - $1s^2\,^1S_0$	z, forbidden line

3.2 Coupling Schemes

The coupling scheme (or vector coupling) is a term describing the interaction of electrons in atoms, ions and molecules. It reveals the vector addition of the orbital ($\boldsymbol{l}_i$) and spin ($\boldsymbol{s}_i$) angular momenta of the individual atomic electrons resulting in the total angular momentum $\boldsymbol{J}$. In the zero-order approximation the energy of an atomic system is defined by the electrostatic interaction of the electrons with the nucleus and by the electrostatic interaction among the electrons themselves. In this approximation the energy level is represented by the configuration $n_1\ell_1, n_2\ell_2, \ldots, n_N\ell_N$ which is g-times degenerate over all possible projections $m_{\ell i}$ and m_{si}:

$$g = 2(2\ell_1 + 1)(2\ell_2 + 2)\ldots(2\ell_N + 1)\,. \tag{3.6}$$

The non-central part of the electrostatic interaction and the magnetic spin-orbit interaction cause a splitting of an atomic level into sublevels the relative position of which can be described by the vector coupling of angular momenta $\boldsymbol{\ell}_i$ and $\boldsymbol{s}_i$. The types of coupling schemes can be easily illustrated by the case

Table 3.5. Notations in the jj-coupling scheme for the transitions in Li- and He-like ions according to [3.1]

Key	Transition
a	$(1s2p^2_{3/2})_{3/2}$ - $1s^22p_{3/2}$
b	$(1s2p^2_{1/2})_{3/2}$ - $1s^22p_{1/2}$
c	$(1s2p_{1/2}2p_{3/2})_{3/2}$ - $1s^22p_{3/2}$
d	$(1s2p_{1/2}2p_{3/2})_{1/2}$ - $1s^22p_{1/2}$
e	$(1s2p_{1/2}2p_{3/2})_{5/2}$ - $1s^22p_{3/2}$
f	$(1s2p_{1/2}2p_{3/2})_{3/2}$ - $1s^22p_{3/2}$
g	$(1s2p_{1/2}2p_{3/2})_{3/2}$ - $1s^22p_{1/2}$
h	$(1s2p^2_{1/2})_{1/2}$ - $1s^22p_{3/2}$
i	$(1s2p^2_{1/2})_{1/2}$ - $1s^22p_{1/2}$
j	$(1s2p^2_{3/2})_{5/2}$ - $1s^22p_{3/2}$
k	$(1s2p_{1/2}2p_{3/2})_{3/2}$ - $1s^22p_{1/2}$
l	$(1s2p_{1/2}2p_{3/2})_{3/2}$ - $1s^22p_{3/2}$
m	$(1s2p^2_{3/2})_{1/2}$ - $1s^22p_{3/2}$
n	$(1s2p^2_{3/2})_{1/2}$ - $1s^22p_{1/2}$
o	$(1s2s^2)_{1/2}$ - $1s^22p_{3/2}$
p	$(1s2s^2)_{1/2}$ - $1s^22p_{1/2}$
q	$(1s2s2p_{3/2})_{3/2}$ - $1s^22s_{1/2}$
r	$(1s2s2p_{1/2})_{1/2}$ - $1s^22s_{1/2}$
s	$(1s2s2p_{3/2})_{3/2}$ - $1s^22s_{1/2}$
t	$(1s2s2p_{3/2})_{1/2}$ - $1s^22s_{1/2}$
u	$(1s2s2p_{1/2})_{3/2}$ - $1s^22s_{1/2}$
v	$(1s2s2p_{1/2})_{1/2}$ - $1s^22s_{1/2}$
w	$(1s2p_{1/2})_1$ - $(1s^2)_0$
x	$(1s2p_{3/2})_2$ - $(1s^2)_0$
y	$(1s2p_{3/2})_1$ - $(1s^2)_0$
z	$(1s2s)_1$ - $(1s^2)_0$

of two nonequivalent electrons with angular momenta $\boldsymbol{\ell}_1$,$\boldsymbol{s}_1$ and $\boldsymbol{\ell}_2$,$\boldsymbol{s}_2$:

$$\begin{array}{llll} \boldsymbol{LS}\text{ coupling}: & \boldsymbol{\ell}_1+\boldsymbol{\ell}_2=\boldsymbol{L}, & \boldsymbol{s}_1+\boldsymbol{s}_2=\boldsymbol{S}, & \boldsymbol{L}+\boldsymbol{S}=\boldsymbol{J} \\ \boldsymbol{LK}\text{ coupling}: & \boldsymbol{\ell}_1+\boldsymbol{\ell}_2=\boldsymbol{L}, & \boldsymbol{L}+\boldsymbol{s}_1=\boldsymbol{K}, & \boldsymbol{K}+\boldsymbol{s}_2=\boldsymbol{J} \\ \boldsymbol{jK}\text{ coupling}: & \boldsymbol{\ell}_1+\boldsymbol{s}_1=\boldsymbol{j}_1, & \boldsymbol{j}_1+\boldsymbol{\ell}_2=\boldsymbol{K}, & \boldsymbol{K}+\boldsymbol{s}_2=\boldsymbol{J} \\ \boldsymbol{jj}\text{ coupling}: & \boldsymbol{\ell}_1+\boldsymbol{s}_1=\boldsymbol{j}_1, & \boldsymbol{\ell}_2+\boldsymbol{s}_2=\boldsymbol{j}_2, & \boldsymbol{j}_1+\boldsymbol{j}_2=\boldsymbol{J}. \end{array} \tag{3.7}$$

In the case of equivalent $n\ell$ electrons due to the Pauli principle the $\boldsymbol{LS}$ and $\boldsymbol{jj}$ couplings are the only ones possible because all electrons participate symmetrically. Each type of coupling scheme is characterized by relative values of different types of interactions. The $\boldsymbol{LS}$ coupling, also called Russel-Saunders coupling, is adequate when the electrostatic interaction V_{es} is much larger than the spin-orbit one V_{so}:

$$V_{\text{es}} \gg V_{\text{so}} \,.$$

The $\boldsymbol{LS}$ coupling is used for not very heavy neutral atoms and low-charged ions in low states of excitation. With increasing nuclear charge $Z \gg 1$ the situation is reversed

$$V_{\mathrm{es}} \ll V_{\mathrm{so}}$$

and the $\boldsymbol{jj}$ coupling is realized which is of particular interest for multicharged ions. In the case $V_{\mathrm{es}} \approx V_{\mathrm{so}}$ the states are described by the so-called *intermediate coupling* scheme. A smooth transformation from the $\boldsymbol{LS}$ to the $\boldsymbol{jj}$ coupling is shown in Fig. 3.1 when the nuclear charge Z increases from $Z = 10$ to $Z = 100$. The example is for the $1s2p$ states of helium-like ions and the numerical results from *Plante* et al. [3.2] were taken to generate the plot. For low values of the atomic number a very small energy splitting of the $^3P_{0,1,2}$ is observed whereas at high atomic numbers the levels are grouped according to their total angular momentum J.

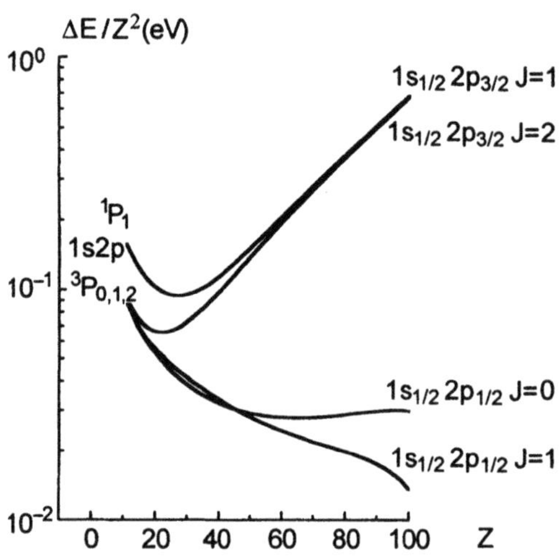

Fig. 3.1. Electronic binding energies of the $1s2p$ states in helium-like ions as a function of the nuclear charge Z. Plotted are the energy differences to the $1s2s\,^3S_1$ state divided by the square of the nuclear charge. The numerical data from *Plante* et al. [3.2] have been used for the plot

The $\boldsymbol{jK}$ coupling is also called the $\boldsymbol{jl}$ coupling. The $\boldsymbol{jl}$ coupling is applied when the spin-orbit interaction of the electrons of the atomic core is larger than the electrostatic interaction of these electrons with the excited electron. The $\boldsymbol{jl}$ coupling appears in the spectra of excited states of ions with noble-gas configuration (Ne-, Ar-like, etc.) and other cases when one of the electrons is, on the average, at a large distance from the atomic core.

In $\boldsymbol{LS}$ coupling the designation $^{2S+1}L_J$ of terms has already been introduced. For example, for the two-electron configuration $npn'p$ one has the singlet ($S = 0$) and the triplet ($S = 1$) states:

$$^1S_0,\ ^1P_1,\ ^1D_2,\ ^3S_1,\ ^3P_{0,1,2},\ ^3D_{1,2,3}\,.$$

The designation of terms in the other coupling schemes is more complicated. For the same configuration $npn'p$ one has:
LK coupling, level designation $L[K]_J$:

$$S[1/2]_{0,1},\ P[1/2]_{0,1},\ P[3/2]_{1,2},\ D[3/2]_{1,2},\ D[5/2]_{2,3}\ ,$$

jK coupling, level designation $j[K]_J$:

$$1/2[1/2]_{0,1},\ 3/2[1/2]_{0,1},\ 1/2[3/2]_{1,2},\ 3/2[3/2]_{1,2},\ 3/2[5/2]_{2,3}\ ,$$

jj coupling, level designation $[j_1 j_2]_J$:

$$[1/2\,1/2]_{0,1},\ [1/2\,3/2]_{1,2},\ [3/2\,1/2]_{1,2},\ [3/2\,3/2]_{0,1,2,3}\ .$$

For a given total angular momentum J, the number of sublevels is the same for all types of the coupling scheme. In many cases the use of a 'pure' scheme is not possible and it is necessary to use the intermediate coupling scheme [3.3].

3.3 Ionization and Transition Energies

An adequate interpretation of x-ray spectra of HCI is possible provided the relativistic and quantum electrodynamic (QED) corrections are accounted for. The role of these effects strongly increases with increasing ion charge. However, even in the case of a one-electron system (H-like ions) this task entails many difficulties.

The nonrelativistic Schrödinger equation does not describe the level fine structure which is caused by relativistic effects such as the dependence of the electron mass on the velocity and the existence of the electron spin $s = 1/2$. The relativistic analogue of the Schrödinger equation for H-like ions is the four-dimensional Dirac equation which has the form [3.4]:

$$H\Psi = E\Psi, \qquad H = -e\varphi + \beta mc_o^2 + \boldsymbol{\alpha}(c_o\boldsymbol{p} + e\boldsymbol{A})\ , \tag{3.8}$$

where φ and $\mathbf{A}$ are the scalar and vector potentials of the external electromagnetic field, respectively, $\mathbf{p} = -i\hbar\nabla_r$ is the electron-momentum operator, $\boldsymbol{\alpha}$ is the 4×4 Dirac matrix associated with the Pauli spin matrix $\boldsymbol{\sigma}^{\mathrm{P}}$,

$$\boldsymbol{\alpha} = \begin{pmatrix} 0 & \boldsymbol{\sigma}^{\mathrm{P}} \\ \boldsymbol{\sigma}^{\mathrm{P}} & 0 \end{pmatrix} \quad \text{and} \quad \boldsymbol{\beta} = \begin{pmatrix} \boldsymbol{I} & 0 \\ 0 & -\boldsymbol{I} \end{pmatrix}, \tag{3.9}$$

with $\boldsymbol{I}$ denoting a unit quadratic two-row matrix. For a point nucleus of charge Z the Coulomb potential is given by

$$\varphi = -Ze^2/r,\ \mathbf{A} = 0\ , \tag{3.10}$$

and (3.8) has an analytical solution for each component of the wavefunction Ψ. The corresponding eigenvalue of the relativistic energy E_{rel} for the bound $n\ell$ state is given by

$$E_{\text{rel}} = mc_o^2 \left[1 + \left(\frac{\alpha Z}{n - |\kappa| + \sqrt{\kappa^2 - \alpha^2 Z^2}} \right)^2 \right]^{-1/2}, \tag{3.11}$$
$$\kappa = (-1)^{j+\ell+1/2}(j+1/2),$$

where j is the total angular momentum including the electron spin and α is the fine structure constant. The energy E_{rel} includes the rest-mass energy of the electron. Expansion of (3.11) in powers of $\alpha^2 Z^2$ and the substitution

$$E_{n\kappa} = E_{\text{rel}} - mc_o^2 \tag{3.12}$$

gives

$$E_{n\kappa} = E_0 + E_1 + \cdots$$
$$E_0 = -\frac{Z^2}{n^2}\text{Ry}, \qquad E_1 = -\left(\frac{1}{j+1/2} - \frac{3}{4n} \right) \frac{\alpha^2 Z^4}{n^3}\text{Ry}, \tag{3.13}$$

where E_0 is the nonrelativistic part and E_1 is the fine-structure splitting (Sect. 3.4). As can be seen from (3.13), the states with the same quantum number j but different ℓ have equal energies. However, this degeneration is removed by QED effects caused by the interaction of the electron with its own electromagnetic radiation field (Sect. 3.5).

The solution of the Dirac equation (3.11) together with (3.12) can be written in the equivalent form [3.3]:

$$E_{n\kappa} = -\frac{2Z^2}{N(N+n+\gamma-|\kappa|)}\text{Ry}, \qquad \gamma = \sqrt{\kappa^2 - \alpha^2 Z^2}$$
$$N = \sqrt{n^2 - 2(n-|\kappa|)(|\kappa|-\gamma)}, \qquad |\kappa| = j + 1/2. \tag{3.14}$$

Relativistic and QED effects are very important for accurate calculations of energy levels for high-Z ions [3.5]. In Table 3.6 the calculated contributions of the Coulomb energy (3.14), reduced-mass and Lamb-shift corrections are given for the ground-state energy of H-like ions. Ionization energies for the 1s electron and energy intervals between $n = 1$ and $n = 2$ states in H-like ions with the nuclear charge $1 \leq Z \leq 110$ are given in Table A.1 in Appendix A.1. The data given in Table 3.6 and in Table A.1 in Appendix A.1 have been calculated by *Johnson* and *Soff* [3.6] with inclusion of the QED corrections to all orders in αZ together with the finite-nuclear-size, reduced-mass and recoil corrections. In Tables A.2,3 given in Appendix A.1 the wavelengths for transitions from higher n states in H-like ions are listed. They have been calculated [3.7] by solving the Schrödinger equation with high-order QED corrections including anomalous magnetic moment, self-energy, vacuum-polarization and nuclear-recoil effects.

For ions having more than one electron (He-, Li-like ions, etc.) the difficulties involved in the effort of calculations of wavefunctions and energy levels increase significantly due to additional interactions between atomic electrons (e.g., the two-electron screened Lamb shift). There have been applied several sophisticated theoretical approaches to highly charged many-electron ions:

Table 3.6. Contributions of the Coulomb energy, reduced-mass correction and the Lamb shift to the total ionization energy in cm^{-1} of H-like ions in the $1s$ ground state, from [3.6]

Z	Coulomb energy	Reduced-mass correction	Lamb shift	Total energy
1	-109738.776164	59.732344	0.272624	-109678.771196
10	-10988379.7	301.2	1201.3	- 10986877.2
26	- 74862320.	730.	32040.	- 74829550.
54	- 3334859×10^2	13×10^2	3799×10^2	- 3331046×10^2
92	- 1066910×10^3	2×10^3	3698×10^3	- 1063210×10^3
110	- 166355×10^4	0.	1495×10^4	- 164860×10^4

the unified method [3.8], the Z-expansion method [3.9], Multiconfiguration Dirac-Fock (MCDF) [3.10–13], the Configuration-Interaction (CI) method [3.14], relativistic Many-Body Perturbation Theory (MBPT) [3.15,16], relativistic generalization of the random phase approximation [3.17], and a semiempirical approach [3.18]. In Table A.4 the calculated energy levels for the ground state n=1 and excited singlet and triplet n=2 states in He-like ions with the nuclear charge $3 \leq Z \leq 100$ are given [3.2]. Calculations have been performed on the basis of the relativistic all-order many-body perturbation theory. Calculated $K\alpha$ x-ray energies for He-like ions with $4 \leq Z \leq 92$ are compared with experimental data in Tables 3.7,8.

A general overview of experimental spectroscopic studies of HCI is given in [3.19,20]. Experimental $2s - 2p$ energy intervals in high-Z He-like ions are reported in [3.19,21–29]; measured wavelengths for transitions $1s^2 - 1sn\ell$, $n > 2$, in He-like ions can be found in [3.19,30].

Relativistic configuration-interaction calculations of energy levels for the $n = 2$ states in Li-like ions with nuclear charges in the range $10 \leq Z \leq 92$ are presented in Table 3.9. The calculations comprise QED and mass-polarization corrections. Calculated $2s - 2p$ transition energies are compared with available experimental data and other calculations in Tables 3.10 and 3.11. Experimental atomic energy intervals for $2s - 2p$ transitions in Li-like ions are reported in [3.19,22,31–35]. Accurate calculations of $2s_{1/2} - 2p_{1/2,3/2}$ intervals in Li-like uranium ions are presented in [3.16,36,37].

The fine structure of highly charged Be-like ions is considered in [3.38–40]. Relativistic many-body calculations of $2p^5 3s$ excited-state energy levels of Ne-like ions with nuclear charges $10 \leq Z \leq 92$ are presented in [3.41]. Energy levels and transition probabilities for Ne-like ions with Z = 27–92 have been calculated in [3.42] using relativistic perturbation theory with a model potential. New observations of the $3s^2 3p\,^2P - 3s3p^2\,^4P$ intercombination transitions in Al-like ions up to a nuclear charge of Z = 42 have been made [3.43] in a sliding-spark light source and in a tokamak plasma.

Wavelengths and energy levels for Na-like ions from Y^{28+} up to U^{81+} from a laser-produced plasma have been measured and identified in [3.44]. Energy levels and transition probabilities for intercombination transitions

Table 3.7. X-ray energies (in eV) for $K\alpha_2$ transitions in He-like ions

Z	Rel. CI [3.45]	Unif. theory [3.8]	All-order [3.46]	MCDF [3.10]	Experiment	Ref.
4	121.9226	121.9222				
6	304.4051	304.4035				
8	568.6443	568.6401	568.6408			
10	914.8109	914.8029	914.8034			
14	1853.801	1853.780	1853.781			
18	3123.574	3123.530	3123.534	3123.567	3123.52(4)	[3.47]
					3123.60(25)	[3.48]
22	4727.007	4726.925	4726.933			
24	5654.938	5654.831	5654.843			
26	6667.692	6667.552	6667.567	6667.669	6667.50(25)	[3.49]
32	10220.98	10220.73	10220.76		10221.80(35)	[3.50]
36	13026.36	13026.00	13026.05	13026.31	13026.8(3)	[3.51]
					13026.30(71)	[3.30]
44	19717.65	19716.98	19717.10			
54	30206.91	30205.58	30205.87	30206.53	30209.6(3.5)	[3.52]
64	43244.63	43242.29	43242.92			
80	70145.32	70140.93	70142.94			
83		76127.12			76085(85)	[3.21]
92	96174.5	96167.2	96172.5		96171(52)	[3.53]

Table 3.8. X-ray energies, in eV, for $K\alpha_1$ transitions in He-like ions

Z	Rel. CI [3.45]	Unif. theory [3.8]	All-order [3.46]	MCDF [3.10]	Experiment	Ref.
4	123.6707	123.6704				
6	307.9038	307.9026				
8	573.9645	573.9612	573.9616			
10	922.0072	922.0006	922.0009			
14	1865.020	1865.000	1865.002			
18	3139.617	3139.577	3139.582	3139.649	3139.55(4)	[3.47]
					3139.57(25)	[3.48]
22	4749.708	4749.630	4749.641		4749.74(17)	[3.54]
24	5682.149	5682.048	5682.064		5682.32(40)	[3.54]
26	6700.539	6700.404	6700.427	6700.603	6700.73(20)	[3.54]
					6700.90(25)	[3.49]
32	10280.39	10280.14	10280.19		10280.70(22)	[3.50]
36	13114.70	13114.34	13114.42	13114.80	13115.31(30)	[3.51]
					13114.78(71)	[3.30]
44	19904.07	19903.40	19903.57			
54	30630.64	30629.28	30629.68	30630.76	30629.1(3.5)	[3.52]
64	44109.08	44106.64	44107.49			
80	72454.43	72449.62	72452.26			
83		78859.86			78825(85)	[3.21]
92	100615.7	100607.3	100614.0		100626(35)	[3.53]
					100649(65)	[3.24]

$4s^2\,{}^1S_0 - 4s4p\,{}^3P_1$ in Zn-like ions with the nuclear charges Z=31-47 have been calculated in [3.55] using the configuration interaction method.

Table 3.9. Calculated atomic energy levels (in a.u.) for $n = 2$ states in Li-like ions, from [3.56]

Z	$2s_{1/2}$	$2p_{1/2}$	$2p_{3/2}$
10	102.79472	102.21080	102.20328
15	238.82049	237.87187	237.82070
20	432.04940	430.72775	430.54164
26	740.56971	738.78368	738.19690
32	1134.5082	1132.2332	1130.7946
42	1988.0137	1984.8492	1980.2231
54	3358.2485	3353.8451	3340.1601
64	4816.9256	4811.2901	4782.4780
74	6601.1075	6594.0028	6538.7795
82	8296.7811	8288.3031	8199.6898
90	10270.053	10260.106	10122.138
92	10812.432	10802.122	10648.563

Table 3.10. $2s_{1/2}$ - $2p_{1/2}$ transition energies (in eV) for Li-like ions, from [3.56]

Z	Rel. CI [3.56]	RMBPT [3.36]	Relativistic [3.57]	MCDF [3.13]	Experiment	Reference
10	15.8888	15.8885			15.8887(2)	[3.58]
15	25.813	25.812		25.806	25.814(3)	[3.59]
20	35.963	35.964		35.957	35.962(2)	[3.60]
26	48.600	48.602		48.597	48.599(1)	[3.61]
					48.602(4)	[3.62]
32	61.907	61.911		61.908	61.902(4)	[3.31]
42	86.11	86.12		86.129	86.10(1)	[3.31]
54	119.82	119.84		119.901	119.97(10)	[3.22]
90	270.80	270.85		272.320		
92	280.74	280.84	280.52	282.568	280.59(9)	[3.32]

Table 3.11. $2s_{1/2} - 2p_{3/2}$ transition energies (in eV) for Li-like ions, from [3.56]

Z	Rel. CI [3.56]	RMBPT [3.36]	MCDF [3.13]	Experiment	Reference
10	16.0933	16.0931		16.0932(2)	[3.58]
15	27.205	27.205	27.197	27.206(3)	[3.58]
20	41.028	41.028	41.021	41.029(2)	[3.58]
26	64.567	64.568	64.560	64.566(2)	[3.61]
				64.567(4)	[3.62]
28			74.954	74.9684(68)	[3.34]
32	101.051	101.055	101.043	101.043(12)	[3.31]
42	211.99	211.99	211.980	211.94(7)	[3.31]
54	492.21	492.22	492.212	492.34(62)	[3.22]
90	4025.11	4025.25	4026.26	4025.23(14)	[3.63]
92	4459.31	4459.70	4460.810	4459.37(35)	[3.33]

3.4 Fine and Hyperfine Structures

In the case of H-like ions the relativistic effects, i.e. the velocity dependence of the electron mass and the spin-orbit interaction, lead to a splitting of an atomic $n\ell$ level into two components: $j = \ell + 1/2$ and $j = \ell - 1/2$. This splitting is called *fine* or *multiplet* splitting. Due to the relativistic effects mentioned each of the components is also shifted by the value $\Delta E_{n\ell j}$ given by

$$\Delta E_{n\ell j} = -\left(\frac{1}{j+1/2} - \frac{3}{4n}\right)\frac{\alpha^2 Z^4}{n^3}\text{Ry}, \qquad j = \ell \pm 1/2, \tag{3.15}$$

where α is the fine-structure constant. A set of spectral lines arising from transitions between the fine-structure components (transitions $n\ell j - n'\ell' j'$) are called *multiplet* with the *selection rule* $\Delta j \equiv j' - j = 0, \pm 1$ (Sect. 4.1). For a hydrogenic $n\ell$ level the fine structure is illustrated in Fig. 3.2.

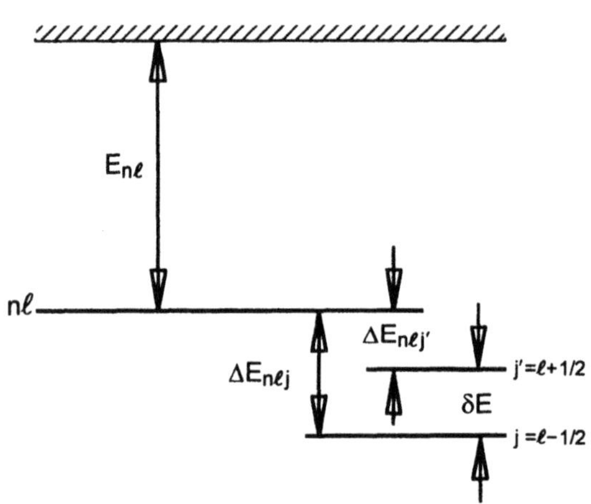

Fig. 3.2. Fine-structure shift and splitting of a hydrogenic $n\ell$ level

The scales of splitting and shifting are of the same order of magnitude ($\sim\alpha^2$). According to (3.15) the splitting between two components $j = l \pm 1/2$ is given by

$$\delta E = \frac{\alpha^2 Z^4}{n^3 \ell(\ell+1)}\text{Ry}. \tag{3.16}$$

For the $2p$ state one has from (3.15,16), in units of $\alpha^2 Z^4$Ry,

$$\Delta E_{2p_{1/2}} = -\frac{5}{64}, \quad \Delta E_{2p_{3/2}} = -\frac{1}{64}, \quad \delta E = \frac{1}{16}. \tag{3.17}$$

In a multielectron system, described in ***LS*** coupling, the spin-orbit interaction plays a dominant role and leads to the dependence of the energy term on the orbital ***L***, spin ***S*** and total angular momentum ***J***. If $L \geq S$ the atomic term splits into $2S+1$ different components. The quantity $2S+1$, determining the number of components, is called the *multiplicity* of the term. If $L < S$ the term splits into $2L+1$ components. In the ***LS***-coupling scheme (Sect. 3.2) the energy splitting between neighboring levels is given by the *Landé* interval rule [3.3]:

$$\Delta E_{j,j-1} \equiv E_j - E_{j-1} = a(LS)\,J, \tag{3.18}$$

i.e., proportional to J. The multiplet splitting constant $a(LS)$ depends only on the quantum numbers L and S.

The interaction of a nucleus with nonzero magnetic dipole moment ($\mu \neq 0$) or nonzero electric quadrupole moment ($Q \neq 0$) with atomic electrons leads to a splitting of a level with an angular momentum J into components. This splitting is called the *hyperfine* splitting (hfs). Each component is characterized by a set of quantum numbers $JIFM$, where $\boldsymbol{J} = \boldsymbol{L} + \boldsymbol{S}$ is the total (orbital plus spin) angular momentum of the atomic electrons, $\boldsymbol{I}$ is the nuclear spin, $\boldsymbol{F} = \boldsymbol{J} + \boldsymbol{I}$ is the total angular momentum of an atom and M is the projection of $\boldsymbol{F}$ on the quantum axis.

The total splitting of the level j of a H-like ion in the nonrelativistic approximation is given by [3.3,64]:

$$\begin{aligned} \Delta E &= \frac{1}{2}AC + BC(C+1) - \frac{4}{3}BI(I+1)j(j+1), \\ C &= F(F+1) - j(j+1) - I(I+1), \end{aligned} \tag{3.19}$$

where A and B are the dipole and quadrupole hyperfine splitting constants, respectively. In general, determination of the constants A and B is a quite complicated problem. In case of H-like ions of nuclear charge $Z \ll 137$ and for an orbital quantum number ℓ one has

$$A = \frac{g_I}{j(j+1)(\ell+1/2)} \frac{m_e}{m_p} \frac{\alpha^2 Z^3}{n^3} \mathrm{Ry}, \tag{3.20}$$

$$B = \frac{3Qa_o^{-2}}{8I(2I-1)j(j+1)(\ell+1)(\ell+1/2)\ell} \frac{Z^3}{n^3} \mathrm{Ry}, \tag{3.21}$$

where a_o is the Bohr radius, m_p is the proton mass and Q is the nuclear electric quadrupole moment. The quantity Q has a dimension of an area and is a measure of the deviation of the charge distribution from spherical symmetry. The dimensionless quantity g_I is called the *gyromagnetic* ratio or nuclear g-factor and is defined by

$$g_I = \frac{\mu}{\mu_N I}, \tag{3.22}$$

where μ is the nuclear magnetic dipole moment, I is the nuclear spin and μ_N is the nuclear magneton. The magnetic interaction (the two terms in (3.19) proportional to the constant B) is nonzero if the nuclear spin $I \geq 1$ and if the electron angular momentum $j \geq 1$. Experimental data on nuclear spin, nuclear dipole and quadrupole moments are given in [3.65].

Usually, the quadrupole terms in (3.19) depending on Q, have been found to be small compared with the magnetic dipole term and the hfs splitting is represented in the form:

$$\Delta E = \frac{1}{2} AC, \tag{3.23}$$

where the constants C and A are given in (3.19,20), respectively. For highly charged ions having one optical electron outside a spherical core, the inclusion of relativistic and QED corrections leads to the following form of the hfs [3.66]:

$$\Delta E = \frac{1}{2} AC \left\{ \left[\Phi(\alpha Z)(1-\delta)(1-\epsilon) + x_{\text{rad}} \right] + \cdots \right\}, \tag{3.24}$$

where $\Phi(\alpha Z)$ is the one-electron relativistic factor, δ and ϵ denote the nuclear charge- and magnetization-distribution corrections and x_{rad} is the one-electron radiative correction. Calculated wavelengths for the hfs of the ground state H- and Li-like ions are given in Tables 3.12,13 together with contributions from different corrections.

According to Table 3.12, the shortest calculated wavelength (241.2 nm) corresponds to Bi^{82+} ions (M1 transition F=5 $\leftrightarrow$ F=4). In fact, this is the only case of highly charged H-like ions where the wavelength of the

Table 3.12. Nuclear spin I, correction factors and wavelengths (in nm) for transitions between hfs components of the ground state in H-like ions [3.66,67]. For ions from In^{48+} up to U^{91+} the data have been calculated without account for the radiative correction

Ion	I	μ/μ_N	$\Phi(\alpha Z)$	δ	ϵ	x_{rad}	λ[nm]
$^{13}C^{5+}$	-1/2	0.702412(1)	1.00288	0.00055	0.00036	0.00050	387400(80)
$^{14}N^{6+}$	+1	0.403761	1.00393	0.00067	0.00004	0.00040	565190(110)
$^{17}O^{7+}$	+5/2	-1.8938(1)	1.00514	0.00082	0.00033	0.00029	100850(20)
$^{113}In^{48+}$	+9/2	5.5289(2)	1.2340	0.0171	0.0039		1349.4
$^{121}Sb^{50+}$	+5/2	3.3634(3)	1.2582	0.0191	0.0046		1791.5
$^{123}Sb^{50+}$	+7/2	2.5498(2)	1.2582	0.0191	0.0016		2473.8
$^{127}I^{52+}$	+5/2	2.8133(1)	1.2843	0.0212	0.0047		1873.7
$^{133}Cs^{54+}$	+7/2	2.5825(5)	1.3125	0.0236	0.0020		1876.2
$^{139}La^{56+}$	+7/2	2.7830	1.3430	0.0262	0.0027		1533.8
$^{141}Pr^{58+}$	+5/2	4.2754(5)	1.3761	0.0291	0.0072		843.0
$^{151}Eu^{62+}$	+5/2	3.4717(6)	1.4509	0.0362	0.0079		815.3
$^{159}Tb^{64+}$	+3/2	2.014(4)	1.4933	0.0404	0.0073		1123.2
$^{165}Ho^{66+}$	-7/2	4.173(27)	1.5395	0.0454	0.0086		563.9
$^{175}Lu^{70+}$	+7/2	2.234(4)	1.6453	0.0560	0.0018		831.7
$^{181}Ta^{72+}$	+7/2	2.3705(7)	1.7061	0.0625	0.0030		701.1
$^{185}Re^{74+}$	+5/2	3.1871(3)	1.7731	0.0698	0.013		448.6
$^{209}Tl^{80+}$	+1/2	1 .6223	2.0217	0.0978	0.020		382.2
$^{205}Tl^{80+}$	+1/2	1.6382	2.0217	0.0191	0.020		378.6
$^{207}Pb^{81+}$	-1/2	0.587(6)	2.0718	0.104	0.036		1017
$^{209}Bi^{82+}$	-9/2	4.1106(2)	2.1250	0.110	0.011		241.2
$^{235}U^{91+}$	-7/2	0.39(7)	2.7978	0.186	0.034		1540

Table 3.13. Nuclear spin I, correction factors and wavelengths (in cm) for transitions between hfs components of the ground state in Li-like ions [3.67,68]. For Bi^{80+} ions the data have been calculated without account for the radiative correction

Ion	I	μ/μ_N	$\Phi(\alpha Z)$	δ	ϵ	x_{rad}	λ[cm]
$^{14}N^{4+}$	+1	0.403761	1.00557	0.00067	0.00004	0.00042	7.072(4)
$^{19}F^{6+}$	+1/2	2.6289	1.00923	0.00099	0.00036	0.00022	0.34102(20)
$^{23}Na^{8+}$	+3/2	2.2176(1)	1.01384	0.00125	0.00035	0.00002	0.30924(19)
$^{25}Mg^{9+}$	+5/2	0.85545(8)	1.01650	0.00142	0.00058	0.00007	0.6679(4)
$^{27}Al^{10+}$	+5/2	3.6415	1.01941	0.00155	0.00048	0.00016	0.12060(7)
$^{29}Si^{11+}$	+1/2	0.55529(3)	1.02257	0.00172	0.00063	0.00025	0.37250(22)
$^{35}Cl^{14+}$	+3/2	0.82187	1.03355	0.00232	0.00026	0.00049	0.20073(12)
$^{57}Fe^{23+}$	-1/2	0.090623	1.08130	0.00452	0.0028	0.00106	0.3079(5)
		0.090764					0.3068(5)
		0.09044(7)					0.3072(5)
$^{209}Bi^{80+}$	-9/2	4.1106(2)	2.790	0.117(2)	0.015(6)		$1.543(11)\times10^{-4}$

ground state hyperfine structure splitting has been measured with a high accuracy: $\lambda_{exp} = (243.87 \pm 0.04)$ nm corresponding to (5.0840 ± 0.0008) eV [3.69]. The related theoretical value was found recently [3.70] by combining the most elaborated calculation of the nuclear-magnetization distribution correction (the Bohr-Weisskopf effect) [3.71] and QED corrections, and constitutes $\lambda_{th} = (5.058 \pm 0.007)$ eV. At present, the reason for the discrepancy between theory and experiment is not clear so far. The authors [3.72] suggested reinvestigating the experimental magnetic dipole moment $\mu(\mathrm{Bi}) = (4.1106 \pm 0.0002)\mu_N$ which includes the diamagnetic correction for a free ion, but no reliable correction for the chemical shift. The latter is difficult to estimate but presently it causes the largest uncertainty of the theoretical value of the hyperfine interaction , see [3.72] for details. An experiment on hydrogen-like $^{207}Pb^{81+}$ is underway for which the nuclear parameters are known with a much higher accuracy than for $^{209}Bi^{82+}$.

3.5 Lamb Shift

In recent years, a significant progress has been achieved in accurate measurements of the energy levels in heavy few-electron HCI including the $1s$ Lamb shift in H-like uranium [3.73–75], the hyperfine structure of H-like bismuth [3.69] and the ground state energies of He-like ions [3.29,20]. These experimental data provide a unique possibility to test the QED effects in strong nuclear fields. The term *Lamb shift* was originally introduced after the famous experiment on hydrogen atoms carried out by *Lamb* and *Retherford* [3.76] which showed that the $2s_{1/2}$ and $2p_{1/2}$ levels are nondegenerate and that the splitting between them (the Lamb shift) is very small but measurable: $\Delta S = 1058$ MHz or 3.380×10^{-5} eV. The fine structure and the Lamb shift of the $n = 2$ level is schematically shown in Fig. 3.3. This splitting is

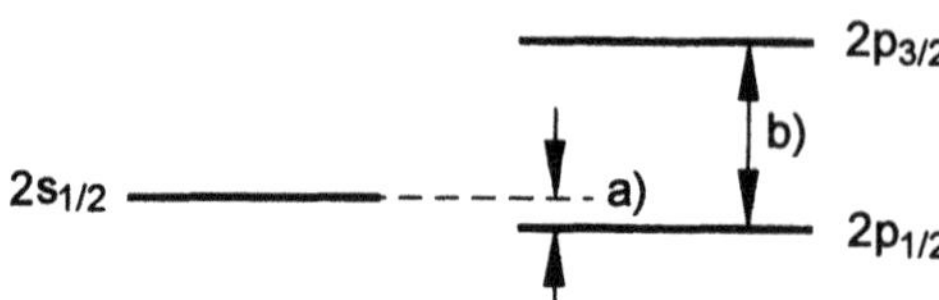

Fig. 3.3. Lamb shift (a) and fine structure (b) in H-like ions

caused by the interaction of the electron with its own electromagnetic radiation field and is well explained by the theory of quantum electrodynamics (QED) e.g., [3.77–79]. Although the term Lamb shift was originally used only for the $2s_{1/2}-2p_{1/2}$ energy splitting in hydrogen, it is now common to extend it to isolated levels, e.g., $1s_{1/2}$, $2s_{1/2}$, $2p_{1/2}$, etc. We will use the term Lamb shift for one-electron systems as the difference between the true (experimental) binding energy and the one calculated for a point nuclear field (Dirac eigenvalue) disregarding all QED effects. In the definition of the Lamb shift for many-electron systems (He-, Li-like ions etc.), one has to be very accurate because there is no well defined analog to the Dirac equation for these ions, even for He-like ones as the simplest multielectron system.

The current status of the QED effects and experimental data for hydrogen-like ions is given in [3.72,80,81]. In the QED theory, the Lamb shift can be written as

$$\Delta S = \frac{\alpha}{\pi}\frac{(\alpha Z)^4}{n^3}F(\alpha Z)m_e c_o^2, \tag{3.25}$$

where $F(\alpha Z)$ is a dimensionless slowly varying function describing the sum of different corrections as explained below. The function $F(\alpha Z)$ is plotted in Fig. 3.4 for the $1s$ Lamb shift as a function of the atomic number Z including a large number of experimental data which will be discussed in more detail below. For states with an angular momentum $\ell \neq 0$, $F(\alpha Z) \approx 1$ whereas $F(\alpha Z) \approx \ln[(\alpha Z)^{-2}]$ for s states. Therefore, the Lamb shift is largest for the ground state and strongly increases ($\propto Z^4$) with the nuclear charge Z.

Various treatments for light ions based on an expansion of the Coulomb interaction between the electron and the nucleus in terms of the parameter αZ, break down for high-Z systems so it is necessary to develop QED methods which are nonperturbative in the parameter (αZ). One of the main difficulties arising in developing such methods consists in adopting the standard renormalization procedure based on the (αZ) expansion of the bound-electron propagator, to the calculations exact in (αZ). At present, the main

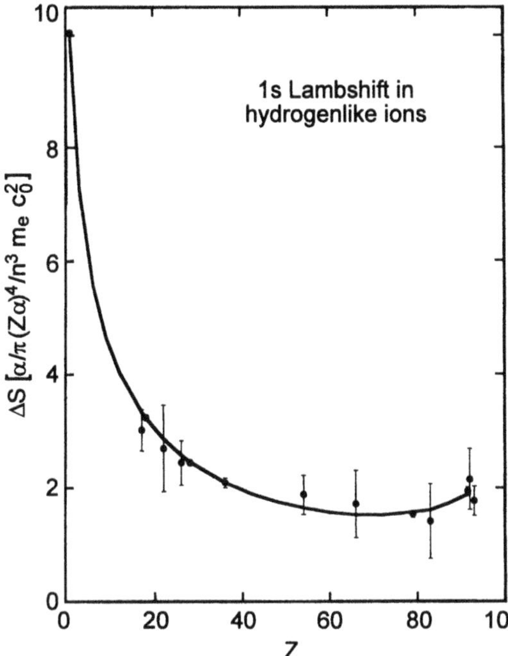

Fig. 3.4. The function $F(\alpha Z)$ for the scaled $1s$ Lamb shift as introduced in (3.25): *solid circles* with uncertainty bars correspond to experimental data listed in Table 3.16, the *curve* represents the theoretical calculation [3.6]

efforts of the different QED approaches for high-Z ions are concentrated on calculations of the various corrections in the first- and second-order of the fine-structure constant α. The leading QED corrections, not included in the Breit equation or the non-pair Hamiltonian formulation, are the self-energy (SE) and vacuum-polarization (VP) corrections which constitute a large part of the Lamb shift in HCI. Both corrections are of the order of $\alpha^5 Z^4 m_e c_0^2$. The SE corrections are related to emission and reabsorption of a virtual photon by an electron in an ion. The VP effect describes the Coulomb interaction of the electron and the nucleus when the exchanged photon excites a virtual electron-positron pair of the vacuum.

Highly Charged One-Electron Ions. Various theoretical contributions to the one-electron Lamb shift in HCI have been discussed in the literature [3.6,72,80,82]. Table 3.14 shows these contributions, taken from [3.72,80], for the $1s$ state in hydrogen-like uranium relative to the electron rest energy.

As can be observed in Table 3.14, the main radiative contributions arise from the self-energy (SE) and vacuum-polarization (VP) corrections, while the main non-QED correction is due to the nuclear finite-size effect. The contributions mentioned are represented by the Feynman diagrams in Fig. 3.5.

The FS correction is usually calculated by solving the Dirac equation in a model potential (e.g., a Fermi potential) describing the nuclear charge distribution with the root-mean-square (rms) nuclear radius R. For H-like uranium, such calculations have been performed [3.83] using the experimental

Table 3.14. Theoretical contributions, in eV, to the 1s ground-state energy of H-like uranium

Contribution	Value [eV]
One-electron Dirac eigenvalue (point nucleus)	-132280.0
One-electron SE	355.0
One-electron VP	-88.6
One-electron nuclear finite size	198.7
Nuclear recoil	0.5
Nuclear polarization	- 0.2
One-electron second-order QED	±2
Total energy	-131815±2

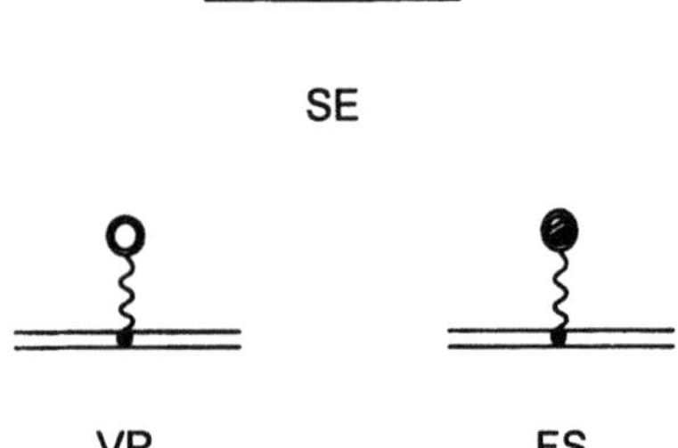

Fig. 3.5. Feynman diagrams for the one-electron Lamb shift: SE – self energy, VP – vacuum polarization and FS – nuclear finite size

[3.84] value: $R = 5.860$ fm. A highly accurate formula for the nuclear-size correction, suitable for an arbitrary model of the nuclear charge distribution, is presented in [3.85]. A complete relativistic treatment of the nuclear recoil effect can be given only within QED theory. In the lowest-order relativistic approximation ($\propto (m_\mathrm{e}/M)(\alpha Z)^4 m_\mathrm{e} c_\mathrm{o}^2$) this correction is given by [3.86]:

$$\Delta E_M = \left(m_\mathrm{e}^2 c_\mathrm{o}^4 - E_\mathrm{rel}^2\right)/(2Mc_\mathrm{o}^2), \tag{3.26}$$

where M is the nuclear mass, and E_rel is the Dirac eigenvalue given in (3.15). Calculations of the nuclear recoil correction in all orders of αZ have been performed in [3.87]. As was mentioned, the leading QED corrections which contribute to the Lamb shift, are the first-order self energy [3.88–90] and the first-order vacuum polarization [3.91–93].

As can be seen in Table 3.14, the total theoretical value of the ground state U^{91+} ion is estimated to be $E(1s) = (131\,815 \pm 2)$ eV with the accuracy mainly limited by the higher-order QED corrections which are expected to be about a few eV. The $E(1s)$ value can be compared with that given by *Johnson* and *Soff* [3.6], $E_\mathrm{th}(1s) = (131821 \pm 3)$ eV. The corresponding 1*s* Lamb shift, as follows from Table 3.14, is $\Delta S = (465 \pm 2)$ eV.

Experimentally, the Lamb shift of the ground state in H-like ions can be derived from a measurement of photon energies of K x-ray lines originating from a one-electron capture by a bare ion into excited states or from a direct transition of a free electron into the $1s$ state. The latter process of radiative recombination will be addressed in greater detail in Chap. 5. The difference between the measured and the Dirac transition energy approximately gives the $1s_{1/2}$ Lamb shift because the shifts of higher states are very small.

A few experiments with high-Z ions ($Z \geq 66$) have been carried out in which an electron is captured either in a thin foil in a single-pass experiment [3.21,53] or in a gas target of the heavy-ion storage ring (ESR) at GSI Darmstadt [3.73,94,95]. The main uncertainty in these measurements arose from the Doppler effect because the x rays were emitted by ions moving at high velocities ($\beta \geq 0.4$). Therefore, besides statistical uncertainties, the main uncertainty was introduced by the uncertainty of the observation angles of the emitted x rays. However, this uncertainty was completely overcome in the experiments with hydrogen-like Au^{78+} [3.96] and U^{91+} [3.74,75] ions which were conducted at the electron cooler at GSI by observing the x rays at an angle near 0° relative to the ion-beam direction. The necessary accuracy in the determination of the beam velocity has been achieved via the calibration of the terminal voltage of the electron cooler.

The process of radiative recombination employed is illustrated in Fig. 3.6. The cooling electrons have a very low electron temperature, typically about 0.2 eV which is small compared to the experimental uncertainties. The line with the highest energy corresponds to a direct transition of cooling electrons into the $1s$ ground state. Because the electrons travel with the same mean

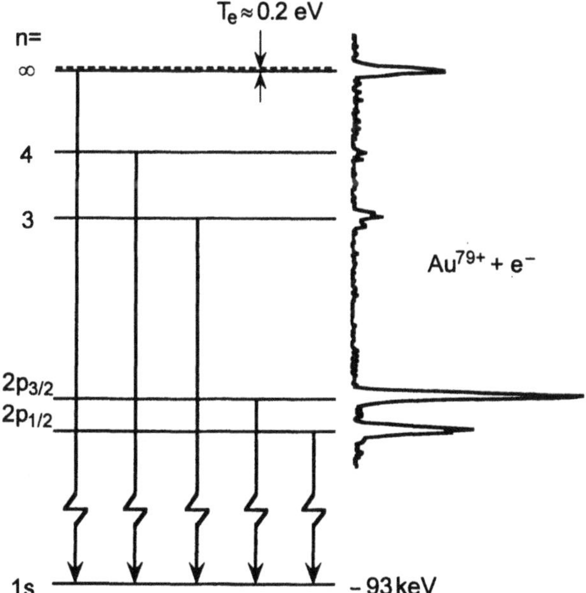

Fig. 3.6. Level scheme and x-ray spectrum measured from bare Au^{79+} ions traversing the electron cooler of the ESR, from [3.74] © 1995 IEEE

velocity as the ions with a very narrow distribution, this transition energy equals the 1s binding energy. Other lines are due to characteristic $n\ell \rightarrow 1s$ transitions induced by radiative recombination into higher $n\ell$ states with subsequent cascade transitions. We note that RR of *free* cooling electrons produce *narrow* x-ray lines in contrast to the radiative capture of *bound* electrons from target atoms (Chap. 6) which reveals considerable broadening and kinematic centroid shifts requiring a careful analysis. The x-ray spectrum resulting from the radiative recombination of stored Au^{79+} ions with free cooling electrons is shown in Fig. 3.7. The lines originate from the radiative capture into a 1s state from the Lyman series in Au^{78+} ions. The 1s Lamb shift measured in this experiment was 202.3 ± 7.9 eV. The results of a similar measurement with bare uranium ions in the electron cooler [3.75] are displayed in Fig. 3.8 where the $Ly\alpha_2$ and $Ly\alpha_1$ lines are shown. The 1s Lamb shift in this experiment with uranium ions was measured with the highest accuracy (at the time of writing this book) resulting in the value 470±16 eV. The $2s_{1/2}-2p_{1/2}$ Lamb splittings in hydrogen-like ions are given in Table 3.15. The available experimental and theoretical data on the 1s Lamb shift in H-like ions with nuclear charge $1 \leq Z \leq 110$ are presented in Table 3.16.

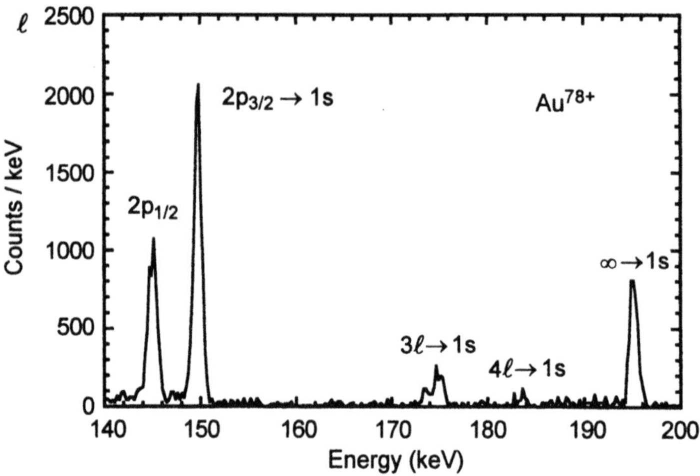

Fig. 3.7. Measured x-ray spectrum of Au^{78+} ions in the electron cooler, from [3.96] © 1994 Elsevier

Two-Electron Ions. The situation of the QED theory of high-Z two-electron ions is more complicated as compared to the one-electron system because of the necessity to include corrections caused by electron-electron interactions. Despite these difficulties, substantial progress in the theory of two-electron systems has been achieved [3.72,97–100] and the theoretical accuracy is now approaching that of one-electron systems.

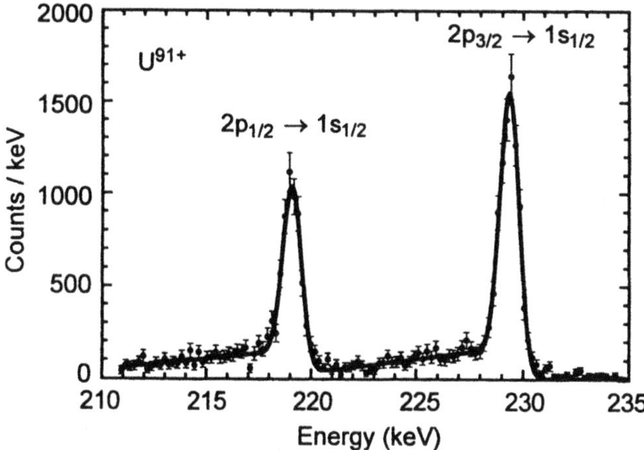

Fig. 3.8. Measured $Ly\alpha$ lines of U^{91+} ions in the electron cooler of the ESR storage ring. The intensity appearing on the low-energy side of the peaks is caused by delayed cascade feeding. From [3.75] © 1995 Springer

Table 3.15. The $2s_{1/2} - 2p_{1/2}$ Lamb splitting, in eV, in H-like ions

Z	Experiment	Uncertainty	Reference	Theory [3.6]
1	4.374897×10^{-6}	3.7×10^{-11}	[3.101]	4.37500×10^{-6}
	4.374923×10^{-6}	7.9×10^{-12}	[3.102]	
	4.374872×10^{-6}	5.0×10^{-11}	[3.103]	
	4.374864×10^{-6}	3.7×10^{-11}	[3.104]	
2	5.80752×10^{-5}	6.6×10^{-10}	[3.105]	5.8074×10^{-5}
	5.80731×10^{-5}	5.0×10^{-9}	[3.106]	
3	2.59575×10^{-4}	8.7×10^{-8}	[3.107]	2.5940×10^{-4}
6	3.22624×10^{-3}	3.3×10^{-5}	[3.108]	3.2340×10^{-3}
8	9.06539×10^{-3}	6.2×10^{-5}	[3.109]	9.0823×10^{-3}
	9.16299×10^{-3}	3.1×10^{-5}	[3.110]	
	9.10674×10^{-3}	4.5×10^{-5}	[3.111]	
9	0.013809	1.4×10^{-4}	[3.112]	0.01383
15	0.08325	0.00083	[3.113]	0.08376
	0.08343	0.00029	[3.114]	
	0.08349	0.00012	[3.115]	
	0.08342	0.00016	[3.116]	
16	0.10397	0.00099	[3.117]	0.10494
	0.10449	0.00026	[3.118]	
17	0.12899	9.9×10^{-4}	[3.119]	0.12964
	0.129	9.1×10^{-4}	[3.120]	
18	0.15716	2.5×10^{-3}	[3.121]	0.15817
	0.15670	1.6×10^{-3}	[3.122]	
92				75.94 [3.123] 75.296 [3.6]

Table 3.16. The 1*s* Lamb shift (in eV) in H-like ions versus nuclear charge Z

Z	Experiment	Uncertainty	Reference	Theory [3.6]
1	3.380006×10^{-5}	2.9×10^{-10}	[3.124]	3.38011×10^{-5}
	3.380024×10^{-5}	2.5×10^{-10}	[3.125]	
	3.380011×10^{-5}	2.07×10^{-10}	[3.126]	
	3.379999×10^{-5}	1.9×10^{-10}	[3.104]	
2				4.45426×10^{-4}
3				1.97829×10^{-3}
4				5.64215×10^{-3}
5				1.26377×10^{-2}
6				2.43257×10^{-2}
7				4.21795×10^{-2}
8				6.77822×10^{-2}
9				0.102759
10				0.148942
16	1.015	0.35	[3.127]	0.762000
17	0.85	0.10	[3.128]	0.938310
	0.84	0.12	[3.129]	
	0.90	0.10	[3.129]	
	0.825	0.12	[3.130]	
	0.807	0.15	[3.130]	
18	1.0	0.5	[3.48]	1.14140
	1.145	0.016	[3.131]	
20				1.63461
22	2.12890	0.6	[3.132]	2.26020
26	4.13	0.7	[3.133]	3.97246
	3.4	0.6	[3.133]	
	4.07	0.7	[3.134]	
	4.29	0.6	[3.134]	
28	5.07	0.10	[3.135]	5.09575
30				6.42983
36	11.95	0.50	[3.136]	11.8579
40				16.8743
50				36.0174
54	54.0	10.0	[3.52]	47.1016
60				68.3649
66	110.0	38.0	[3.94]	97.4020
70				122.831
79	212	15	[3.96]	205.300
	202.3	7.9	[3.75]	
80				217.840
82	290	75	[3.137]	244.745
83	215	100	[3.21]	259.870
90				401.833
92	520	130	[3.53]	458.490
	470	16	[3.75]	463.4 [3.138]
	429	63	[3.73]	464.6 [3.80]
	508	98	[3.139]	465.5 [3.100]
100				808.998
110				1853.57

The energy of a He-like system can be presented as the sum of two independent one-electron contributions and a two-electron contribution. The latter includes effects from the electron-electron interaction such as one- and two-photon exchange diagrams, self-energy and vacuum-polarization screening diagrams, and higher-order corrections. The leading corrections for the two-electron Lamb shift are presented in Fig. 3.9a–f by their Feynman diagrams.

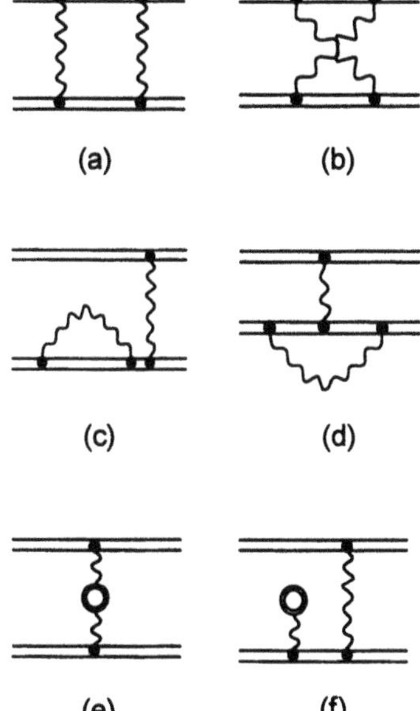

Fig. 3.9a–f. Feynman diagrams representing one- and two-photon contributions to the Lamb shift of two-electron ions: (a) and (b) – two-photon exchange diagrams; (c) and (d) – self energy; (e) and (f) – vacuum polarization

The main part of the two-electron contribution is given by the one-photon exchange diagram. Theoretical contributions to the ground state of He-like uranium compiled from [3.72,80,87,100] are given in Table 3.17. The theoretical value for the $1s$ state in He-like uranium is $E_{\mathrm{th}} = (261\,383 \pm 4)$ eV which can be compared to the ionization energy: $I_{\mathrm{th}}(1s) = 261\,383$ eV - $131\,815$ eV $= 129\,568(4)$ eV. According to calculations by Drake [3.8], $I_{\mathrm{th}}(1s) = 129\,560(9)$ eV.

A differential method established at the SuperEBIT (LLNL) which exploits radiative recombination (RR) transitions into the vacant $1s$ shell of bare and H-like ions [3.20,29], has allowed one to carry out direct measurements of the two-electron contributions to the ground state binding energies of He-like ions with nuclear charge $32 \leq Z \leq 83$. The experimental data are in good agreement with the available theoretical predictions (Table 3.18): the unified method [3.8], the Multi-Configuration Dirac-Fock treatment (MCDF) [3.10,29], the Relativistic Many-Body Perturbation Theory (RMBPT) [3.100]

Table 3.17. Theoretical contributions to the 1s ground-state energy of He-like uranium

Contribution	Value [eV]
One-electron Dirac eigenvalue (point nucleus)	-264560.0
One-electron SE	710
One-electron VP	-177
One-electron nuclear finite size	397
Nuclear recoil	1
Nuclear polarization	-0.4
One-electron second-order QED	±4
One-photon exchange	2266
Two-photon exchange	-13
SE screening	-10
VP screening	3
Total energy	-261383±4

Table 3.18. Two-electron contribution to the ionization potentials of He-like ions, in eV, from [3.20,100]. For the theoretical results, the numbers in brackets give the difference between experimental and calculated data

Z	Experiment	Relativistic MBPT	Relativistic all order	MCDF	Unified
32	562.5(1.6)	562.0(+0.5)	562.1(+0.4)	562.1(+0.4)	562.1(+0.4)
54	1027.2(3.5)	1028.2(-1.0)	1028.4(-1.2)	1028.2(-1.0)	1028.8(-1.6)
66	1341.6(4.3)	1338.6(+5.0)	1337.2(+4.4)	1336.5(+5.1)	1338.2(3.4)
74	1568(15)	1573.9(-5.0)	1574.8(-7)	1573.6(-6)	1576.6(-9)
76	1608(20)		1639.2(-31)	1637.8(-30)	1641.2(-33)
83	1876(14)	1885.8(-9)		1880.8(-5)	1886.3(-10)

and the all-order MBPT [3.2]. For comparison, the one-electron contribution was subtracted by using the hydrogen-like energies from [3.6].

In future, the experimental uncertainty in the SuperEBIT experiments is expected to be improved by up to an order of magnitude. This would provide for the first time an experimental sensitivity for the ground state QED of H- and He-like ions which is beyond the lowest-order Lamb shift.

4. Transition Probabilities

The interpretation and identification of x-ray spectra emerging from highly charged ions, the development of short wavelength lasers, x-ray plasma diagnostics and many other physical applications require the knowledge of transition probabilities and lifetimes of excited states which are considered in this chapter.

4.1 Selection Rules

Selection rules qualify the change of quantum numbers of an atom or ion undergoing a transition under the interaction with the electromagnetic radiation field. Exact and approximate selection rules are distinguished. The first ones follow from the conservation laws and the properties of the angular parts of the operators for electric and magnetic transitions. The *exact* selection rules are independent of the coupling scheme of the angular momenta and are formulated for exact quantum numbers – parity P, the total angular momentum J and its projection M – in the form:

$$\begin{aligned} \Delta P &= \begin{cases} (-1)^\kappa & \text{for E}\kappa \text{ transitions} \\ -(-1)^\kappa & \text{for M}\kappa \text{ transitions} \end{cases} \\ \Delta J &= 0, \pm 1, \ldots, \pm\kappa, \quad J + J' \geq \kappa \\ \Delta M &= 0, \pm 1, \ldots, \pm\kappa, \end{aligned} \tag{4.1}$$

where κ is the multiplicity of the transition. The *approximate* selection rules are formulated for the quantum numbers of the transition described in a certain coupling scheme.

In the ***LS*** (Russel-Saunders) coupling scheme one has the additional selection rules:
for electric Eκ transitions:

$$\begin{aligned} \Delta L &= 0, \pm 1, \ldots, \pm\kappa, \quad L + L' \geq \kappa \\ \Delta S &= 0\,, \end{aligned} \tag{4.2}$$

for magnetic Mκ transitions:

$$\begin{aligned} \Delta L &= 0, \pm 1, \ldots, \pm(\kappa - 1), \quad L + L' \geq \kappa - 1 \\ \Delta S &= 0, \pm 1, \ldots, \pm(\kappa - 1), \quad S + S' \geq \kappa - 1\,. \end{aligned} \tag{4.3}$$

For E1 transitions one has:

$$\Delta L = 0, \pm 1, \quad \Delta S = 0, \tag{4.4}$$

and $L = 0 \rightarrow L = 0$ transitions are forbidden.
For E2 transitions:

$$\Delta L = 0, \pm 1, \pm 2, \quad L + L' \geq 2\,, \tag{4.5}$$

i.e. transitions between S terms ($L = L' = 0$) and between S and P terms ($L = 0$, $L' = 1$) are ruled out.

According to (4.1) the exact selection rules for the *electric dipole* (E1) radiative transition $LSJM - L'S'J'M'$ in a multielectron system are:

$$\begin{aligned} \Delta J \equiv J' - J &= 0, \pm 1, \quad J + J' \geq 1 \\ \Delta M \equiv M' - M &= 0, \pm 1, \\ \Delta P \equiv P' - P &= \pm 1. \end{aligned} \tag{4.6}$$

For H-like ions the selection rules for dipole transitions $n\ell m - n'\ell' m'$ are:

$$\Delta \ell \equiv \ell' - \ell = \pm 1, \quad \Delta m \equiv m' - m = 0, \pm 1, \tag{4.7}$$

and for transitions $nlj - n'l'j'$ between fine-structure components, respectively,

$$\Delta j \equiv j' - j = 0, \pm 1, \quad j = \ell \pm 1/2\,, \tag{4.8}$$

where j is the total angular momentum. There are no limitations on the principal quantum numbers n and n'. Electric dipole transitions between the different hyperfine-structure levels with quantum numbers F and F' obey the following selection rules:

$$\text{E1 transitions}: \ \Delta F \equiv F - F' = 0, \pm 1, \quad F + F' \geq 1\,. \tag{4.9}$$

E1 transitions between two F-components of the same level J are forbidden by the parity selection rule, but M1 and E2 transitions are allowed. For these two cases one has:

$$\begin{aligned} &\text{M1 transitions}: \quad \Delta F = 0, \pm 1, \qquad F + F' \geq 1 \\ &\text{E2 transitions}: \quad \Delta F = 0, \pm 1, \pm 2, \quad F + F' \geq 2\,. \end{aligned} \tag{4.10}$$

A transition is called *forbidden* if the selection rules are violated. A violation of the selection rules can be spawned by electromagnetic effects such as the spin-orbit interactions. The probabilities for forbidden transitions in HCI strongly increase with the ion charge. Selection rules also exist for transitions between states described by other types of the coupling schemes ($\boldsymbol{LK}, \boldsymbol{jK}, \boldsymbol{jj}$ and others) [4.1].

4.2 Transition Probabilities

There are two main types of radiative transitions: electric multipole Eκ and magnetic multipole Mκ transitions. Within the framework of quantum electrodynamics, the probability per unit of time of a one-photon transition from an initial excited state $|a\rangle$ to a final state $|b\rangle$ is given in terms of the matrix element e.g., [4.2,3]

$$\mathrm{d}A = \frac{e^2\omega}{2\pi\hbar c_\mathrm{o}} \left|\langle b|\boldsymbol{\alpha}\boldsymbol{\epsilon}^*\mathrm{e}^{-\mathrm{i}\boldsymbol{kr}}|a\rangle\right|^2 \mathrm{d}\Omega, \tag{4.11}$$

where $\omega = (E_a - E_b)/\hbar$ is the frequency of the emitted photon, $\boldsymbol{\epsilon}$ its polarization vector, $\boldsymbol{\alpha}$ are the usual Dirac matrices, $\boldsymbol{k}$ is the photon momentum and $\mathrm{d}\Omega$ denotes the solid-angle element for the radiation. Here, $|a\rangle$ and $|b\rangle$ denote the relativistic wavefunctions obtained from the Dirac equation with a certain effective potential.

In the nonrelativistic electric-dipole approximation describing E1 transitions, the transition probability is simply obtained by utilizing the replacement

$$\mathrm{e}^{-\mathrm{i}\boldsymbol{kr}} \to 1, \quad \boldsymbol{\alpha}\boldsymbol{\epsilon}^*\mathrm{e}^{-\mathrm{i}\boldsymbol{kr}} \to \boldsymbol{p}\boldsymbol{\epsilon}^*/mc_o, \quad \boldsymbol{p} \to -\mathrm{i}m\boldsymbol{r},$$

where $\boldsymbol{p}$ is the electron momentum. Now $|a\rangle$ and $|b\rangle$ denote the nonrelativistic wavefunctions obtained from the Schrödinger equation. The total probability for E1 transition has the form

$$A(E1) = \frac{4}{3}\frac{e^2\omega^3}{c_\mathrm{o}^2} \left|\langle b|\boldsymbol{r}|a\rangle\right|^2, \tag{4.12}$$

i.e., defined by the electric-dipole matrix element. The transition probabilities for the higher multipole transitions, i.e., the electric multipoles Eκ and the magnetic multipoles Mκ, with $\kappa > 1$, are obtained from (4.11) using the higher terms in the power-series expansion

$$\mathrm{e}^{-\mathrm{i}\boldsymbol{kr}} \approx 1 - \mathrm{i}\boldsymbol{kr} + k^2r^2 + \cdots .$$

In the general case, the formulae for the corresponding transition probabilities $A^{\mathrm{E}\kappa}$ and $A^{\mathrm{M}\kappa}$ are quite complicated, they are derived in the monograph [4.2]. Let $A_{ki}^{\mathrm{E}\kappa}$ and $A_{ki}^{\mathrm{M}\kappa}$ denote the probabilities for a transition $k \to i$ of the respective Eκ or Mκ modes. Then we can introduce the dimensionless oscillator strengths $f_{ik}^{\mathrm{E}\kappa}$ and $f_{ik}^{\mathrm{M}\kappa}$ through the relations

$$\begin{aligned} g_k A_{ki}^{\mathrm{E}\kappa} &= g_i A_\mathrm{o} (\Delta E/\mathrm{Ry})^2 f_{ik}^{\mathrm{E}\kappa}, \\ g_k A_{ki}^{\mathrm{M}\kappa} &= \alpha^2 g_i A_\mathrm{o} (\Delta E/\mathrm{Ry})^2 f_{ik}^{\mathrm{M}\kappa}, \\ A_\mathrm{o} &= \alpha^3 \mathrm{Ry}/\hbar = 8.033 \times 10^9\,\mathrm{s}^{-1}, \end{aligned} \tag{4.13}$$

where ΔE is the transition energy, g is the statistical weight and α is the fine-structure constant.

For both Eκ and Mκ transitions, an order-of-magnitude estimate of the oscillator strength is

$$f^{\kappa} \simeq (\omega/c)^{2\kappa-2} \simeq (\alpha \Delta E/\mathrm{Ry})^{2\kappa-2}, \tag{4.14}$$

where ω denotes the transition frequency. The probability of two- and more-photon radiation is much more complicated than (4.11) and can also be found in [4.2].

H-like Ions. In the case of H-like ions, the transition probabilities of the lowest excited states can be expressed in a closed analytical form. The $2s$ state has the same parity as the $1s$ ground state, therefore it can decay to the ground state either by a magnetic dipole (M1) transition or by a two-photon electric-dipole (2E1) transition. For M1 transitions the leading term of the relativistic transition probability is [4.4]

$$\mathrm{M1}: \; A(2s_{1/2} - 1s_{1/2}) = \frac{\alpha^9 Z^{10}}{972} \frac{m_e e^4}{\hbar^3} \simeq 2.46 \times 10^{-6} Z^{10} \; [\mathrm{s}^{-1}], \tag{4.15}$$

where Z is the nuclear-charge number of the ion.

Fully relativistic calculations for the 2E1 transitions in H-like ions have been carried out [4.5–8]. The corresponding probability can be approximated by a simple analytical expression [4.7]

$$\begin{aligned} \mathrm{2E1}: \; A(2s_{1/2} - 1s_{1/2}) &= 8.22943\, Z^6 \\ &\times \frac{1 + 3.9448\,(\alpha Z)^2 - 2.040\,(\alpha Z)^4}{1 + 4.6019\,(\alpha Z)^2} \; [\mathrm{s}^{-1}] \end{aligned} \tag{4.16}$$

with an accuracy of 0.05% for nuclear charges $1 \leq Z \leq 92$.

Measured probabilities for the decay of the $2s_{1/2}$ states in H-like ions are compared with theoretical data in Table 4.1.

Table 4.1. The decay rates, in s^{-1}, of the $2s_{1/2}$ state in H-like ions

Z	Experiment (from [4.9,10])	Theory [4.7]	Theory [4.6]
2	491^{+95}_{-140}	526.61	526.62
	520(21)		
	525(5)		
8	$2.21(22)\times10^6$	2.1552×10^6	2.1553×10^6
9	$4.22(28)\times10^6$	4.3699×10^6	4.3072×10^6
16	$1.37(13)\times10^8$	1.3964×10^8	1.3966×10^8
18	$2.82(20)\times10^8$	2.8590×10^8	2.8594×10^8
28	$4.606(38)\times10^9$	4.6368×10^9	4.6372×10^9

Two-photon transitions take place via intermediate virtual states. We note that both probabilities A(2E1) and A(M1) for the $2s$ state increase with the nuclear charge Z more rapidly than the probability A(E1) for the electric dipole transition $2p - 1s$, which approximately scales as $A(\mathrm{E1}) \sim Z^4$. For ions with $Z > 40$ the M1 transitions represent the main contribution to the radiative decay of the $2s$ state. The ratio $A(\mathrm{M1})/A(\mathrm{E1})$ strongly increases with Z and for H-like uranium ($Z = 92$) reaches the value of about 4×10^{-3}. Calculated transition probabilities in H-like ions from the $n = 2$ states are given in Table 4.2. Experimental and theoretical values for A in H-like ions can be also found in [4.4,7,9–11].

Table 4.2. Calculated M1, 2E1 and E1 transition probabilities (in s^{-1}) for transitions originating from $n = 2$ states in H-like ions

Z	$2s_{1/2}$-$1s_{1/2}$ M1 [4.11]	$2s_{1/2}$-$1s_{1/2}$ 2E1 [4.6]	$2p_{1/2} - 1s_{1/2}$ E1 [4.11]	$2p_{3/2} - 1s_{1/2}$ E1 [4.11]
1	2.4946×10^{-6}	8.2291	$6.2649\times10^{+8}$	$6.2648\times10^{+8}$
2	2.5559×10^{-3}	$5.2660\times10^{+2}$	$1.0028\times10^{+10}$	$1.0027\times10^{+10}$
3	1.4744×10^{-1}	$5.9973\times10^{+3}$	$5.0772\times10^{+10}$	$5.0764\times10^{+10}$
4	$2.6192\times10^{+0}$	$3.3689\times10^{+4}$	$1.6048\times10^{+11}$	$1.6043\times10^{+11}$
5	$2.4406\times10^{+1}$	$1.2847\times10^{+5}$	$3.9181\times10^{+11}$	$3.9163\times10^{+11}$
6	$1.5121\times10^{+2}$	$3.8348\times10^{+5}$	$8.1252\times10^{+11}$	$8.1198\times10^{+11}$
7	$7.0694\times10^{+2}$	$9.6657\times10^{+5}$	$1.5054\times10^{+12}$	$1.5041\times10^{+12}$
8	$2.6895\times10^{+3}$	$2.1526\times10^{+6}$	$2.5684\times10^{+12}$	$2.5654\times10^{+12}$
9	$8.7423\times10^{+3}$	$4.3614\times10^{+6}$	$4.1146\times10^{+12}$	$4.1084\times10^{+12}$
10	$2.5100\times10^{+4}$	$8.2015\times10^{+6}$	$6.2721\times10^{+12}$	$6.2604\times10^{+12}$
12	$1.5580\times10^{+5}$	$2.4453\times10^{+7}$	$1.3009\times10^{+13}$	$1.2975\times10^{+13}$
14	$7.3003\times10^{+5}$	$6.1554\times10^{+7}$	$2.4110\times10^{+13}$	$2.4022\times10^{+13}$
16	$2.7845\times10^{+6}$	$1.3688\times10^{+8}$	$4.1145\times10^{+13}$	$4.0950\times10^{+13}$
18	$9.0777\times10^{+6}$	$2.7686\times10^{+8}$	$6.5935\times10^{+13}$	$6.5538\times10^{+13}$
20	$2.6149\times10^{+7}$	$5.1965\times10^{+8}$	$1.0054\times10^{+14}$	$9.9797\times10^{+13}$
22	$6.8154\times10^{+7}$	$9.1804\times10^{+8}$	$1.4728\times10^{+14}$	$1.4596\times10^{+14}$
24	$1.6357\times10^{+8}$	$1.5427\times10^{+9}$	$2.0872\times10^{+14}$	$2.0648\times10^{+14}$
28	$7.7347\times10^{+8}$	$3.8637\times10^{+9}$	$3.8719\times10^{+14}$	$2.8154\times10^{+14}$
30	$1.5525\times10^{+9}$	$5.8231\times10^{+9}$	$5.1061\times10^{+14}$	$2.0206\times10^{+14}$
34	$5.5106\times10^{+9}$	$1.2238\times10^{+10}$	$8.4378\times10^{+14}$	$3.2559\times10^{+14}$
38	$1.7050\times10^{+10}$	$2.3636\times10^{+10}$	$1.3190\times10^{+15}$	$1.2834\times10^{+15}$
42	$4.7295\times10^{+10}$	$4.2657\times10^{+10}$	$1.9723\times10^{+15}$	$1.9073\times10^{+15}$
46	$1.2003\times10^{+11}$	$7.2824\times10^{+10}$	$2.8443\times10^{+15}$	$2.7317\times10^{+15}$
50	$2.8303\times10^{+11}$	$1.1868\times10^{+11}$	$3.9800\times10^{+15}$	$3.7935\times10^{+15}$
54	$6.2741\times10^{+11}$	$1.8595\times10^{+11}$	$5.4291\times10^{+15}$	$5.1318\times10^{+15}$
58	$1.3198\times10^{+12}$	$2.8162\times10^{+11}$	$7.2458\times10^{+15}$	$6.7872\times10^{+15}$
62	$2.6541\times10^{+12}$	$4.1409\times10^{+11}$	$9.4897\times10^{+15}$	$8.8017\times10^{+15}$
66	$5.1345\times10^{+12}$	$5.9328\times10^{+11}$	$1.2225\times10^{+16}$	$1.1218\times10^{+16}$
70	$9.6037\times10^{+12}$	$8.3060\times10^{+11}$	$1.5521\times10^{+16}$	$1.4079\times10^{+16}$
74	$1.7443\times10^{+13}$	$1.1391\times10^{+12}$	$1.9453\times10^{+16}$	$1.7427\times10^{+16}$
78	$3.0881\times10^{+13}$	$1.5331\times10^{+12}$	$2.4100\times10^{+16}$	$2.1301\times10^{+16}$
82	$5.3459\times10^{+13}$	$2.0287\times10^{+12}$	$2.9547\times10^{+16}$	$2.5740\times10^{+16}$
86	$9.0760\times10^{+13}$	$2.6427\times10^{+12}$	$3.5884\times10^{+16}$	$3.0776\times10^{+16}$
90	$1.5151\times10^{+14}$	$3.3931\times10^{+12}$	$4.3202\times10^{+16}$	$3.6434\times10^{+16}$
92	$1.9468\times10^{+14}$	$3.8251\times10^{+12}$	$4.7260\times10^{+16}$	$3.9502\times10^{+16}$

He-like Ions. Helium-like ions are the simplest many-electron systems. However, the situation of transition probabilities is complex because the atomic structure in He-like ions changes drastically with an increasing ion charge. The energy-level pattern of He-like ions is characterized by two independent terms: singlet and triplet. Figure 4.1 shows the energy-level diagrams for the low-lying states in a one- and in a two-electron heavy ion. Excitations to $2s$ and $2p$ states are regarded. For the ion with two electrons the states are described by the $\boldsymbol{jj}$-coupling scheme. In the pure $\boldsymbol{LS}$ coupling all intercombination transitions ($\Delta S = 1$) are forbidden. However, in highly charged ions the selection rule $\Delta S = 0$ is violated through influences of relativistic effects (electromagnetic interactions) which rapidly increase with increasing Z. Consequently, the intensities of the intercombination lines increase as well. In this case one has to use the intermediate coupling scheme introduced in Sect. 3.2. For example, for the intercombination transition $2\,^3P_1 - 1\,^1S_0$, the intermediate coupling mixes the $2\,^3P_1$ and $2\,^1P_1$ states and the transition probability $A(2\,^3P_1 - 1\,^1S_0)$ increases with increasing ion charge: $A(\text{He}) = 1.79 \times 10^2\,\text{s}^{-1}$, $A(\text{Fe}^{24+}) = 4.42 \times 10^{13}\,\text{s}^{-1}$, and $A(\text{U}^{90+}) = 2.99 \times 10^{16}\,\text{s}^{-1}$, respectively (Table A.5).

Table 4.3 summarizes the approximate Z dependence of the transition probabilities A in the form $A \sim Z^a$ for transitions between the $n = 1$ and $n = 2$ states in He-like ions. For highly charged ions the decay of triplet states increases dramatically due to relativistic effects and the excited L states decay promptly to the ground state, except the $2\,^3P_0$ state.

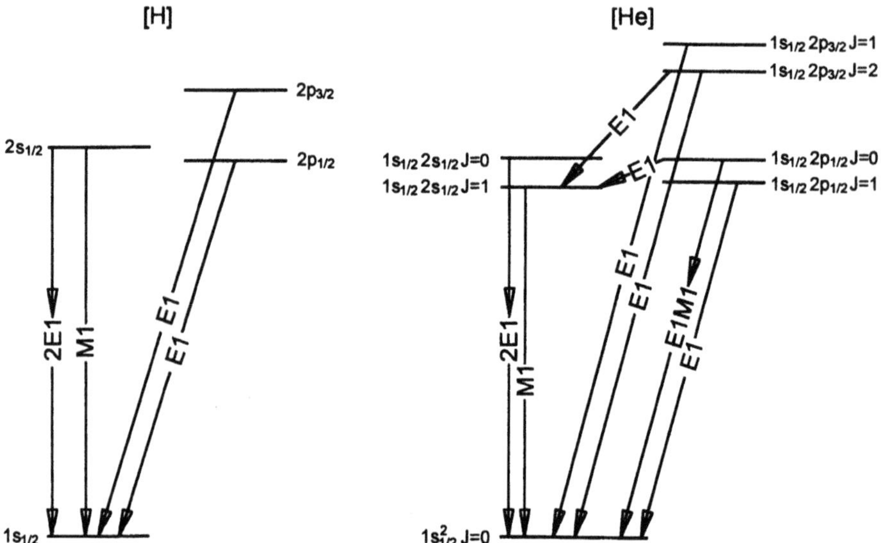

Fig. 4.1. Level diagrams for the low-lying states of hydrogen-like and helium-like heavy ions

Table 4.3. Approximate scaling of the transition probabilities A with the nuclear charge number Z, expressed as a power law $A \propto Z^a$, for the transitions $2 \to 1$ and $2 \to 2$ in He-like ions

Transition	Type	Range of Z	a
$2\,^1P_1 \to 1\,^1S_0$	E1	$Z < 80$	4
$2\,^1P_1 \to 2\,^1S_0$	E1	$Z < 30$	3
$2\,^1P_1 \to 2\,^3S_1$	E1	$Z > 4$	6
$2\,^1S_0 \to 2\,^3S_1$	M1	$Z < 50$	7
$2\,^1S_0 \to 2\,^3P_1$	E1	$10 < Z < 20$	2
$2\,^1S_0 \to 1\,^1S_0$	2E1	$Z > 1$	6
$2\,^3S_1 \to 1\,^1S_0$	M1	$4 < Z < 80$	10
$2\,^3S_1 \to 1\,^1S_0$	2E1	$Z < 25$	10
$2\,^3P_1 \to 1\,^1S_0$	E1	$6 < Z < 20$	10
$2\,^3P_1 \to 1\,^1S_0$	E1	$20 < Z < 40$	8
$2\,^3P_1 \to 1\,^1S_0$	E1	$Z > 40$	4
$2\,^3P_1 \to 2\,^3S_1$	E1	$Z < 50$	1
$2\,^3P_1 \to 2\,^3S_1$	E1	$Z > 50$	2
$2\,^3P_2 \to 2\,^3S_1$	E1	$15 < Z < 20$	1
$2\,^3P_2 \to 2\,^3S_1$	E1	$Z > 20$	4
$2\,^3P_2 \to 1\,^1S_0$	M2	$Z < 80$	8
$2\,^3P_0 \to 2\,^3S_1$	E1	$Z < 50$	1

Accurate nonrelativistic calculations of transition probabilities in He-like ions have mainly been made for $2 \to 1$ and for $2 \to 2$ transitions, e.g., [4.12–14]. Relativistic calculations of transition probabilities of high-Z ions are based on the Relativistic Random-Phase Approximation (RRPA) [4.15,16], the Multiconfiguration Dirac-Fock (MCDF) [4.17], Configuration Interaction (CI) expansions [4.18], relativistic many-body theory [4.19], and the QED perturbation theory [4.4,20,21]. Sophisticated relativistic calculations of transition amplitudes and probabilities have been performed in the helium isoelectronic sequence up to the nuclear charge $Z = 100$ [4.22].

The E1 Decay Rates of the $1s2p\,^1P$ and ^3P States. Accurate relativistic calculations of atomic probabilities A(E1) for electric-dipole transitions from singlet and triplet $2p$ states to the ground and $2s$ states in He-like ions have been performed by *Johnson* et al. [4.22].

All three $2\,^3P_{0,1,2}$ levels decay to the $2\,^3S_1$ state by optically allowed E1 transitions. This is the dominant decay mode for low ionization stages. The $2\,^3P_1$ state is mixed with the $2\,^1P_1$ and higher $1snp\,^1P_1$ states by the spin-orbit interaction. The mixing is very small in helium but increases rapidly with increasing nuclear charge Z. The allowed decay mode for the $2\,^3P_1$ state is the electric-dipole E1 transition to the $2\,^3S_1$ state. The $2\,^1P_1$ level, however, can directly decay to the ground state by electric-dipole E1 radiation, and has a very short lifetime. The $2\,^3P_{0,2}$ levels do not mix with the singlet system and they have lifetimes ~ 10 times higher as compared to the $2\,^3P_1$ level.

The $2\,^3P_0$ level can decay to the ground state only by two- (E1M1) or three-photon (3E1) transitions which are ineffective in the whole range of Z. The $2\,^3P_2$ state can also decay to the ground state by the magnetic-quadruple M2 transition. Theoretical analysis shows that the transition probabilities are small for the low-Z systems but since A(M2) scales as Z^8 it becomes the dominating depopulation mode of the $2\,^3P_2$ state for ions with $Z > 20$. The magnetic-quadrupole M2 transition $2\,^3P_2 \to 2\,^1S_0$ is also possible but its probability remains low for all Z values.

The $2\,^3P_2$ state decays to the ground state by an M2 transition for which the probability can be estimated by the formula [4.4]:

$$\begin{aligned} \text{M2: } & A(2\,^3P_2 - 1\,^1S_0) \\ & = \frac{\alpha^7 Z^8}{1215} \left(\Delta E/Z^2\text{Ry}\right)^5 \frac{me^4}{\hbar^3} \left[1 + 0.28\,(\alpha Z)^2\right] \\ & \simeq 0.037\, Z^8 \left[1 + 0.28\,(\alpha Z)^2\right] \left(\Delta E/Z^2\text{Ry}\right)^5 \;[\text{s}^{-1}], \quad Z \geq 30, \end{aligned} \tag{4.17}$$

The total decay of the $2\,^3P_2$ state consists of two branches, the $2\,^3P_2 \to 2\,^3S_1$ (E1) and the $2\,^3P_2 \to 1\,^1S_0$ (M2) modes. Experimental and theoretical data for the total decay rates of the $2\,^3P_2$ state are given in Table 4.4.

Table 4.4. The decay rates (in s^{-1}) of the $2\,^3P_2$ state in He-like ions (from [4.22])

Z	Experiment	Theory [4.22]	Unit
6	5.99(22)	5.705	10^7
7	6.71(32)	6.901	10^7
8	8.26(14)	8.182	10^7
9	10.0(1.0)	9.598	10^7
	9.58(14)	9.598	10^7
13	1.852(69)	1.877	10^8
15	2.778(77)	2.904	10^8
	2.94(26)	2.904	10^8
16	4.00(32)	3.743	10^8
17	5.38(29)	4.933	10^8
18	5.9(1.0)	6.618	10^8
	6.17(30)	6.618	10^8
22	2.27(15)	2.326	10^9
	2.48(25)	2.326	10^9
24	4.65(76)	4.347	10^9
26	9.1(1.6)	7.929	10^9
	8.9(1.0)	7.929	10^9
	8.3(1.0)	7.929	10^9
28	1.429(61)	1.404	10^{10}
29	2.13(23)	1.849	10^{10}
47	8.06(72)	9.140	10^{11}

The $2\,^1S_0 \to 1\,^1S_0$ Transitions. The $1s2s\,^1S_0$ level decays to the ground state by a two-photon emission 2E1. The corresponding transition probabilities have been calculated in [4.13,23]. An accurate extrapolation formula for the decay rate including relativistic corrections has the form [4.13]

$$\begin{aligned}
2\text{E1} : A(2\,^1S_0 - 1\,^1S_0) &= 16.458762\, Z^6 \\
&\times \left[\left(\frac{Z - 0.806389}{Z} \right)^6 \left(1 + \frac{1.539}{(Z+2.5)^2} \right) \right. \\
&\left. -(\alpha Z)^2 \, \frac{0.6571 + 2.040\,(\alpha Z)^2}{1 + 4.6019\,(\alpha Z)^2} \right] \; [\text{s}^{-1}],
\end{aligned} \tag{4.18}$$

where Z is the nuclear charge. A nonrelativistic power-series expansion of (4.18) yields:

$$\begin{aligned}
2\text{E1} :\; A(2\,^1S_0 - 1\,^1S_0) &\simeq 16.458762\, Z^6 \\
&\times \left(1 - \frac{4.8383}{Z} + \frac{11.293}{Z^2} - \frac{25.63}{Z^3} + \frac{87.15}{Z^4} + \cdots \right) [s^{-1}] \,.
\end{aligned} \tag{4.19}$$

Experimental and calculated two-photon 2E1 decay probabilities for the $2\,^1S_0$ state are given in Table 4.5. In general, the relativistic effects decrease the decay rates although they increase the $2\,^1S_0 - 1\,^1S_0$ energy separation.

Table 4.5. The two-photon decay rates (in s^{-1}) of the $2\,^1S_0$ state in He-like ions

Z	Experiment (from [4.9,24])	Theory [4.13]
2	50.0(2.5)	50.943(73)
3	1988(98)	1938.6(1.8)
18	$4.31(34)\times10^8$	$4.215(10)\times10^8$
28	$6.410(66)\times10^9$	$6.482(21)\times10^9$
36	$2.934(30)\times10^{10}$	$2.993(12)\times10^{10}$
92		$7.238(77)\times10^{12}$

The $2\,^1S_0$ level can also decay to the $2\,^3P_1$ state by a spin-forbidden E1 transition but with much lower probability because the $2\,^1S_0$ level is close to the $2\,^3P_1$ level in energy. We note that for ions with $Z \leq 6$ the $2\,^1S_0$ level lies lower than the $2\,^3P_1$ level (Table A.4).

The $2\,^3S_1 \rightarrow 1\,^1S_0$ Transitions. The $2\,^3S_1$ level can decay to the ground state by a relativistically induced M1 transition or by a 2E1 two-photon emission. The M1 transition probability is 3–4 orders of magnitude higher than that for 2E1 decay for all Z and can be estimated by [4.4]:

$$\begin{aligned}
\text{M1}: A(2\,^3S_1 - 1\,^1S_0) &= \frac{2^5}{3^9} \alpha^9 Z^{10} (\Delta E / Z^2 \text{Ry})^3 \frac{me^4}{\hbar^3} \simeq \\
&\simeq 0.397\, Z^{10} (\Delta E/Z^2\text{Ry})^3 \; [\text{s}^{-1}], \quad Z \geq 10,
\end{aligned} \tag{4.20}$$

Experimental and calculated M1 decay rates are compared in Table 4.6. Relativistic calculations of intense M1 decay rates are also reported in [4.4,15,22].

Table 4.6. The M1 decay rates (in s^{-1}) of the $2\,^3S_1$ state in He-like ions (from [4.22])

Z	Experiment	Theory [4.22]	Unit
2	1.11(33)	1.266	10^{-4}
3	1.71(38)	2.035	10^{-2}
6	4.857(11)	4.860	10^{1}
7	2.561(31)	2.537	10^{2}
10	1.105(18)	1.092	10^{4}
16	1.41(17)	1.426	10^{6}
17	2.82(19)	2.661	10^{6}
18	4.93(29)	4.787	10^{6}
22	3.88(20)	3.750	10^{6}
23	5.92(24)	5.913	10^{6}
26	2.08(26)	2.075	10^{8}
35	4.46(14)	4.360	10^{9}
36	5.848(75)	5.822	10^{9}
41	2.1954(68)	2.217	10^{10}
47	9.01 (16)	9.083	10^{10}
54	3.92(12)	3.846	10^{11}

Experimental transition probabilities for the intercombination $2\,^3P_1 \to 1\,^1S_0$ (E1) decay are given in Table 4.7 in comparison with calculated values. In general, forbidden transitions like Mκ and intercombination transitions are very important for the identification of x-ray lines in Solar and laboratory plasma spectra.

Table 4.7. The E1 decay rates (in s^{-1}) of the $2\,^3P_1$ state in He-like ions, from [4.22]. The calculated rate represents the sum of the rates for the decay to the $2\,^3S_1$ and to the $1\,^1S_0$ states

Z	Experiment	Theory [4.22]	Unit
6	8.85(39)	8.478	10^{7}
7	2.04(12)	2.071	10^{8}
8	6.58(35)	6.292	10^{8}
9	1.88(7)	1.924	10^{9}
12	3.45(18)	3.388	10^{10}
13	7.81(43)	7.533	10^{10}
14	1.57(8)	1.572	10^{11}
16	6.37(73)	5.820	10^{11}

For He-like uranium one has the following theoretical values for transition probabilities A (in s^{-1}):

$$\begin{aligned}
\text{2E1}: A(2\,^1S_0 - 1\,^1S_0) &= 7.24 \times 10^{12}, \\
\text{M1}: A(2\,^3S_1 - 1\,^1S_0) &= 1.21 \times 10^{14}, \\
\text{E1}: A(2\,^1P_1 - 1\,^1S_0) &= 5.00 \times 10^{16},
\end{aligned}$$

$$\begin{aligned} \text{M2}: A(2\,^3P_2 - 1\,^1S_0) &= 2.06 \times 10^{14}, \\ \text{E1}: A(2\,^3P_1 - 1\,^1S_0) &= 2.99 \times 10^{16}, \\ \text{E1M1}: A(2\,^3P_0 - 1\,^1S_0) &= 5.61 \times 10^{9}. \end{aligned} \tag{4.21}$$

Theoretical and experimental data on transition probabilities for other sequences (Li-, Be-, B-like, etc.) have been the subject of reviews [4.25–27]. The information on transition probabilities and oscillator strengths can also be found in several papers [4.28–35].

4.3 Lifetimes

The radiative lifetime τ_k of an excited state k is defined as

$$\tau_k = \left[\sum_{i<k} A_{ik}\right]^{-1}, \tag{4.22}$$

where the A_{ik} are transition probabilities (Sect. 4.2) of all possible radiative transitions including forbidden and intercombination transitions.

Data and bibliography on radiative lifetimes for neutral atoms can be found in the monograph [4.36]. Recent progress in lifetime measurements of intercombination transitions in different isoelectronic sequences is reviewed in [4.37]. Most of the data on lifetimes for HCI have been obtained employing the beam-foil time-of-flight method covering a wide range of ion charges. For high-charge few-electron systems, traps are used including ion storage rings [4.38,39] and electron-beam ion traps (EBIT) [4.40]. Precision lifetime measurements of HCI permit systematic checks of theoretical approaches but only for ions having one or two electrons outside a closed shell.

Beam-foil time-of-flight measurements of forbidden transitions (2E1, M1, M2) in H- and He-like ions have been carried out with accuracies ranging from 0.1% for light ions up to about a few percent for heavy ions. Experimental data on lifetimes of the $2s_{1/2}$ states of H-like ions are given in [4.4,11,41]. Forbidden transitions from the $n = 2$ levels in He-like ions ranging from $Z = 18$ up to $Z = 64$ in the x-ray region can be found in [4.37,41–43]. The lifetimes of the metastable $2\,^3S_1$ state in He-like ions are given in Table 4.8 and some recent data on lifetimes of HCI are presented in Table 4.9.

Measured lifetimes involving the $2s^2\,^1S_0 - 2s2p\,^{1,3}P_1$ resonance and intercombination transitions in Be-like ions with the nuclear charge $Z < 55$ are critically evaluated and systematized in [4.31]. Accurate calculations of radiative lifetimes of HCI are given in [4.6,11,13,44–46].

Radiative lifetimes of excited ions with noble-gas configuration are of particular interest because these levels are employed in certain short-wavelength laser schemes, astrophysics and thermonuclear fusion studies. Corresponding beam-foil measurements of lifetimes have been analyzed in [4.28,32,37,47] and some results are reproduced in Table 4.10. Figure 4.2 represents results [4.47]

Table 4.8. Theoretical and experimental lifetimes τ (in s) of the $2\,^3S_1$ state in He-like ions

Z	Experiment (from [4.4])	Theory [4.11]	Unit
2	0.909(27)	0.939	10^4
	0.9(3)		
16	0.706(83)	0.700	10^{-6}
17	0.354(24)	0.375	10^{-6}
18	0.172(12)	0.208	10^{-6}
	0.202(2)		
22	0.258(13)	0.266	10^{-7}
23	0.169(7)	0.169	10^{-7}
26	0.50(5)	0.481	10^{-8}
	0.48(6)		
36	0.20(6)	0.171	10^{-9}
54	0.2554(76) [4.48]		10^{-12}

Table 4.9. Theoretical and experimental lifetimes τ (in s) of excited states in H- and in He-like ions

Ion	State	Experiment	Reference	Theory	Reference	Unit
O^{6+}	$2\,^3P_2$	12.10(20)	[4.43]	12.2	[4.44]	10^{-9}
	$2\,^3P_1$	1.52(8)	[4.43]	1.6	[4.44]	10^{-9}
F^{7+}	$2\,^3P_2$	10.44(15)	[4.43]	10.4	[4.44]	10^{-9}
	$2\,^3P_1$	0.531(20)	[4.43]	0.52	[4.44]	10^{-9}
	$2\,^3P_0$	9.48(20)	[4.43]	11.0	[4.44]	10^{-9}
Ar^{17+}	$2\,^2S_{1/2}$	3.487(36)	[4.49]	3.497	[4.6]	10^{-9}
Ni^{26+}	$2\,^1S_0$	156.1(1.6)	[4.10]	154.3(0.5)	[4.13]	10^{-12}
Ni^{27+}	$2\,^2S_{1/2}$	217.1(1.8)	[4.10]	215.45	[4.13]	10^{-12}
Kr^{34+}	$2\,^1S_0$	34.08(34)	[4.42]	33.41(14)	[4.13]	10^{-12}
Br^{33+}	$2\,^1S_0$	39.32(32)	[4.10]	39.63(16)	[4.13]	10^{-12}
Nb^{39+}	$2\,^1S_0$	15.33(60)	[4.41]	15.245	[4.13]	10^{-12}
Ag^{45+}	$2\,^3P_2$	1.24(11)	[4.41]	1.08	[4.45]	10^{-12}
Ag^{46+}	$2s_{1/2}$	4.49(8)	[4.41]	4.35	[4.6]	10^{-12}
U^{90+}	$2p_{1/2}\,^3P_0$	54.4(3.4)	[4.50]	57.31	[4.12,22]	10^{-12}

for lifetimes of the $3p\,^3D_3$ and $3d\,^3F_3$ levels in Ne-like ions. Both experimental and theoretical scaled lifetimes $(Z-9)\tau$ seem to approach a constant value as predicted by *Westerlind* et al. [4.51].

The lifetimes for forbidden M1 transitions $3d^4\,^5D_2-{}^5D_3$ between finestructure levels of the ground state have been measured in Ti-like Xe^{32+} and Ba^{34+} ions using the EBIT source [4.52]. Besides those results, the experiments clearly demonstrated that the EBIT can also be used as a valuable tool for optical spectroscopy of highly charged ions.

Experimental data on lifetimes of highly charged ions with many-electrons are given in various papers, e.g., [4.31,33–35,53–57].

Table 4.10. Experimental and theoretical lifetimes (in ps) of $2p^5 3s\,^3P$ and ^1P states in Ne-like ions (from [4.28])

Ion	$^1P_1$ Experiment	$^1P_1$ Theory	$^3P_1$ Experiment	$^3P_1$ Theory
Ne^0	1470±100	1547	29600±1000	
	1870±180		31700±1600	
	1300±100		29800±2000	
Na^+	320±50	346	6000±1200	5623
	580±60		10600±500	
Mg^{2+}	100±15	131	1900±190	1820
	110±15			
Al^{3+}	46±10	63		637
Si^{4+}	28±8	36		248
P^{5+}	18±7	21	130±30	137
S^{6+}	14.5±1.0	15	52±2	52
	12±3		49±13	
Cl^{7+}	13±1	11	27±1	27
	10±2		34±12	
	8±2		30±5	
Ar^{8+}	6.5±2.0		19±4	16
Se^{24+}		0.60		0.28
Br^{25+}		0.53		0.24
Kr^{26+}		0.47		0.21
Rb^{27+}		0.42		0.19
Sr^{28+}		0.38		0.16
Y^{29+}		0.34		0.14

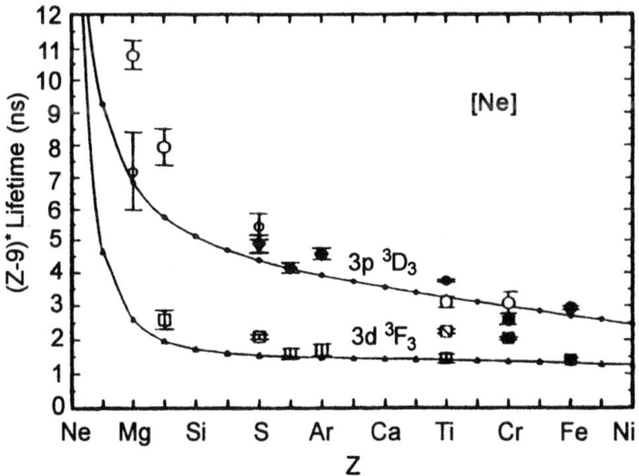

Fig. 4.2. Scaled lifetimes $(Z-9)\tau$ for the $3p\,^3D_3$ and $3d\,^3F_3$ levels of Ne-like ions as a function of the nuclear charge Z. *Symbols*: experimental data from *Ando* et al. [4.47]; *solid lines*: theoretical values of *Hibbert* et al. [4.35], [4.47] © 1996 Academic Press

4.4 Autoionizing States

The autoionizing states of highly charged ions – or the states lying above the ionization limit – take a significant part in many processes important for x-ray spectroscopy and diagnostics of hot laboratory and astrophysical plasmas [4.58–60] and in electron-ion collisions, especially in resonance scattering such as resonant excitation and single and multiple ionization [4.61], dielectronic recombination and photoionization [4.62].

In the context of x-ray spectra of HCI, one of the most striking features of the autoionizing states is the formation of satellite lines (or dielectronic satellites) arising from a radiative decay of autoionizing states (Sect. 6.1). In the case of a doubly excited autoionizing state $X^{q+**}(a_1 n\ell)$ there can be two decay channels:

$$\text{autoionization}: \; X^{q+**}(a_1 n\ell) \;\rightarrow\; X^{(q+1)+}(a_0) + e \tag{4.23}$$

$$\text{radiation}: \; X^{q+**}(a_1 n\ell) \;\rightarrow\; X^{q+*}(a_0 n\ell) + \hbar\omega \tag{4.24}$$

where n and ℓ are the principal and orbital quantum numbers, respectively. A transition $a_1 n\ell \rightarrow a_0 n\ell$ associated with the emission of a photon $\hbar\omega$ is called a *satellite* line because it appears in the vicinity of the parent line $a_1 \rightarrow a_0$. The intensities of satellites strongly increase with increasing ion charge and are very sensitive to the plasma density and temperature. To a large extent, x-ray diagnostics of hot plasmas is based on the spectroscopy of autoionizing states. A general discussion of autoionization processes can be found in [4.60,63,64]. Dielectronic satellites to lines of He- and Li-like ions have been analyzed and calculated [4.65–71] and Be-like ions have also been studied [4.72].

If the energy of a discrete atomic state is greater than the first ionization potential, then an atom (or ion) can make a radiationless transition to a continuum state of the same energy, so in the final state one has a free electron and an ion in its ground or excited state:

$$X^{q+**}(a_1 n\ell) \;\rightarrow\; X^{(q+1)+}(a_0) + e(\epsilon\ell'). \tag{4.25}$$

This process is called Auger or autoionization decay. Here ϵ and ℓ' are the kinetic energy and angular momentum, respectively, of the free electron and

$$\epsilon = E(a_1 n\ell) - E(a_0) > 0 \,, \tag{4.26}$$

where E are the corresponding atomic energies.

The type of the initial state may vary appreciably. In particular, it can be a state with a hole in an inner shell or a state with one or more simultaneously excited electrons. For example, excitation of an inner $1s$ electron of a Li-like ion, with configuration $1s^2 2s$, into a $2p$ state may form the metastable autoionizing state $1s2s2p\,^4P_{5/2}$. In Be-like ions (4 electrons), the autoionizing states $1s2s^2 2p$, $1s2s2p^2$, $1s2p^3$, or generally, $1sn\ell n'\ell' n''\ell''$ are also possible.

In first-order nonrelativistic perturbation theory, the autoionization probability A_a is given by

$$A_a = \frac{2\pi}{\hbar} \left|\langle a_1 n\ell|V|a_0\epsilon\ell'\rangle\right|^2 , \tag{4.27}$$

where V is the interaction operator, coupling the discrete and continuum states and may be either electrostatic $(1/r_{12})$ or magnetic.

An excited metastable state may also decay by a radiative transition to a lower-lying state with a probability A_r. The total lifetime τ of the autoionizing state is given by

$$\tau = (A_a + A_r)^{-1} , \tag{4.28}$$

where A_a and A_r denote the probabilities for autoionization and for radiative decay, respectively. The selection rules for nonradiative transitions via different interactions are summarized in Table 4.11. There, the notation $\Delta L \equiv L' - L$, $\Delta S \equiv S' - S$, etc. is used where the angular quantum numbers LSJ and $L'S'J'$ correspond to the discrete and continuum states, respectively. The parity of a state is defined by the sum $\Sigma_i \ell_i$ taken over all electrons.

Table 4.11. Selection rules for autoionization transitions (from [4.73]); α is the fine-structure constant, m_e and m_p are electron and proton masses, respectively

Interaction	ΔL	ΔS	ΔJ	ΔP	Order of magnitude
Coulomb	0	0	0	0	1
Spin-orbit	0, ±1	0, ±1	0	0	α^4
Spin-other-orbit	0, ±1	0, ±1	0	0	α^4
Spin-spin	0, ±1, ±2	0, ±1, ±2	0	0	α^4
Hyperfine	0, ±1, ±2	0, ±1	0, ±1	0	$\alpha^4(m_e/m_p)^2$

In the relativistic approximation, the autoionization probability can be written in the form:

$$\Gamma = \frac{2\pi}{\hbar} \left| \sum_{i<f} \langle \Psi_1 | V_{if} | \Psi_0 \rangle \right|^2 \rho(\epsilon), \tag{4.29}$$

where $\Psi_{0,1}$ are the initial bound Dirac and final free states, respectively, and the two-electron operator V_{12} is a sum of the Coulomb and Breit interactions of the type

$$V_{12} = \frac{1 - \boldsymbol{\alpha}_1 \boldsymbol{\alpha}_2}{r_{12}} - \frac{(\boldsymbol{\alpha}_1 \nabla_1)(\boldsymbol{\alpha}_2 \nabla_2) r_{12}}{2}, \quad r_{12} = |r_1 - r_2|, \tag{4.30}$$

Here $\boldsymbol{\alpha}_1$ and $\boldsymbol{\alpha}_2$ are the four-dimensional Dirac matrices and $\rho(\epsilon)$ is the energy density of the outgoing continuum electron states. The wavefunction Ψ_0 is usually normalized to the Dirac delta function.

The following computer codes are used for calculation of energy levels and transition probabilities of autoionizing states: Multi-Configuration Hartree-Fock method (MCHF) [4.74], SUPERSTRUCTURE [4.75], Z-expansion Method (MZ) [4.76,77], AUTOLSJ [4.78] and Multi-Configuration Dirac-Fock method (MCDF) [4.79].

Table A.7 in Appendix A.2, shows calculated absolute energies for doubly excited He-like ions using the Z-expansion method (MZ-method) with relativistic corrections and (a) the intermediate coupling scheme or (b) with the use of the $\boldsymbol{LS}$-coupling scheme. The calculated autoionization probabilities of the satellites to the resonance line in H-like ions are given in Table A.8. Table A.9 presents the calculated transition probabilities for doubly excited states in He-like ions.

5. Radiative Processes

The elementary processes photoionization, radiative recombination and Bremsstrahlung will be treated as being responsible for the ionization-recombination balance and for the emission of characteristic short-wavelength radiation of highly charged ions in plasmas. We will also briefly summarize the development of x-ray lasers as monochromatic sources of VUV and x-ray radiation employing plasmas of highly charged ions as active media.

5.1 Photoionization and Radiative Recombination

5.1.1 General Relations

Photoionization and the inverse process, radiative recombination, belong to the basic radiative processes characterizing the interaction of electromagnetic radiation with atoms and ions. Radiative emission resulting from electron-ion collisions are responsible for the appearance of prominent spectral features in the VUV and x-ray regions, which are frequently employed for the investigation of fundamental atomic collisional and radiative interactions and for the spectroscopic determination of basic plasma properties, such as temperature, density, polarization of radiation and charge-state distribution [5.1,2]. The properties of photoionization and radiative recombination have been considered in many reviews and monographs [5.3–8].

The photoionization process consists of photon absorption and ejection of a bound electron into continuum

$$X^{q+} + \hbar\omega \rightarrow X^{(q+1)+} + e(\epsilon, \lambda), \tag{5.1}$$

where ϵ and λ are the energy and the angular momentum of the photoelectron. The inverse process to photoionization is the photorecombination or radiative recombination (RR)

$$X^{(q+1)+} + e \rightarrow X^{q+} + \hbar\omega. \tag{5.2}$$

Cross sections $\sigma_k(\omega)$ of multielectron photoionization

$$X^{q+} + \hbar\omega \rightarrow X^{(q+k)+} + k\,e, \quad k \geq 1, \tag{5.3}$$

are measurable quantities (e.g., [5.9,10]). The total or *photoabsorption* cross section is given by

$$\sigma_{\mathrm{abs}}(\omega) = \sum_{k=1}^{N}\sum_{\gamma}\sigma_k^{(\gamma)}(\omega), \tag{5.4}$$

where the sum over γ runs over all subshells of the target, and N is the total number of target electrons.

Photoionization σ_ν and photorecombination σ_{r} cross sections are mutually related through the principle of detailed balance (the Milne relation, e.g., [5.11]):

$$g_q\,\sigma_\nu(\omega) = \frac{2m_{\mathrm{e}}c_{\mathrm{o}}^2\epsilon}{\hbar^2\omega^2}\,g_{\mathrm{q}+1}\,\sigma_{\mathrm{r}}(\epsilon), \tag{5.5}$$

where g denotes the statistical weight. The total probability, per unit of time, for photoionization is give by

$$W = \int_I^\infty c_{\mathrm{o}}N_\omega\sigma_\nu(\omega)\,\mathrm{d}\omega, \tag{5.6}$$

where N_ω is the photon density per frequency interval at a frequency ω and I is the threshold energy.

In the relativistic approximation the photoionization cross section, similar to the radiative transition probability (Sect. 4.2), is determined by the matrix element

$$\sigma_\nu \propto |\langle\Psi_f|\boldsymbol{\alpha}\boldsymbol{a}^*\exp(\mathrm{i}\boldsymbol{k}\boldsymbol{r})|\Psi_i\rangle|^2\,, \tag{5.7}$$

where $\boldsymbol{\alpha}$ is the usual Dirac matrix, $\boldsymbol{a}$ is the polarization vector of the photon, $\boldsymbol{k}$ is the photon momentum, $\Psi_{i,f}$ are initial (bound) and final (continuum) atomic wavefunctions obtained from the Dirac equation with the self-consistent-field electrostatic potential.

In the nonrelativistic approximation σ_ν is obtained as

$$\sigma_\nu \propto |\langle\Psi_f|\boldsymbol{a}^*\boldsymbol{p}|\Psi_i\rangle|^2\,, \tag{5.8}$$

where $\boldsymbol{p}$ is the momentum of the photoelectron and $\Psi_{i,f}$ now are solutions of the Schrödinger equation with the effective potential.

In the nonrelativistic dipole approximation the photorecombination cross section has the form

$$\sigma_\nu = \frac{4}{3\times 137}\frac{Q}{2l+1}\frac{\hbar\omega}{\mathrm{Ry}}\left[\ell R_{\ell-1}^2 + (\ell+1)R_{\ell+1}^2\right]\pi a_0^2, \tag{5.9}$$

$$R_{\ell\pm1} = \int_0^\infty P_{n\ell}(r)P_{\epsilon,\ell\pm1}(r)\mathrm{d}r, \tag{5.10}$$

where $P_{n\ell}(r)$ and $P_{\epsilon\lambda}$ are the radial wavefunctions of the electron in the discrete and continuum states, respectively, with the normalization

$$\int_0^\infty P_{n\ell}^2(r)\mathrm{d}r = 1, \tag{5.11}$$

$$\int_0^\infty P_{\epsilon\lambda}P_{\epsilon'\lambda}(r)\mathrm{d}r = \pi\delta(\epsilon-\epsilon'). \tag{5.12}$$

The angular coefficients Q depend on the type of transition. For a transition $n\ell^N$ - $n\ell^{N-1}$ they are given by

$$Q = N,$$

where N is the number of equivalent electrons. For a transition $n\ell^N S_0 L_0$ - $n\ell^{N-1} S_1 L_1$ they read

$$Q = N\left|G_{S_1L_1}^{S_0L_0}\right|^2,$$

where G is the Racah fractional parentage coefficient [5.12].

In the case of photoionization (and RR) of H-like ions from the $n\ell$ states with $1s \le n\ell \le 5g$ the cross sections can be written in a closed analytical form [5.3,13,14]. Extensive numerical calculations [5.15,16] on bound-free transitions in H-like ions were performed and the asymptotic behavior [5.17–19] of the photoionization cross sections was evaluated. For H-like ions in the ground state the cross section has the form

$$\sigma_\nu(1s) = \frac{2^9\pi^2}{3}\frac{\alpha}{Z^2}\left(\frac{\omega_0}{\omega}\right)^4\frac{\mathrm{e}^{-4\kappa\,\mathrm{arccot}\kappa}}{1-\mathrm{e}^{-2\pi\kappa}}\,a_0^2, \tag{5.13}$$

$$\sigma_r(1s) = \frac{2^8\pi^2}{3}\frac{\alpha^3}{Z^2}\frac{\mathrm{e}^{-4\kappa\,\mathrm{arccot}\kappa}}{1-\mathrm{e}^{-2\pi\kappa}}\,a_0^2, \tag{5.14}$$

where $\hbar\omega_0 = Z^2\mathrm{Ry} \approx I_{1s}$, $\kappa^2 = \omega_0/(\omega-\omega_0)$, and α denotes the fine-structure constant. With increasing frequency ω, the photoionization cross section σ_ν decreases first according to the $\omega^{-8/3}$ law, and then at $\hbar\omega - I_{n\ell} \gg I_{n\ell}$ it varies as

$$\sigma_\ell^\nu \propto \omega^{-(3+\ell+1/2)}, \quad \sum_\ell \sigma_\ell^\nu \propto \omega^{-7/2}. \tag{5.15}$$

Here $I_{n\ell}$ is the absolute value of the binding energy of the $n\ell$ state. In the relativistic region $\hbar\omega \gg 50Z^2\mathrm{Ry}$, one has

$$\sum_\ell \sigma_\ell^\nu \propto \omega^{-3}. \tag{5.16}$$

The RR cross section $\sigma_\mathrm{r} \to \infty$ at threshold $\hbar\omega = I_{n\ell}$ and decreases as $\omega^{-5/2}$ at $\hbar\omega - I_{n\ell} \gg I_{n\ell}$.

Photoionization cross sections and radiative rate coefficients α_r, i.e., the RR cross sections averaged over a Maxwell electron-velocity distribution with a temperature T_e, scale in proportion to the following quantities

$$Z_\mathrm{eff}^2\,\sigma(\omega/Z_\mathrm{eff}^2), \quad Z_\mathrm{eff}^{-1}\,\alpha_\mathrm{r}(T_\mathrm{e}/Z_\mathrm{eff}^2),$$

where ω is the photon frequency and Z_eff is the effective nuclear charge of the ion (the spectroscopic symbol).

The Kramers Formulae. The photoionization σ_ν and RR cross section σ_r are quite often presented in the form

$$\sigma_\nu(n\ell) = Q_\nu g(n\ell)\,\sigma_\nu^{\rm Kr}(n), \quad \sigma_{\rm r}(n\ell) = Q_{\rm r} g(n\ell)\,\sigma_{\rm r}^{\rm Kr}(n), \tag{5.17}$$

where Q_ν and Q_r are the angular factors and $g(n\ell)$ is a dimensionless quantity called the bound-free *Gaunt* factor. The quantities $\sigma_{\nu,\rm r}^{\rm Kr}(n)$ are the Kramers cross sections, obtained from classical considerations [5.20], describing the ℓ-averaged cross sections

$$\sigma_\nu^{\rm Kr}(n) = \frac{64\pi\,\alpha}{3\sqrt{3}\,Z^2\,n^5}\left(\frac{\omega_0}{\omega}\right)^3 a_0^2, \tag{5.18}$$

$$\sigma_{\rm r}^{\rm Kr}(n) = \frac{32\pi\,\alpha^3\omega_0^2}{3\sqrt{3}\,\omega\,n^3(\omega-\omega_0/n^2)}\,a_0^2, \quad \hbar\omega_0 = Z^2\,{\rm Ry}. \tag{5.19}$$

Through the detailed balance (5.5), the cross sections $\sigma_{\nu,\rm r}^{\rm Kr}(n)$ are related to the statistical weights $g_q = 2n^2$ and $g_{q+1} = 1$, respectively. The Kramers cross sections contain the main dependencies on atomic parameters such as the ion charge Z, the photon frequency ω and the principal quantum number n but do not describe the dependence on the angular-momentum quantum number ℓ. The formulae (5.18,19) are often used with the so-called Stobbe correction [5.21] which is especially important for low-lying n levels (Fig. 5.1).

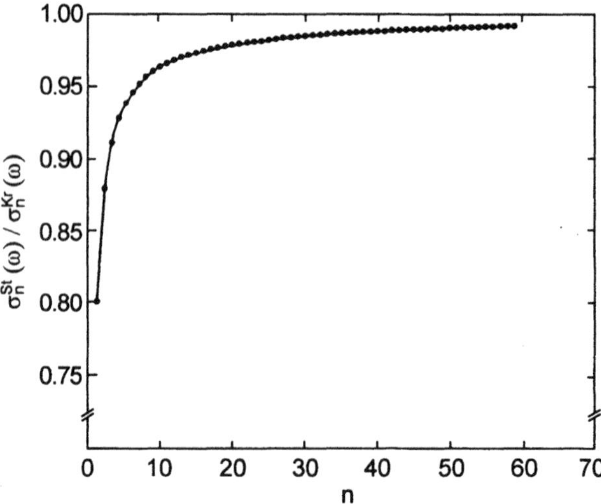

Fig. 5.1. The ratio of the Stobbe photoionization cross section to the Kramers cross section as a function of the principal quantum number n at the low-energy limit

The RR rate coefficients corresponding to the cross sections (5.19) have the form

$$\alpha_{\rm r}^{\rm Kr}(n) = 2KZ\beta^{3/2}e^\beta\,|Ei(-\beta)|, \tag{5.20}$$

$$\beta = \frac{Z^2{\rm Ry}}{n^2T_{\rm e}}, \quad K = \frac{32\sqrt{\pi}\,a_0^2c_0}{3\sqrt{3}\,137^4} = 2.60\times10^{-14}\,{\rm cm^3s^{-1}},$$

where T_e is the electron temperature and $Ei(x)$ denotes the exponential integral. According to (5.20), $\alpha_r^{Kr}(n) \propto n^{-1}$ at $\beta \gg 1$ and $\alpha_r^{Kr}(n) \propto n^{-3}$ at $\beta \ll 1$. The Kramers formula (5.20) is very practical for estimates of the contribution from highly excited states to the total rate of radiative recombination. A classical (Kramers) approach for describing, in general, radiative and collisional processes, caused by electrons of low and moderate energies, has been developed by *Kogan* et al. [5.22].

5.1.2 Photoionization

Photoionization processes play an important role in applied problems of quantum radiophysics, atomic physics, solid-state physics and astrophysics. In particular, many problems of diagnostics of laser-produced plasmas, on its radiation in VUV and X-ray regions, energy transport problems and radiative cooling in laser-thermonuclear targets, the use of EXAFS (EXtended Absorption Fine Structure) and XANES (X-ray Absorption Near-Edge Structure) spectroscopical methods for investigation of solid-state structure [5.23] require the knowledge of high-accuracy photoionization cross sections and of photorecombination rates of atoms and ions.

Experimental data on photoionization cross sections of positive ions are very scarce and, to our knowledge, are available only for ions with a charge state $q < 5$ (see [5.7,24–27]), because the photoabsorption spectra of multiply ionized atoms are quite difficult to observe. In the last few years some techniques of absorption spectroscopy of light ions have been developed. In particular, two methods have to be mentioned. In the first, a dense column of atomic vapors irradiated by a saturating pulse of radiation, resonant on a suitable line of the neutrals, is ionized almost completely. By using dye lasers, singly and doubly charged ions can be produced. The second technique employs two laser-produced plasmas, one acting as a background continuum-radiation source, and the other as an absorbing medium.

At present, absorption spectra of several multiply charged ions have been observed in various experiments, in particular, for B^+ and Be^{q+}, $q = 1, 2, 3$ ions including both discrete and continuous transitions in the soft x-ray range [5.24], and for C^{4+} [5.25], K^+ and Ca^{2+} [5.27] ions. Typical experimental photoabsorption spectra are shown in Fig. 5.2. The spectrum clearly shows the discrete series of lines followed by the photoionization jumps of the $1s$ electron, and the continuum. The dashed curves represent calculations by *Reilman* and *Manson* [5.28].

As a rule, at threshold the measured photoionization cross sections are in good agreement with calculations performed in the nonrelativistic dipole approximation, e.g., [5.28–31]. At higher photon energies, the behavior of photoionization cross sections is strongly affected by the presence of large and wide resonances spanning more than an order of magnitude as can be observed in Fig. 5.3. These resonances are formed from the autoionization of

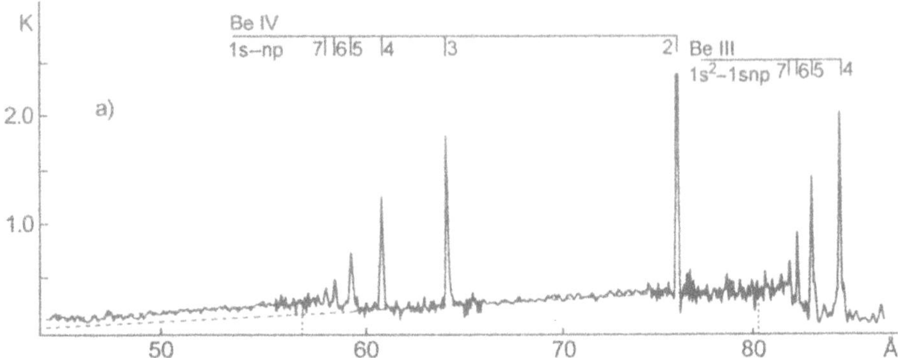

Fig. 5.2. Experimental absorption coefficient [5.24] of Be^{3+} and Be^{2+} ions as a function of the photon energy in Å. *Dashed curve* – calculation from [5.28]. Figure from [5.24] © 1987 Royal Swedish Academy of Sciences

intermediate Rydberg states of the target and strongly enhance the background cross sections. The resonance series converge onto their ionization limits, e.g., to the states $3p^6 3d^9 4snp$, $3p^5 3d^{10} 4sns, \ldots$, $n = 5, 6, \ldots, \infty$ in case of photoionization of Cu-like Kr^{7+} ions.

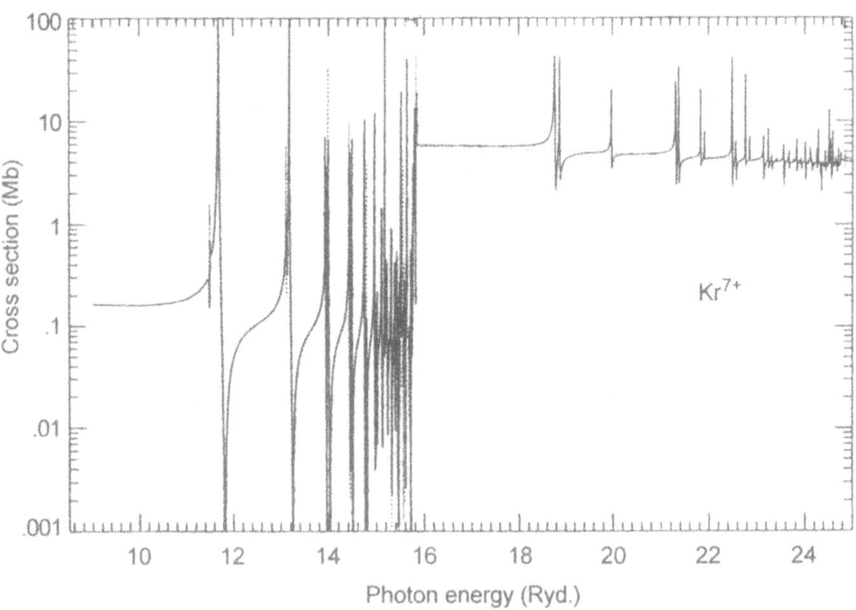

Fig. 5.3. Cross section for photoionization of Cu-like Kr^{7+} ions from their $4s$ ground state, calculated [5.31] using the R-matrix method: *dotted curve* – calculated with 3 final states accounted for, *solid curve* – 5-state calculations. [5.31] © 1992 Royal Swedish Academy of Sciences

Photoionization from excited states is followed by Photo Excitation of Core Electrons, (the PEC processes, [5.32]) which leads to the appearance of large resonances, followed by autoionization, and can strongly modify the photoionization cross section over a wide range of frequencies as demonstrated in Fig. 5.4a,b. PEC resonances are especially important for photoionization

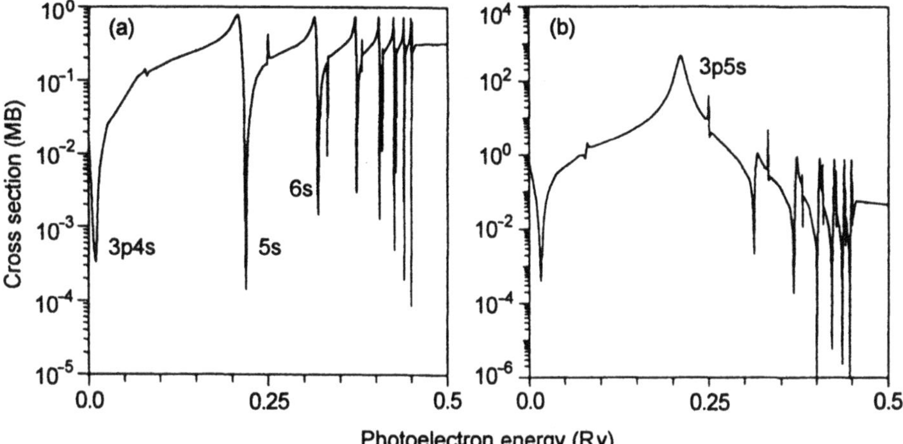

Fig. 5.4a,b. The importance of PEC resonances $3pns$ in the case of photoionization of Al^+ ions shown by close-coupling calculations of *Butler* et al. [5.33]: (a) photoionization from the $3s^2$ ground state, (b) photoionization from the $3s5s$ excited state. [5.33] © 1992 IOP

from excited states of low-Z ions. A simple example of a PEC process is

$$2sn\ell + \hbar\omega \rightarrow (2pn\ell)^{**} \rightarrow 1s2s + e, \quad (5.21)$$

where $(2pnl)^{**}$ is an autoionizing state lying above the $2s$ series limit.

Recently, an ion source has been exposed to synchrotron radiation to perform a merged-beam ion-spectrometry experiment on singly charged ions. Some attempts are also in progress to study inner-shell photoionization processes in ions trapped in a Penning ion trap, using the white synchrotron radiation from a bending magnet. Nevertheless, various experiments performed during the last years in the different synchrotron radiation centers, showed that electron spectrometry is the most suitable technique for the study of electron correlations [5.34] and, in particular, it allows one to carry out separately the measurement of photoionization cross sections in various subshells of atoms or ions, in contrary to photoabsorption or ion spectrometry which only measure total photoionization or photoabsorption cross sections. It is also the best way to study the nonradiative de-excitation (autoionization and Auger decays) of inner-shell vacancies created by photoionization.

A big volume of accurate calculations of atomic radiative data (oscillator strengths and photoionization cross sections) have been performed in the framework of the international collaboration known as the *Opacity Project* aiming at computing extensive sets of accurate data required for re-estimating stellar envelope opacities [5.35–37]. Calculations have been carried out for astrophysically abundant elements from H to Ni in all their ionization stages, using the R-matrix technique. Recent advances in the field of opacity calculations for stellar envelopes are summarized in [5.38]. The elements considered are listed in Table 5.1 together with their relative abundance in Solar photosphere.

Table 5.1. Solar photospheric abundance used in work for the Opacity Project. The values of the logarithm of the abundance are normalized so that log(abundance) = 12.00 for hydrogen. From [5.38]

Element	Z	log(abundance)
H	1	12.00
He	2	11.00 ±0.02
C	6	8.55 ±0.05
N	7	7.97 ±0.07
O	8	8.87 ±0.07
Ne	10	8.07 ±0.06
Na	11	6.33 ±0.03
Mg	12	7.58 ±0.05
Al	13	6.47 ±0.07
Si	14	7.55 ±0.05
S	16	7.21 ±0.06
Ar	18	6.52 ±0.10
Ca	20	6.36 ±0.02
Cr	24	5.67 ±0.03
Mn	25	5.39 ±0.03
Fe	26	7.51 ±0.01
Ni	28	6.25 ±0.04

Calculations of photoionization cross sections of atoms and ions are often performed in the nonrelativistic dipole approximation. Relativistic, retardation and multipole effects in photoionization cross sections have been considered in [5.39]. The relative importance of various effects change with photon energy and with nuclear-charge number Z. Many-electron correlation effects are important for low photon energies and low-Z atoms [5.40,41], while relativistic effects are important for high-energy photoionization processes [5.42].

The nonrelativistic dipole approximation breaks down at photon energies higher than $\hbar\omega > 20$ keV when the factor kr in the exponent $\exp(i\boldsymbol{kr})$ is no longer small. However, as was shown in [5.39], a comparison between the nonrelativistic dipole photoionization cross sections and those calculated in the fully relativistic multipole approach is remarkably good for atomic s subshells even at the highest energies where the nonrelativistic approximation is not valid. This phenomenon does not occur for subshells with $\ell \neq 0$ and in angular distributions. The authors [5.39] explain their results in terms of an approximate cancellation of relativistic, retardation and multipole effects which occurs only for s subshells (Sect. 6.2.3).

5.1.3 Radiative Recombination

Radiative recombination processes (RR) play an important role in the determination of the ionization-recombination balance of HCI in both colliding electron-ion beam experiments and in high-temperature laboratory and astrophysical plasmas. Motivations for studying these processes include an understanding of fundamental processes in reactions of free electrons with ions, determination of the corresponding cross sections and rates for use in models of energy transport phenomena and plasma diagnostics. Furthermore, x rays from RR of bare ions and subsequent x-ray transitions originating from deexcitation cascades, offer new possibilities to study the QED effects in very heavy H-like ions [5.43,44].

Ion-accelerator facilities and ion-storage-ring technology have opened a new era of electron-ion collision studies about a decade ago [5.45]. Using heavy-ion storage rings supplied with electron coolers, a very high energy resolution in electron-ion recombination-rate and cross-section measurements could be accomplished with energy spreads as small as about 10 meV at low center-of-mass energies.

Two major technical developments have facilitated this enormous progress: the design of cold and yet dense electron targets and the construction of ion storage rings providing extremely intense high-quality ion beams. By the technique of adiabatic expansion of electron beams in a decreasing magnetic field [5.46] (Sects. 2.3.2, 6.1) new recombination measurements with unprecedented quality, both in energy resolution and statistics, have become feasible.

Only recently [5.47,48] did it become possible to determine the characteristic energy spread of the electron beam by an independent method and to measure the RR cross sections or, more precisely, the RR rate coefficients $\langle \sigma v \rangle$ averaged over the electron-velocity distribution function $f(\boldsymbol{v})$ in a beam, which is the Maxwell two-temperature function (2.34) displayed on page 50. Usually, $T_\perp \approx 0.1$ eV and $T_\parallel \approx 0.001$ eV, so $T_\parallel \ll T_\perp$ which corresponds to the flattened distribution and is typical for modern electron coolers (Sect. 2.3.2). The typical behavior of the RR rate is shown in Fig. 5.5 for recombination of Li-like Bi^{80+} ions. For ions with one valence electron, experimental data are well described by the Kramers formula with the Stobbe correction. For many-electron ions the situation is much more complicated, but the semiclassical Kramers formula for the total RR cross sections is still used with an appropriate effective charge Z_{eff} which depends on the electron energy, the charge of the ion and the total number of the target electrons, e.g., [5.49,50]. In most cases, this procedure leads to an agreement with experimental data within 50% but in some cases it fails completely.

The ionized and recombined HCI were detected at the electron cooler in the Heidelberg test storage ring TSR with stored Li-like Si^{11+}, Cl^{14+} ions

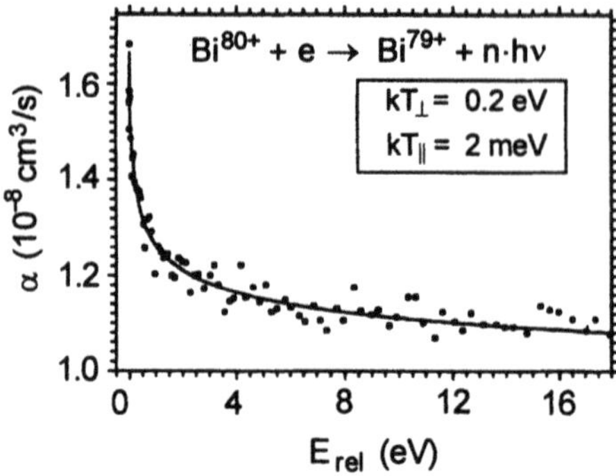

Fig. 5.5. Rates for radiative recombination of Li-like Bi^{80+} ions as a function of the relative energy E_{rel}. *Dots*: experiment [5.51]; *solid curve*: Kramers formula (5.19) with Stobbe correction [5.21] convoluted with the experimental electron distribution function with the two temperatures indicated in the figure. Reprinted from [5.51] with permission

as well as Na-like Cl^{6+}, Fe^{15+} and Se^{23+} ions [5.52]. Absolute cross sections and rates of radiative and dielectronic recombination (Chap. 6) have been measured for electron-ion center-of-mass energies E_{cm} ranging from 0 eV up to several keV. In separate experiments at the GSI accelerator facilities absolute rates for recombination of highly charged ions such as Au^{76+} and U^{28+} were determined with an energy spread as low as 10 meV. The RR rates for Ne^{10+} and Ar^{13+} have been measured at very low electron energies (< 1 meV) using the ion storage ring CRYRING at the Manne Siegbahn Laboratory [5.53].

The initial recombination studies in merged-beam single-pass experiments showed good agreement between RR theory and experimental results. However, in the storage-ring experiments, such a good agreement was only observed for low-Z ions. Increasing deviations from theoretical RR predictions for heavier ions have been found. In some of the nonbare heavy ions, deviations from RR theory by one to two orders of magnitude have been reported. For example, in the case of RR in U^{28+} ions, the observed cross section is a factor 20 to 50 higher than the RR expectation. For highly charged Ar^{13+} ions with several active electrons, the enhancement amounts to one order of magnitude compared with the expected RR rate [5.53]. Very low-lying DR resonances were shown to contribute the majority of this enhancement. However, even for a bare ion of similar charge, Ar^{18+}, a rate coefficient enhanced by a factor of 10 was measured near zero relative energy (Fig. 5.6). A convincing explanation of these phenomena does not exist yet.

The theory of electron-ion recombination processes in plasmas has been reviewed by *Hahn* [5.50]. Various analytical expressions for the photoionization cross sections and radiative rate coefficients have been considered in many papers, e.g., [5.3,18,20,21,54–56]. Extensive tabulations of recombination cross sections and of rate coefficients can be found in the literature [5.50,55,57–59]. A formalism based on projection operators was employed for electron-ion photorecombination processes which takes into account the interference between dielectronic and radiative recombination [5.60].

The photon-energy spectrum for radiative recombination has been calculated in analytical form based on the nonrelativistic dipole approximation

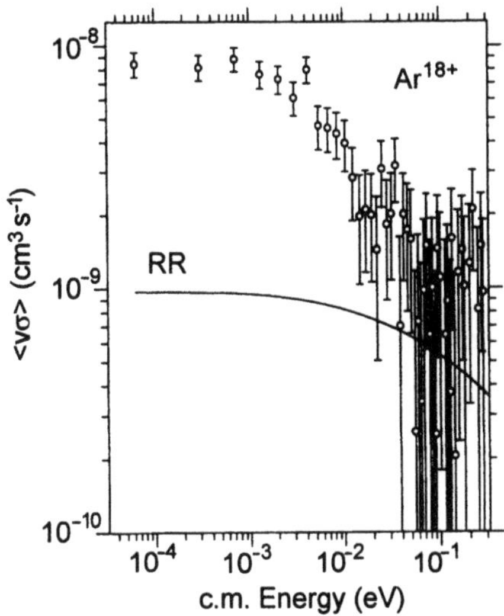

Fig. 5.6. Measured recombination rate of bare Ar^{18+} ions [5.61] as a function of the electron-ion center-of-mass energy. The solid curve represents the RR rate calculated with the Kramers formula (5.19) and convoluted with the electron distribution ($T_\perp = 0.2$ eV, $T_\parallel = 2$ meV). Reprinted from [5.61] with permission

[5.62] (also [5.18,63]). In case of a final $1s$ state and for the situation at an electron cooler with an anisotropic velocity distribution, the corresponding recombination rate differential in solid angle and photon energy can be expressed as

$$\begin{aligned}\frac{\partial^2 \alpha_{\mathrm{r}}}{\partial \Omega \partial \hbar\omega} &= 6.7 \times 10^{-16}\, \frac{I_{1s}[\mathrm{eV}]^3/(\hbar\omega)^2}{T_\perp T_\parallel} \exp\left(-\frac{\hbar\omega - I_{1s}}{T_\perp}\right) \\ &\times \int_0^1 \left[1 + \cos^2\theta + \left(1 - 3\cos^2\theta\right) x^2\right] \\ &\times \exp\left[-\left(\hbar\omega - I_{1s}\right)^2 \left(T_\parallel^{-1} - T_\perp^{-1}\right)^2 x^2\right] \mathrm{d}x \; \mathrm{cm^3 s^{-1} eV^{-1}}.\end{aligned} \tag{5.22}$$

The most striking feature of the photon-energy spectrum for the RR into the ground state is the existence of a sharp edge corresponding exactly to the electron binding energy I_{1s} [5.62]. This makes the photons from RR very attractive for precise x-ray spectroscopy, especially, in light of the recent progress in producing cold electron beams, with the transverse temperature as low as $T_\perp = 1$–10 meV [5.46]. For illustration, the calculated energy spectrum of photons originating from radiative recombination of U^{92+} into the K-shell is shown in Fig. 5.7. Such photons are detected in a collinear geometry at the GSI electron cooler where the small intrinsic width of the x-ray lines is used for an accurate determination of K-shell binding energies of very heavy few-electron ions.

The method has been used for an accurate measurement of the $1s$ Lamb shift in highly charged H-like ions (Sect. 3.5). Using free cooling electrons with a low temperature of about 0.2 eV and a dense ion beam, the binding energy of the $1s$ state is measured which, after subtracting the calculated

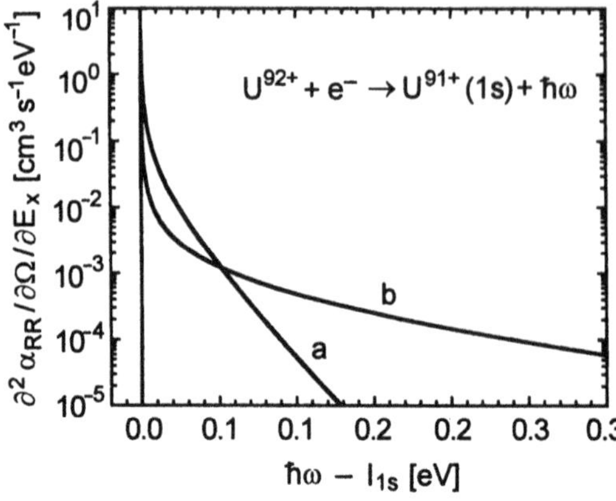

Fig. 5.7. Photon-energy- and angle-differential recombination coefficients according to the analytical expression (5.22) for U^{92+} ions, shown in the CM system at $\theta_{CM} = 0°$. The two *curves* are for (*a*) $T_\perp = 0.02$ eV, $T_\parallel = 1 \times 10^{-4}$ eV and (*b*) $T_\perp = 0.2$ eV, $T_\parallel = 1 \times 10^{-4}$ eV

Dirac energy, gives the $1s_{1/2}$ Lamb shift with a high accuracy. Because of the small temperatures of the cooling electron beam and of the cooled ion beam, RR in the electron cooler produces x-ray spectra with very narrow spectral lines. The spectra can have very low background levels when the lines are Doppler shifted beyond the Bremsstrahlung limit corresponding to the cooler voltage and when the x-ray photons are measured in coincidence with ions suffering a decrement of their charge state. This way x-ray spectra were obtained for H-like Au^{79+} and U^{91+} ions [5.43,64,65]. Figure 5.8 shows a coincident x-ray spectrum for 321 MeV/u U^{92+} ions stored in the ESR. The main intensity is distributed among the characteristic $Ly\alpha_{1,2}$ radiation and the radiative-recombination line labeled $\infty \rightarrow 1s$. The former are induced by radiative recombination populating high-n states with subsequent cascades [5.44]. This leads to a characteristic lineshape which is well understood in terms of the RR population mechanism and the atomic lifetimes involved. A general overview of the physics of highly charged heavy ions revealed by storage-cooler rings is presented in several papers [5.66,67].

Photorecombination cross sections for highly charged uranium ions have been calculated in [5.68] using the manybody perturbation theory. The cross sections are resolved with regard to the final recombined level, corresponding to different x-ray light emitted during radiative stabilization of the dielectronic capture resonances. Interference between the radiative and dielectronic recombination processes has been found more important for the weaker x-ray-resolved partial photorecombination cross sections than for the total cross section.

Recent experimental observations of photorecombination processes in highly charged ions [5.2,52,69] have challenged our theoretical understanding of strong-field quantum electrodynamics. The energy positions of the observed resonance structures provide sensitive tests of the Lamb-shift calculations in multielectron systems [5.70]. Information on the detailed shape of the nuclear charge distribution may also be extracted from the resonance

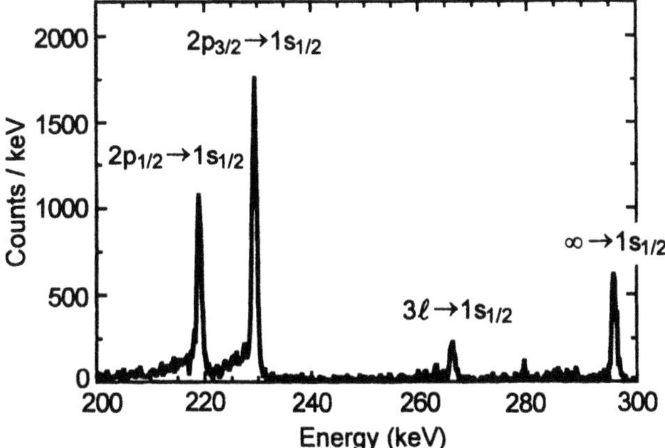

Fig. 5.8. True coincident x-ray spectrum of H-like U^{91+} ions at an energy of 321 MeV/u matching the electron energy [5.65] © 1995 Springer

energies [5.71]. The cross section heights of the observed resonance features provide good tests of the Auger-rate calculations in which static, magnetic, and retardation effects have to be included in the electron-electron interaction [5.72,73]. Finally, the detailed shape of the observed resonance structures provides tests of theoretical studies which include the quantum interference between the processes of radiative and dielectronic recombination [5.60,74,75].

5.2 Bremsstrahlung

Bremsstrahlung (BS) or deceleration radiation is the process in which one photon is radiated in the scattering of an electron from an atom or ion, the internal state of which remains unchanged,

$$X^{q+} + e(\boldsymbol{p}_0) \to X^{q+} + e(\boldsymbol{p}_1) + \hbar\omega, \tag{5.23}$$

where $\boldsymbol{p}_{0,1}$ are the electron momenta before and after the collision. In the nonrelativistic approximation the photon and electron energies are related through

$$\hbar\omega = \frac{1}{2m}\left(p_0^2 - p_1^2\right) = \hbar k c_o, \tag{5.24}$$

where k is the momentum of the radiated photon and c_o is the speed of light.

BS is closely related to elastic electron scattering, radiative recombination and pair production. BS is also a fundamental process for a variety of applications in astrophysics, biology, fusion-plasma problems, transport and energy loss in charged particle beams.

In general, Bremsstrahlung represents a quite complicated theoretical problem, because it involves bound and continuum states. The latter are usually described by continuum Coulomb wavefunctions. The problem is solved only in the nonrelativistic limit and for some special cases, for which numerical calculations were performed. Experiments are also limited in number, especially for positive ions. The general theory of BS can be found in the literature [5.76–83] and the relativistic theory of BS is discussed in various original papers and review articles [5.84–87].

Basic Formulae. There are several analytical expressions for the BS cross section in the pure Coulomb case. The BS cross section in the frequency interval $\omega, \omega + \mathrm{d}\omega$ is usually written in the form:

$$\mathrm{d}\sigma = g(\eta_0, \eta_1)\,\mathrm{d}\sigma_{\mathrm{Kr}}\,, \tag{5.25}$$

$$\mathrm{d}\sigma_{\mathrm{Kr}} = \frac{16}{3\sqrt{3}}\alpha^3\eta_0^2\,\frac{\mathrm{d}\omega}{\omega}\,[\pi a_0^2]\,, \tag{5.26}$$

$$\eta_0 = \frac{Ze^2}{\hbar v_0}, \quad \eta_1 = \frac{Ze^2}{\hbar v_1}, \tag{5.27}$$

where $\mathrm{d}\sigma_{\mathrm{Kr}}$ is the Kramers BS cross section [5.20], g is a dimensionless quantity called the *Gaunt factor* and Z is the nuclear charge of the ion and v_0 and v_1 denote the initial and final velocity of the electron, respectively. So, in general, the BS cross section is a function of two variables, η_0 and η_1.

Nonrelativistic quantum mechanics in dipole approximation results in the well-known Sommerfeld formula [5.76]

$$\begin{aligned} g(\eta_0, \eta_1) &= \frac{\sqrt{3}\pi\, x_0}{[\exp(2\pi\eta_0) - 1][1 - \exp(-2\pi\eta_1)]} \\ &\quad \times \frac{\mathrm{d}}{\mathrm{d}x_0}\,|{}_1F_2(i\eta_0, i\eta_1, 1; x_0)|^2\,, \\ x_0 &= -\frac{4\eta_0\eta_1}{(\eta_1 - \eta_0)^2}, \end{aligned} \tag{5.28}$$

where ${}_1F_2$ is the confluent hypergeometric function.

In the quasiclassical limit $\eta_1 > \eta_0 \gg 1$, (5.28) reduces to [5.78,20]:

$$g = \frac{\pi\sqrt{3}}{4}\,\mathrm{i}\nu\, H_{\mathrm{i}\nu}^{(1)}(\mathrm{i}\nu)\, H_{\mathrm{i}\nu}^{(1)\prime}(\mathrm{i}\nu), \quad \nu = \frac{Ze^2\omega}{mv_0^3}, \tag{5.29}$$

where $H_{\mathrm{i}\nu}^{(1)}(\mathrm{i}\nu)$ is the Hankel function of the complex argument and index and $H_{\mathrm{i}\nu}^{(1)\prime}(\mathrm{i}\nu)$ is its derivative with respect to the argument. The function (5.29) is monotonic and has the limiting values

$$\begin{aligned} g &= 1, \quad \nu \gg 1, \\ g &= \frac{\sqrt{3}}{\pi}\ln\left(\frac{2}{\gamma_{\mathrm{E}}\nu}\right), \quad \nu \ll 1 \end{aligned} \tag{5.30}$$

where $\gamma_{\mathrm{E}} \approx 0.57216$ denotes the Euler constant.

The BS spectrum for a point Coulomb potential has a logarithmic divergence for soft photons ($\nu \ll 1$) and it becomes flat for harder photons

($\nu \gg 1$). In case of a large initial electron velocity ($\eta_0 \ll 1$) one obtains from (5.28) the Born-Elwert approximation

$$g = \frac{\sqrt{3}}{\pi} f_{\mathrm{E}} \ln \frac{\eta_1 + \eta_0}{\eta_1 - \eta_0}\,, \tag{5.31}$$

where f_{E} is the Elwert factor

$$f_{\mathrm{E}} = \frac{\eta_1}{\eta_0} \frac{1 - \exp(-2\pi\eta_0)}{1 - \exp(-2\pi\eta_1)}\,. \tag{5.32}$$

If the final electron energy is also large, $2\pi\eta_1 \ll 1$, (5.31) is equal to the nonrelativistic Born approximation

$$g = \frac{\sqrt{3}}{\pi} \ln \frac{\eta_1 + \eta_0}{\eta_1 - \eta_0}, \quad f_{\mathrm{E}} = 1, \quad 2\pi\eta_0 < 2\pi\eta_1 \ll 1\,. \tag{5.33}$$

The general approximate formula for the Gaunt factor g can be obtained by a semiclassical method [5.22] as

$$g = \frac{\sqrt{3}\pi}{4} \mathrm{i}\nu \left(1 + \frac{1}{\gamma\eta_0}\right) H^{(1)}_{\mathrm{i}\nu}\left[\mathrm{i}\nu\left(1 + \frac{1}{\gamma\eta_0}\right)\right] H^{(1)\prime}_{\mathrm{i}\nu}\left[\mathrm{i}\nu\left(1 + \frac{1}{\gamma\eta_0}\right)\right] \tag{5.34}$$

from which both limiting cases – the Born approximation and the classical limit – may be derived. At $\eta_0/\nu \gg 1$ (5.34) approaches (5.29). For the limiting case $\omega \to 0$, ($\nu n \ll 1$) and arbitrary η_0, (5.34) may be written as

$$g = \frac{\sqrt{3}}{\pi} \ln \frac{2}{\nu(\gamma + 1/\eta_0)} \tag{5.35}$$

approaching (5.30) in the classical limit $\nu \ll 1$. In the Born approximation and $\omega \to 0$, one can obtain from (5.33) or (5.34)

$$g = \frac{\sqrt{3}}{\pi} \ln \frac{2\eta_0}{\nu}, \quad \eta_1 \simeq \eta_0\,. \tag{5.36}$$

In general, the semiclassical result (5.34) is a good approximation to the Sommerfeld formula (5.28) for all values of the parameters, except the so-called short wavelength limit

$$\nu \to 0, \quad \eta_0 \to 0, \quad \nu/\eta_0 \to 1\,. \tag{5.37}$$

This case corresponds to the large electron velocities, for which the Sommerfeld formula (5.28) is not valid and one must take into account the relativistic and retardation effects.

Numerical Results. Theoretical values of the shape function S, i.e., the ratio of the BS cross section differential in photon energy $\hbar\omega$ and its angle θ, have been obtained by *Kissel* et al. [5.88] in the Born approximation for 24 atoms with the nuclear charge $2 \le Z \le 92$ and for incident electron energies 1 keV $\le E_0 \le$ 500 keV. The shape functions S are fitted in the form suggested by the angular distribution of the first Born approximation

$$S \equiv \frac{\partial^2\sigma/\partial\omega\partial\Omega}{\partial\sigma/\partial\omega} = \frac{A}{4\pi}(1-\beta\cos\theta)^{-9/2}\sum_{i=0}^{5} B_i P_i(\cos\theta), \tag{5.38}$$

where A and B_i are the fitting parameters, β is the velocity of the incident electron in units of the speed of light c_o, and $P_i(\cos\theta)$ are the Legendre polynomials. The factor β is given by

$$\beta = v/c_o = [1 - 1/(1 + E_0[\mathrm{keV}]/510.999\,06)^2]^{1/2}, \tag{5.39}$$

where E_0 is the incident electron energy in keV units.

The BS spectra arising from interactions of electrons with screened atomic nuclei and with orbital electrons were considered by *Zeltzer* and *Berger* [5.89]. They present a comprehensive set of BS cross sections, differential in photon energy, for electron energies ranging from 1 keV to 10 GeV and for target atoms with nuclear charge $1 \leq Z \leq 100$. The set of differential BS cross sections based on the synthesis of various theoretical results, is compared with the available experimental data.

Bremsstrahlung emitted in fast collisions of electrons and hydrogen atoms has been studied by *Dubois* and *Maquet* [5.90] employing the Born approximation for the description of the electron-atom interaction and the Coulomb Green's function for the atomic spectrum. The main features of the BS cross sections are discussed cogitating the screening by the atomic electrons and the influence of atomic structure. Differences in the cross sections between spontaneous BS and stimulated free-free transitions are discussed as well.

An improved cross-section formula applicable for thin targets was derived by *Al-Beteri* and *Raeside* [5.91] by replacing the screening functions in the Bethe-Heitler formula [5.84] by a fitting factor obtained from experimental BS cross sections. The semiempirical formula obtained is in good agreement with experimental data over a wide range of electron energies 1 MeV $\leq E_{\mathrm{kin}} \leq$ 250 MeV and atomic targets with nuclear charge $4 \leq Z \leq 82$. Typical examples of BS cross sections for electrons on gold and lead atoms are shown in Fig. 5.9.

The theory for calculating differential BS cross sections was outlined by *Maximon* et al. [5.92] including an accurate high-energy small-angle approximation and considering polarization of the BS. The numerical results cover electron energies ranging from 50 MeV to 1 GeV.

Different types of the x-ray Bremsstrahlung arising in collisions of heavy charged particles with atoms such as quasi-free electron BS, secondary electron BS, atomic BS, radiative ionization, nuclear BS and others are considered in various papers [5.93–99].

5.3 Polarization of X-Ray Lines

X-ray radiation of highly charged ions is an important source of information on the processes occurring in hot laboratory and astrophysical plasmas. The

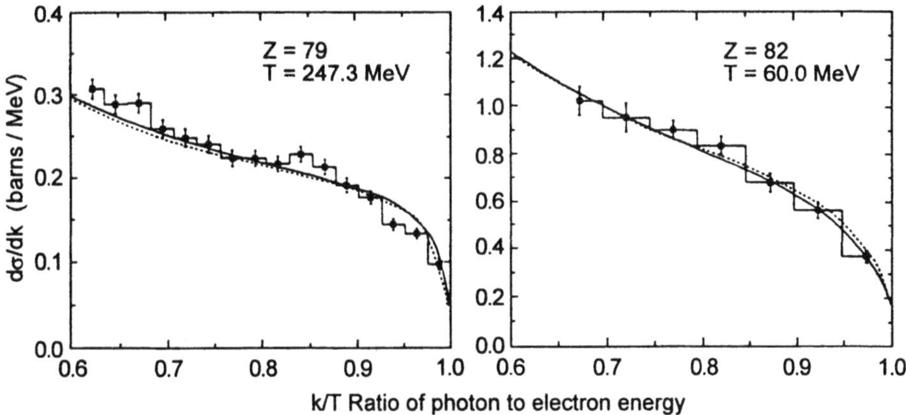

Fig. 5.9. Bremsstrahlung cross sections for electrons incident on thin gold ($Z = 79$) and lead ($Z = 82$) targets, with the electron energy E_0 shown in the figure, as a function of the ratio $\hbar\omega/E_0$ where $\hbar\omega$ is the photon energy. Experiment: $Z = 79$ [5.100], $Z = 82$ [5.101]. Theory: *dotted curves* – [5.89], *solid curves* – proposed BS cross-section formula [5.91]. From [5.91] © 1989 Elsevier

polarization of x-ray lines and of Bremsstrahlung is quite sensitive to the presence of electron beams in plasmas (*nonthermal* or *suprathermal* electrons), or more generally, to the deviation of the electron-velocity distribution function from a Maxwellian one [5.102–105]. Electron beams, in turn, strongly influence the energy balance, heat conductivity, plasma oscillations and polarization of emission spectra from plasmas.

There are many physical mechanisms for creation of nonthermal electrons in a plasma, for example, the presence of electric fields, temperature gradients, parametric instabilities and others, e.g., [5.1,103] and references therein. Such runaway non-Maxwellian electrons were directly registered in laser-produced plasmas [5.106,107], tokamak [5.108,109], vacuum spark [5.110] as well as in the Solar corona [5.102,111]. Most of these measurements have been carried out by observing the Bremsstrahlung.

A significant role of non-Maxwellian electrons in formation of shortwavelength spectra and, therefore, their importance in theoretical models for spectroscopic plasma diagnostics was pointed out in a series of papers devoted to the investigations of the Solar corona [5.112–114] and of a pinch plasma [5.115].

The polarization of continuous radiation (Bremsstrahlung) caused by inelastic scattering of electrons on highly charged positive ions was theoretically studied in many papers, e.g., [5.103,108,116,117]. Wide interest to this problem was stimulated by astrophysical investigations, and observations of the Solar flares hard x-ray emission, in particular. The measurements of polarization of x-ray lines in multicharged ions have been carried out in the laboratory [5.106,118–121] and the Solar plasmas [5.122,123]. There is a compilation of atomic data related to polarization plasma spectroscopy [5.124].

The theory of the polarization of emission lines, first developed for neutral atoms [5.125,126] was later extended to radiation of multicharged ions [5.102,105,127–130]. A theoretical description of the subject is based on the photon polarization density matrix formalism [5.131].

The main mechanism leading to polarization of x-ray lines in highly charged ions is excitation of magnetic sublevels by anisotropic electrons. A directivity of electrons in a plasma leads to an alignment of ions excited by these electrons over the angular momentum L which means a preferential population of certain magnetic sublevels (M-sublevels) belonging to an excited atomic level. Line emission from such levels is polarized and anisotropic. Measurements of polarization of spectral lines can therefore be used for diagnostics of the non-Maxwellian electron distribution function. The degree of polarization is expressed in terms of the electron-impact excitation cross sections of magnetic sublevels of the ion target under the assumption that the population of magnetic sublevels is proportional to the corresponding excitation rate coefficients $\langle \sigma v \rangle$. The nonrelativistic calculations of the x ray line polarization have been reported in various papers [5.127,128,132–134]. Relativistic effects can influence the polarization of line radiation emitted by multicharged ions [5.135,136].

Fortunately, the methods of spectroscopic diagnostics based on the analysis of line spectra or continuous radiation are, to some extent, supplementary, i.e. different electron energies are responsible for discrete or continuous radiation of highly charged ions in a plasma. The high energy photons with $E > 50$ keV are emitted due to Bremsstrahlung of the non-Maxwellian electrons and provide the best tool for a study of high energy tails of the distribution function. At the same time, the main contribution to the polarization of x ray lines is given by electrons having an energy E close to the corresponding transition energy in an ion, i.e., $E = 1$–10 keV.

For the case of an electron beam impinging on a target, the degree of polarization $P(\theta)$ is defined by

$$P(\theta) = \frac{I_{\perp} - I_{\parallel}}{I_{\perp} + I_{\parallel}} , \tag{5.40}$$

where $I_{\perp}$ and $I_{\parallel}$ denote the intensities of the emitted photons, with the polarization vector perpendicular and parallel to the meridian plane, respectively. The meridian plane is the plane formed by the crossing direction of the incident electron beam and of the direction of the radiation. The general quantum-mechanical expression for the linear polarization of photons emitted by ions is given by *Inal* and *Dubau* [5.127].

Calculations of the degree of polarization of x ray lines (including dielectronic satellites, Chap. 6) have been performed mainly for H-, He- and Li-like ions including, first of all, the following transitions:

$2p_{3/2} \to 1s_{1/2}$ — $Ly\alpha_1$ line,
$1s_{1/2}2p_{1/2}, J=1 \to 1s^2_{1/2}, J=0$ — the resonance line (w)
$1s_{1/2}2p_{3/2}, J=1 \to 1s^2_{1/2}, J=0$ — the intercombination line (y)
$1s_{1/2}2p_{3/2}, J=2 \to 1s^2_{1/2}, J=0$ — the magnetic quadrupole line (x)
$1s_{1/2}2s_{1/2}, J=1 \to 1s^2_{1/2}, J=0$ — the forbidden line (z)
$1s2p(^1P)2s\,^2P_{3/2} \to 1s^22s\,^2S_{1/2}$ — q-satellite
$1s2s2p\,^4P_{3/2} \to 1s^22s\,^2S_{1/2}$ — u-satellite.

The degree of polarization $P(\theta)$ of a line corresponding to the electric dipole (E1) or magnetic quadrupole (M2) transitions as a function of observation angle θ relative to the incident electron beam is given by

$$P^{\mathrm{E1}}(\theta) = \frac{P_0 \sin^2\theta}{1 - P_0\cos^2\theta}, \quad P^{\mathrm{M2}}(\theta) = \frac{P_0\sin^2\theta}{1 + P_0\cos^2\theta}, \tag{5.41}$$

where P_0 is the degree of polarization at angle $\theta = 90°$ which can be expressed through the excitation rates $\langle\sigma_0 v\rangle$ and $\langle\sigma_1 v\rangle$ of the corresponding M_L components:

$$P_0 = \frac{a_1[\langle\sigma_0 v\rangle - \langle\sigma_1 v\rangle]}{a_2\langle\sigma_0 v\rangle + a_3\langle\sigma_1 v\rangle}\,. \tag{5.42}$$

Here $\langle\sigma_0 v\rangle$ and $\langle\sigma_1 v\rangle$ denote the excitation rate coefficients, respectively, for $M_L = 0, 1$ (w and y lines), $M_L = 1, 2$ (x line) and $M_L = 1/2, 3/2$ ($Ly\alpha_1$ line). The coefficients a_1, a_2 and a_3 for some lines are given in Table 5.2.

Table 5.2. Coefficients a_1, a_2 and a_3 determining the degree of polarization according to (5.42)

Transition	Line	a_1	a_2	a_3
$1s_{1/2}2p_{1/2}, J=1 \to 1s^2_{1/2}, J=0$	w	1	1	1
$1s_{1/2}2p_{3/2}, J=2 \to 1s^2_{1/2}, J=0$	x	1	1	3
$2p_{3/2} \to 1s_{1/2}$	$Ly\alpha_1$	3	5	7

First calculations of the degree of polarization P for $Ly\alpha$ lines in H-like ions performed for the Solar flare plasma [5.102,125] showed relatively low values of the polarization degree $P < 10\%$. Calculations of the P factor for lines in He-like ions, excited by a monoenergetic electron beam, carried out for iron ions showed 2–3 times higher polarization as compared to $Ly\alpha$ lines. In general, sophisticated calculations of the degree of polarization for the main diagnostically important x-ray lines [5.127–129] yield quite high P values (up to 40–60%) depending on the incident electron energy.

Calculated degrees of polarization for the x, y and w lines in He-like Fe and Ca ions as a function of the energy of the monoenergetic electron beam are shown in Figs. 5.10,11. The polarization of x and w lines as a function of energy (in threshold units) has a weak dependence on the nuclear

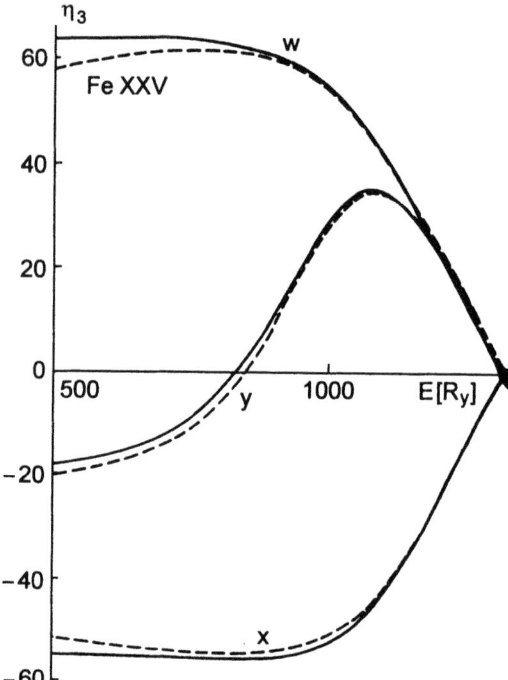

Fig. 5.10. Calculated degree of linear polarization η in % at $\theta = 90°$ for x, y and w lines of He-like Fe^{24+} ions as a function of the incident electron energy. *Dashed curves:* [5.127], *solid curves:* [5.129]

charge while the polarization of the y line reveals a strong dependence on the nuclear charge because the relativistic coupling between $1s_{1/2}2p_{1/2}, J=1$ and $1s_{1/2}2p_{3/2}, J=1$ levels rapidly falls off with decreasing nuclear charge and strongly influences the excitation cross sections of the $1s_{1/2}2p_{3/2}, J=1$ level.

The polarization of w and x lines for Fe^{24+} ions were calculated in the model in which both thermal and non-thermal (anisotropic) electrons were taken into account [5.127]. The lines show a measurable amount of linear polarization when the part of the anisotropic electron density is about 1% or more. For the satellite lines $1s^2 2p\,^2P_{3/2}$ - $^2D_{5/2}$ and $1s^2 2p\,^2P_{1/2}$ - $^2D_{3/2}$ similar calculations were carried out [5.127]. For He-like iron ions the q- and u-satellites are produced almost entirely by inner-shell excitation of Li-like ions in the ground state [5.127]. The degree of polarization is shown in Fig. 5.12.

The polarization of x-ray lines supplementing data on the continuum polarization carry some features which are important for diagnostics of the electron distribution function. The degree of polarization, its sign (the direction of the electrical vector with respect to the beam) and the energy dependence are determined by two factors: the properties of the radiating medium - the anisotropy of the electron distribution function - and the type of atomic transition in the radiation line (multipolarity, electrical or magnetic type, transition from a singly or doubly excited autoionizing state). In this respect, a change in the relative line intensity measured by a polarization-

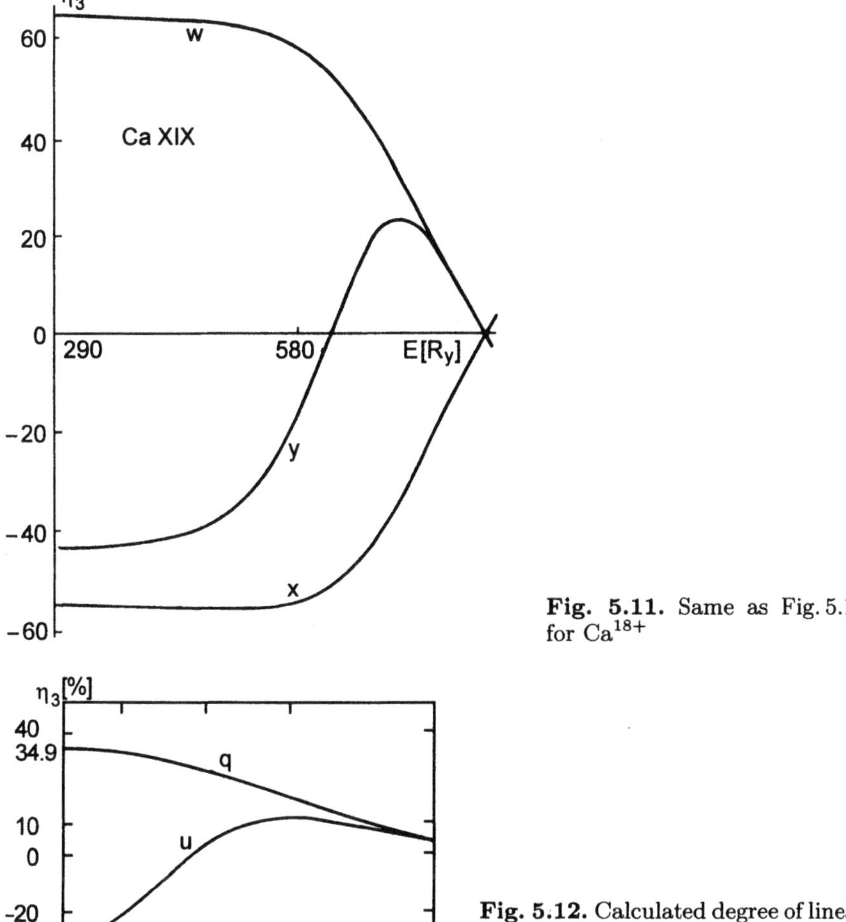

Fig. 5.11. Same as Fig. 5.10 for Ca^{18+}

Fig. 5.12. Calculated degree of linear polarization η in % for q and u satellites in Fe^{24+} ions at $\theta = 90°$ [5.127] © 1989 IOP

sensitive instrument is different for different lines compared to an unpolarized spectrum. This makes it possible to reliably verify the radiating plasma model and to derive the electron-beam parameters.

We note that polarization of x-ray radiation is also possible in radiative recombination processes [5.18,137], synchrotron radiation [5.138], radiative electron capture [5.139], excitation [5.140] and radiative-transfer excitation processes [5.141] in heavy-ion atom collisions.

5.4 X-Ray Lasers

An x-ray laser represents a special type of short-wavelength radiation source. Employing a plasma of highly charged ions as an active media, it has unique properties such as monochromatic radiation and highest brightness in comparison with all other x-ray sources. It makes it possible to use x-ray lasers in many important physical applications such as x-ray lithography, soft x-ray laser microscopy of biological objects, interferometry of laser fusion plasma, medicine and many others.

Since their experimental demonstration in 1985 [5.142,143], x-ray lasers are the object of high interest both in development and applications, e.g., [5.144–148]. Our aim here is not to discuss, in detail, the operational principles of x-ray lasers but to consider, in general, some spectroscopical properties of their short wavelength radiation.

Typical parameters of x-ray lasers are summarized in Table 5.3. The main pumping mechanisms – electron collisional excitation and three-body recombination – have been mainly used with Ne-like ions. Experimental wavelengths for transitions $3p-3s$ in Ne-like ions with nuclear charge $Z \leq 47$ are given in [5.149–152] and are summarized in Table A.10.

At present, soft x-ray lasers exist that operate in the spectral region $30\ \text{Å} < \lambda < 350\ \text{Å}$. A list of x-ray lasers at various institutions is given in Table 5.4 together with wavelengths, ionic species and the gain-length factors [5.145].

The water-window range $23\ \text{Å} < \lambda < 44\ \text{Å}$ has been reached with Ni-like ions. Future x-ray lasers with wavelengths $\lambda < 2\ \text{Å}$ are very prospective. Photons in this wavelength range are not strongly absorbed by air and other matter and their wavelength is of the same order as the size of molecules, interatomic and interplanar distances in crystals and of bond lengths in chemical and biological complexes. Basic physical and optical properties of the dense laser-produced plasmas, which are the main active medium of x-ray lasers, were reckoned by *Jamelot* [5.153].

X-ray optics used in x-ray lasers have to meet special requirements. They should withstand the ultrahigh brightness and provide focusing and manipulation of the x-ray laser beam. The x-ray optics used for x-ray lasers are described in several review articles, e.g., [5.154–157].

Usually, x-ray lasers require a huge facility for laser pumping such as powerful lasers or Z-pinch devices [5.144]. A good alternative greatly reducing the complexity and cost of these devices can be an x-ray lasing in capillaries by direct discharge excitation [5.158,159]. A special geometry of the capillary (length to diameter ratio $l/d > 100$) provides a rapid cooling of the created plasma by electron heat conduction of the walls during the decay of the excitation pulse resulting in a large population inversion. The first experimental demonstration [5.158] of such a device was making use of the 468.75 Å $J=0 \rightarrow J=1$ lasing transition in Ne-like Ar ions. The experimental

Table 5.3. Typical parameters of x-ray lasers

Wavelength	35–350 Å
Active media	High-temperature plasma with T = 0.1–10 keV, $N > 10^{20}\text{cm}^{-3}$
Media dimensions:	
length	0.5–5 cm
diameter	0.01–0.1 cm
Beam divergence	1–10 mrad
Efficiency	10^{-5}
Pulse duration	0.02–10 ns
Peak power	100 MW
Peak brightness	10^{23}–10^{26} photons $\text{mm}^{-2}\text{mrad}^{-2}\text{s}^{-1}$
Photons/pulse	$\leq 10^{15}$
Gain-length factor	3–16
Relative linewidth ($\Delta\lambda/\lambda$)	5×10^{-5}
Pumping devices	Power lasers of visible and infrared ranges with wavelength $\lambda = 0.35$–1.05 µm, impulse longitudinal Z-pinch and transverse discharges, capillary discharge
X-ray optics	Grazing incidence microscopes, normal incidence mirrors, beam splitters, diffractional-limited lenses
Principal pumping mechanisms	More than 10 atomic mechanisms; excitation by electron impact, three-body (ternary) recombination, resonance photoexcitation, inner-shell photoionization, multiphoton ionization
Applications	Atomic and nuclear physics, study of atomic structure of complex clusters, plasma probing by x-ray laser interferometry, x-ray lithography, ESCA spectroscopy, biological objects, holography, crystallography, medicine, radiation chemistry

parameters were: $l = 12$ cm, $d = 4$ mm, gain $g = 0.6\ \text{cm}^{-1}$, the gain-length factor $gl = 7.2$.

Table 5.4. Performance of systems exhibiting soft x-ray gain at laboratories around the world as of November, 1990 (from [5.145])

Wavelength (Å)	Element	Gain-length factor	Institution
35.6	Au^{52+} (Ni-like)	3.0	LLNL
43.2	W^{46+} (Ni-like)	7.0	LLNL
44.8	Ta^{45+} (Ni-like)	8.1	LLNL
47	Ti^{19+} (Li-like)	3.8	LLE
50.2	Yb^{42+} (Ni-like)	2.0	LLNL
54	Na^{10+} (H-like)	3.9	ILE
65.8	Eu^{35+} (Ni-like)	2.1	LLNL
69	Sm^{34+} (Ni-like)	5.4	RAL
71	Eu^{35+} (Ni-like)	3.8	LLNL
73	Cu^{18+} (Na-like)	2.0	MPQ
75	Sm^{34+} (Ni-like)	5.4	RAL
81	F^{8+} (H-like)	2.6	ILE
81	Ni^{17+} (Na-like)	2.6	MPQ
81	F^{8+} (H-like)	2.2	RAL
99	Ag^{37+} (Ni-like)	4.4	CEL
99	Ag^{37+} (Ne-like)	10.1	LLNL
100	Ag^{37+} (Ne-like)	6.1	LLNL
102	O^{7+} (H-like)	3.0	ILE
104	Cu^{18+} (Na-like)	1.9	MPQ
105	Al^{10+} (Li-like)	2.0	ILE
105	Al^{10+} (Li-like)	2.1	LLE
105	Al^{10+} (Li-like)	3.1	LLE
105	Al^{10+} (Li-like)	3.0	LULI
106	Mo^{32+} (Ne-like)	3.8	LLNL
106	Al^{10+} (Li-like)	2.0	RAL
111	Cu^{18+} (Na-like)	1.4	MPQ
129	Si^{11+} (Li-like)	2.0	PPPL
131	Mo^{32+} (Ne-like)	7.1	LLNL
133	Mo^{32+} (Ne-like)	7.3	LLNL
139	Mo^{32+} (Ne-like)	5.0	LLNL
150	Al^{10+} (Li-like)	2.7	LLE
154	Al^{10+} (Li-like)	2.5	LLE
154	Al^{10+} (Li-like)	2.8	LLE
154	Al^{10+} (Li-like)	4.0	PPPL
154	Al^{10+} (Li-like)	3.3	RAL/LSAI
155	Y^{29+} (Ne-like)	11.0	LLNL
157	Y^{29+} (Ne-like)	11.0	LLNL
164	Sr^{28+} (Ne-like)	9.7	LLNL/CEL
165	Sr^{28+} (Ne-like)	8.8	LLNL/CEL
182	C^{5+} (H-like)	4.1	LLE
182	Se^{24+} (Ne-like)	9.6	LLNL
182	Se^{24+} (Ne-like)	3.1	NRL
182	C^{5+} (H-like)	8.0	PPPL
182	C^{5+} (H-like)	2.4	PPPL

Table 5.4. continued

Wavelength (Å)	Element	Gain-length factor	Institution
182	C^{5+} (H-like)	3.4	PPPL
182	C^{5+} (H-like)	3.9	RAL
196	Ge^{22+} (Ne-like)	2.3	LLE
196	Ge^{22+} (Ne-like)	2.8	NRC
196	Ge^{22+} (Ne-like)	4.7	NRL
196	Ge^{22+} (Ne-like)	8.1	RAL/NRL
206	Se^{24+} (Ne-like)	16.0	LLNL
206	Se^{24+} (Ne-like)	5.9	NRL
209	Se^{24+} (Ne-like)	15.2	LLNL
209	Se^{24+} (Ne-like)	5.9	NRL
212	Zn^{20+} (Ne-like)	3.5	NRL
220	Se^{24+} (Ne-like)	9.2	LLNL
221	Cu^{19+} (Ne-like)	3.0	NRL
222	As^{23+} (Ne-like)	8.1	NRL
232	Ge^{22+} (Ne-like)	5.5	LLE
232	Ge^{22+} (Ne-like)	5.0	NRC
232	Ge^{22+} (Ne-like)	6.2	NRL
232	Ge^{22+} (Ne-like)	15.0	RAL
236	Ge^{22+} (Ne-like)	5.5	LLE
236	Ge^{22+} (Ne-like)	5.0	NRC
236	Ge^{22+} (Ne-like)	6.2	NRL
236	Ge^{22+} (Ne-like)	15.0	RAL
247	Ge^{22+} (Ne-like)	2.2	LLE
247	Ge^{22+} (Ne-like)	3.6	NRS
247	Ge^{22+} (Ne-like)	6.6	RAL/NRL
262	Se^{24+} (Ne-like)	11.8	LLNL
262	Zn^{20+} (Ne-like)	3.0	NRL
267	Zn^{20+} (Ne-like)	3.0	NRL
279	Cu^{19+} (Ne-like)	2.6	NRL
284	Zn^{20+} (Ne-like)	2.6	NRL
286	Ge^{22+} (Ne-like)	4.0	LLE
286	Ge^{22+} (Ne-like)	2.4	NRC
286	Ge^{22+} (Ne-like)	9.0	RAL/NRL
326	Ti^{12+} (Ne-like)	5.9	LLE/LLNL

LLNL: Lawrence Livermore National Laboratory, California
LLE: Laboratory for Laser Engineering, University of Rochester, Rochester, New York
ILE: Institute of Laser Engineering, Osaka University, Japan
RAL: Rutherford Appleton Laboratory, Didcot, Oxon, England
MPQ: Max-Plank-Institut fur Quantenoptik, Garching, Germany
CEL: Centre d'Etudes de Limeil-Valenton, Villeneuve St.Georges, France
LULI: National Facility for Use of Intense Lasers, Palaiseau, France
PPPL: Princeton Plasma Physics Laboratory, Princeton, New Jersey
LSAI: Laboratorie de Spectroscopie Atomique et Ionique, Orsay, France
NRL: Naval Research Laboratory, Washington, D.C.
NRC: National Recearch Council of Canada, Ottawa, Canada

6. Collisional Processes

In this chapter some of the collisional processes are covered that play an important role in the population of excited states of highly charged ions. The processes considered are directly connected to or followed by the emission of x rays. With this respect the division of subjects among Chaps. 5,6 may appear somewhat arbitrary. In Sect. 6.1 the resonant *Dielectronic-Recombination* processes and in Sect. 6.2 *Radiative Electron Capture* in fast collisions of heavy ions with atoms are discussed. Section 6.3 deals with the *Resonant Transfer and Excitation* which is related to the two preceding subjects. We close with a concise account on *Three-body Recombination* in Sect. 6.1.

6.1 Dielectronic Recombination

6.1.1 Classification of the Process

Dielectronic Recombination (DR) is a resonant recombination of a free electron with an ion X^{q+} bearing one or more core electrons. Schematically it may be identified with one of the following two-step processes:

$$X^{q+} + e \rightarrow X^{(q-1)+^{**}} \rightarrow \left\{ \begin{array}{l} X^{(q-1)+^{*}} + \hbar\omega \\ \text{or} \\ X^{q+} + e \end{array} \right. . \qquad (6.1)$$

In a first step, often referred to as *dielectronic capture*, a free electron is radiationless transferred to a bound state of the ion and a core electron is simultaneously excited forming a doubly excited state $X^{(q-1)+^{**}}$. Because of the quantization of the ionic energy levels the kinetic energy of the electron has to fulfill a resonance condition for the process to become possible (Sect. 6.1.3) The dielectronic capture, schematically illustrated in Fig. 6.1, is the inverse of the Auger process. In a second step, the doubly excited ion can radiatively stabilize by decaying to a lower level or to its ground state in which case a *dielectronic recombination* has occurred. Alternatively the doubly excited ion may autoionize thus returning to its original charge state in which case a resonance feature in the elastic scattering will occur (Chap. 5).

The same notation is commonly used for the DR and for the Auger transitions. For instance KLL refers to a resonant capture of a free electron into

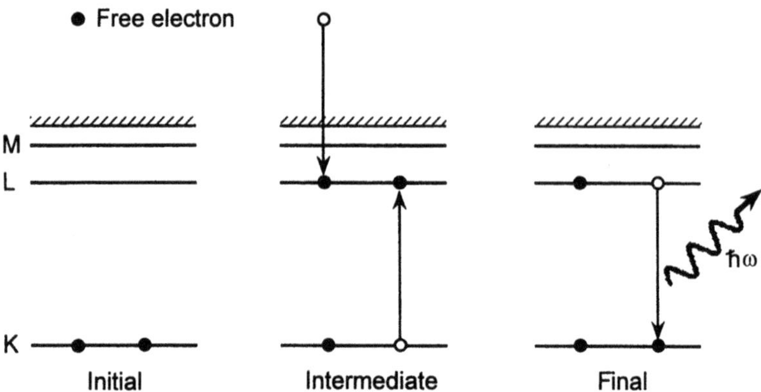

Fig. 6.1. Schematic representation of the dielectronic recombination process. The example is for a KLL resonance on a He-like ion

the L shell of an ion simultaneously exciting a K-shell electron into the L shell. The x-ray spectra associated with the stabilizing decay of the doubly excited states are composed of *dielectronic satellites* [6.1,2] (see also Sect. 6.1.2) appearing in the vicinity of the corresponding singly excited states.

Further decay channels can be opened if the core electron which became excited was an inner electron. Then there might be enough energy in the doubly excited state leading to simultaneous or sequential *double* autoionization. In this case marked structures in the electron-impact ionization [6.3] are found. *Resonant excitation* takes place if an autoionization of the doubly excited state occurs leaving the X^{q+} ion in an excited rather than in its ground state.

The DR process has been considered since the early 1940s [6.4] and its importance for the ionization balance in sources of highly charged ions [6.5–9] has been well recognized. In astrophysical and plasma sources for thermonuclear fusion it is important for the formation of excited states and for plasma cooling by radiation losses. Despite its importance in plasma sources, it is difficult to study DR in such environments because this would require a reliable model of the plasma and controlling of parameters like temperature and density. First measurements of DR cross sections were reported in 1983 [6.10–12] employing crossed electron and ion beams. These early measurements suffered from low counting rates and large backgrounds. Many subsequent measurements have been carried out mainly with ions of low charge states. Problems encountered in the past were due to limited energy resolution in measurements involving intra-shell electron transitions and many closely spaced resonances which could not be resolved. Another complication arose from the fact that measured cross sections were largely enhanced [6.13–15] by electric fields present in the apparatus affecting unresolved resonances near the series limit.

6.1.2 Dielectronic Satellites

Rich spectra of closely spaced x-ray lines originate from the radiative decay of autoionizing states where two or more electrons are excited. Such states lie above the ionization limit and are created either by direct excitation, of an inner-shell electron:

$$X^{q+} + e \rightarrow X^{q+^*} + e \tag{6.2}$$

or by capture of a free electron via dielectronic recombination

$$X^{q+} + e \rightarrow X^{(q-1)+^{**}} \rightarrow X^{(q-1)+^*} + \hbar\omega \,. \tag{6.3}$$

The lines arising from dielectronic recombination (6.3) are called *dielectronic satellites.* Let a_0 and a_1 denote the sets of quantum numbers characterizing an ion X^{q+} in its initial and final state, respectively. A dielectronic satellite to the transition $a_0 \rightarrow a_1$ can occur if simultaneously ionic states are present with a spectator electron in the state $n\ell$ of the ion $X^{(q-1)+}$. Hence the satellite transition can be designated as $a_0 n\ell \rightarrow a_1 n\ell$. For example, $1s2pn\ell \rightarrow 1s^2 n\ell$ transitions in Li-like ions are the satellites to the resonance line $1s2p \rightarrow 1s^2$ in He-like ions. The number of excited electrons in the autoionizing states can be more than two; for example, in Be-like ions the following autoionizing states are possible:

$1s2s^2 2p$, $1s2s2p^2$, $1s2p^3$, ..., $1sn\ell n'\ell' n''\ell''$, ...

The notations for basic spectral lines in H- and He-like ions and corresponding satellites have been compiled in Table 3.5 on page 65 corresponding to Gabriel's notation.

In highly charged ions the satellites are very close, in wavelength, to the *parent* line and their intensities increase approximately as Z_{eff}^4 so that some satellite lines have intensities comparable with the intensity of the resonance line. Therefore, one observes a large number of spectral lines of comparable intensity in a narrow spectral interval. Figure 6.2 represents a typical example for dielectronic-satellite spectra of He-like Kr^{34+} ions observed at the EBIT of the Lawrence Livermore National Laboratory [6.16]

At first glance, it seems that the presence of a large number of the spectral lines makes the identification and the analysis of the VUV and x-ray spectra quite difficult. However, with high-resolution detection techniques ($\delta\lambda/\lambda \simeq 10^{-5}$ in the spectral interval $\lambda = 1$–10 Å), the situation turns out to be most advantageous for x-ray spectroscopy and x-ray plasma diagnostics. Beneficial is that a limited spectral interval contains quantitative information not only about the ionic structure such as wavelengths, radiative and autoionization transition probabilities, rates of electron excitation and ionization – the *micro* parameters – but also about plasma *macro* parameters such as electron and ion temperatures, densities, ion-state distributions and the presence of electron beams in plasmas. The absolute and relative intensities of the dielectronic satellites and their degree of polarization are very sensitive to the plasma macro parameters. This fact is the basis for x-ray

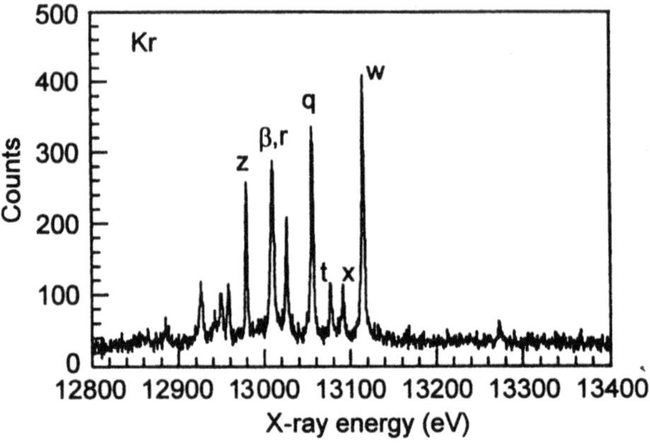

Fig. 6.2. X-ray spectrum of He-like Kr^{34+} ions obtained at the EBIT at LLNL; Li-like and Be-like satellites are labeled r, q, t and β [6.16] © 1992 APS

diagnostics of high-temperature laboratory and astrophysical plasmas, e.g., [6.17–22]. Accurate calculations of wavelengths and intensities of dielectronic satellites have been reported in various papers, see Chaps. 3,4 of this book and [6.23–28].

6.1.3 DR Cross Sections and Rates

We consider the DR process (6.1) for a particular resonance

$$e + |i\rangle \;\rightarrow\; |d\rangle \;\rightarrow\; |f\rangle + \hbar\omega \tag{6.4}$$

involving the initial state $|i\rangle$ of the target ion X^{q+}, the state $|d\rangle$ of the doubly excited ion $X^{(q-1)+^{**}}$ and the final state $|f\rangle$ of the stabilized ion $X^{(q-1)+}$. The reaction is possible when the electron energy satisfies the resonance condition

$$E_{\rm r} \;=\; E_d - E_i \tag{6.5}$$

within the natural width of the level $|d\rangle$, where E_i and E_d denote the total electronic binding energies of the initial and intermediate state, respectively. Within the isolated-resonance approximation [6.5,29–31] the DR cross section can be expressed as

$$\sigma_{\rm DR}(E_{\rm e}) \;=\; \frac{2\pi^2\hbar^3}{p_{\rm e}^2}\, V_{\rm a}(i \to d)\, \tilde{\delta}(E_{\rm e})\, \omega_d\,, \tag{6.6}$$

where $E_{\rm e}$ and $p_{\rm e}$ denote the initial electron energy and momentum, respectively. The quantity $V_{\rm a}(i \to d)$ denotes the total probability per time for the dielectronic capture $|i\rangle \to |d\rangle$ and $\tilde{\delta}(E_{\rm e})$ is the Breit-Wigner profile of the resonance whereas ω_d denotes the fluorscence yield of the doubly excited state. The dielectronic capture probability $V_{\rm a}(i \to d)$ is related to the reverse autoionization probability $A_{\rm a}(d \to i)$ via

$$V_{\rm a}(i \to d) \;=\; \frac{g_d}{2g_i}\, A_{\rm a}(d \to i), \tag{6.7}$$

where g_i and g_d denote the statistical weights. The function $\tilde{\delta}(E_e)$ is defined as

$$\tilde{\delta}(E_e) = \frac{\Gamma_d/2\pi}{(E_e - E_r)^2 + \Gamma_d^2/4}, \qquad \int \tilde{\delta}(E_e)\mathrm{d}E_e = 1 \tag{6.8}$$

and for the fluorescence yield we write

$$\omega_d = \frac{\Gamma_r}{\Gamma_a + \Gamma_r} = \frac{\Gamma_r}{\Gamma_d} \tag{6.9}$$

using the radiative (Γ_r) and Auger (Γ_a) widths and the total width (Γ_d) of the doubly excited state $|d\rangle$.

Combining (6.6–9) we arrive at the following expression for the cross section

$$\sigma_{DR}(E_e) = \frac{\pi\hbar^2}{p_e^2}\frac{g_d}{2g_i}\frac{\Gamma_a(d \to i)\,\Gamma_r(d \to f)}{(E_e - E_r)^2 + \Gamma_d^2/4}. \tag{6.10}$$

Often the natural levelwidth Γ_d is small compared to experimental linewidths. In that case one can use a delta function for the theoretical line profile resulting in the following cross section

$$\sigma_{DR}(E_e) = \frac{2\pi^2\hbar^2}{p_e^2}\frac{g_d}{2g_i}\frac{\Gamma_a(d \to i)\,\Gamma_r(d \to f)}{\Gamma_d}\delta(E_e - E_r). \tag{6.11}$$

This suggests the definition of a resonance strength S_d as the integrated cross section

$$S_d \equiv \int \sigma_{DR}(E)\mathrm{d}E = \frac{2\pi^2\hbar^2}{p_e^2}\frac{g_d}{2g_i}\frac{\Gamma_a(d \to i)\,\Gamma_r(d \to f)}{\Gamma_d}. \tag{6.12}$$

The DR rate coefficient, which sometimes is a more convenient quantity, is obtained by integrating over the velocity distribution of the electrons

$$\alpha_{DR} = \langle v_e \sigma_{DR}(E_e)\rangle. \tag{6.13}$$

For plasma diagnostics a Maxwell distribution is used for the integration [6.32] of (6.10) yielding

$$\alpha_{DR}(T_e) = (2\pi)^{3/2}\hbar^2 m_e^{-3/2}\frac{g_d}{2g_i}\frac{\Gamma_a\Gamma_r}{\Gamma_d}T_e^{-3/2}\exp(-E_r/T_e). \tag{6.14}$$

Within the theoretical framework outlined above the calculation of DR cross sections and rate coefficients entails the evaluation of the rates for autoionization and radiative decay discussed in Chap. 4. Also, the electronic binding energies, determined with a high accuracy, are required for the prediction of the location of the resonances. Depending on the specific electronic configurations embraced, a tremendous computational effort can be involved especially if the number of states within a resonance feature is large. Theoretical results have been reported in [6.33–41].

6.1.4 Experiments

A typical experimental approach to the DR process (6.1) monitors one or more of the reaction products while scanning the electron-ion collision energy over a resonance feature. The signals monitored, embrace the number N_{q-1} of ions that changed their charge state and the number N_x of x rays as a consequence of the stabilizing transitions along with their energy $\hbar\omega$. The resolution is usually limited by the velocity spread of the electron beam. The x-ray energy can be used to sort out a resonance or a group of resonances thus reducing the background. When the x rays are measured with high spectral resolution the dielectronic resonances may be resolved from the photon spectrum rather than from the scanned electron energy. If the corresponding currents and densities along with overlap integrals of crossed or merged beams are determined [6.42], the measured excitation functions can be transformed into rate coefficients or into cross sections suitable for comparison with theory.

Achievements in the experimental techniques of ion sources and storage rings have improved on the following parameters considerably:

- energy resolution
- signal rate
- background
- high charge states.

This is also reflected in the increasing number of high-resolution experiments. Table 6.1 gives an overview of available DR measurements performed at various experimental facilities.

Merged electron and ion beams have been used in single-path experiments [6.43–46] at Århus with the electron cooling device built for the ASTRID storage ring. Dielectronic resonances of metastable helium-like carbon and oxygen involving $\Delta n = 0$ transitions were found to be in agreement with theoretical calculations. Another single-path experiment [6.47] has been conducted with a dedicated high-current electron target using ion beams of the UNILAC at Darmstadt. With a beam of U^{28+} ions unexpectedly high recombination rates were found at very small relative energies. At the same apparatus DR cross sections for $\Delta n = 0$ and $\Delta n = 1$ excitation on lithium-like argon has been measured [6.48] obtaining a good overall agreement with theory.

The other merged-beam experiments listed in Table 6.1 were performed at storage rings with the ionic charge ranging from 7+ up to 90+. At storage rings, the experiments make use of the gain in signal by repeated circulation of the ions and of the electron cooler used as a cold electron target. The signal monitored there, is the change in charge state.

X-ray spectroscopy was used at the ion sources, the EBIS at KSU and the EBIT and SuperEBIT at LLNL. At the EBIS [6.49,50] the charge states of extracted ions was used as a signal together with the number of x rays detected at zero degree. Cross sections for dielectronic recombination with

Table 6.1. High-resolution measurements of dielectronic recombination arranged according to the ionic charge of the heavy ion. *Excitation* refers to the dielectronic transition populating the doubly excited states

Target	Excitation	Signal	Facility	Reference
C^{3+}	$1s^22s \to 1s^22pn\ell$	N_{q-1}	Tandem	[6.43]
C^{4+}	$1s2s \to 1s2pn\ell$	N_{q-1}	Tandem	[6.43]
N^{4+}	$1s^22s \to 1s^22pn\ell$	N_{q-1}	Tandem	[6.43]
C^{5+}	$1s \to 2\ell n\ell'$	N_{q-1}	TSR	[6.51]
N^{5+}	$1s2s \to 1s2\ell n\ell'$	N_{q-1}	Tandem	[6.44]
O^{5+}	$1s^22s \to 1s^22pn\ell$	N_{q-1}	Tandem	[6.43]
O^{6+}	$1s2s \to 1s2pn\ell$	N_{q-1}	Tandem	[6.43,45]
F^{6+}	$1s^22s \to 1s^22pn\ell$	N_{q-1}	Tandem	[6.46]
Cl^{6+}	$2p^63s \to 2p^53sn\ell n'\ell'$	N_{q-1}	TSR	[6.52]
O^{7+}	$1s \to 2\ell n\ell'$	N_{q-1}	TSR	[6.53]
F^{7+}	$1s2s \to 1s2\ell n\ell'$	N_{q-1}	Tandem	[6.44]
Ne^{7+}	$1s^22s \to 1s^22pn\ell$	N_{q-1}	CRYRING	[6.54,55]
Si^{11+}	$1s^22s \to 1s2s2\ell n\ell', 1s2s3\ell n\ell'$	N_{q-1}	TSR	[6.56]
Si^{12+}	$1s2s \to 1s2\ell n\ell'$	N_{q-1}	Tandem	[6.44]
Ar^{13+}	$2s^22p \to 2s2p3\ell3\ell', 2s^23\ell3\ell'$	N_{q-1}	CRYRING	[6.54,55]
Cl^{14+}	$1s^22s \to 1s2s2\ell n\ell', 1s2s3\ell n\ell'$	N_{q-1}	TSR	[6.56]
Fe^{15+}	$2p^63s \to 2p^53sn\ell n'\ell'$	N_{q-1}	TSR	[6.57]
Ar^{15+}	$1s^22s \to 1s^22pn\ell$	N_{q-1}	CRYRING	[6.54,55]
Ar^{15+}	$1s^22s \to 1s^22pn\ell,\ 1s^23\ell n\ell'$	N_{q-1}	UNILAC	[6.48]
Ar^{16+}	$1s^2 \to 1s2\ell n\ell'$	N_{q-1}, N_x	EBIS	[6.49,50]
Ti^{20+}	$1s^2 \to 1s2\ell n\ell'$	$\hbar\omega, N_x$	EBIT	[6.58]
Se^{23+}	$2p^63s \to 2p^53sn\ell n'\ell'$	N_{q-1}	TSR	[6.52]
Fe^{24+}	$1s^2 \to 1s2\ell2\ell'$	$\hbar\omega, N_x$	EBIT	[6.59]
Co^{25+}	$1s^2 \to 1s2\ell2\ell'$	$\hbar\omega, N_x$	EBIT	[6.60]
Cu^{26+}	$1s^22s \to 1s^22pn\ell, 1s^23\ell n\ell'$	N_{q-1}	TSR	[6.36]
Ni^{26+}	$1s^2 \to 1s2\ell n\ell', 1s3\ell n\ell'$	N_x	EBIT	[6.61,62]
U^{28+}	$6s^2 \to 6sn\ell n'\ell'$	N_{q-1}	UNILAC	[6.47]
Mo^{40+}	$1s^2 \to 1s2\ell n\ell', 1s3\ell n\ell'$	N_x	EBIT	[6.62]
Ba^{54+}	$1s^2 \to 1s2\ell n\ell', 1s3\ell n\ell'$	N_x	EBIT	[6.62]
Au^{76+}	$1s^22s \to 1s^22pn\ell$	N_{q-1}	ESR	[6.63,64]
$U^{(87-90)+}$	KLL	N_x	EBIT	[6.65]
U^{89+}	$1s^22s \to 1s^22p5\ell$	N_{q-1}	ESR	[6.66]

$\Delta n = 1$ on helium-like argon were measured and found to agree with Hartree-Fock calculations. At the EBIT, DR was studied [6.59,60,62] by monitoring the x rays measured with solid-state germanium detectors or with a Bragg spectrometer at 90 degrees relative to the electron beam.

Specific examples of DR measurements are given in Figs. 6.3–5. The study of $K\alpha$ radiation of highly charged iron already has a long history in tokamak and solar-flare plasmas [6.67–72] where the dielectronic satellites sample [6.1,2] the electron-energy distribution. Therefore, satellite line radiation can serve as an important diagnostic of the hot plasmas. Within this context, the $K\alpha$ x-ray spectra of iron measured by *Beiersdorfer* et al. [6.59] at the EBIT are particularly interesting. Figure 6.3 demonstrates the ability of the EBIT to select a particular excitation mechanism by choosing the collision energy. For an energy of 4.68 keV the KLL dielectronic resonances of helium-like Fe^{24+} are excited and the lines observed are due to the stabilization of the autoionizing states of the kind $1s2\ell2\ell'$. The peaks in Fig. 6.3 are labeled in *Gabriel*'s [6.1] notation as introduced in Sect. 3.1. Because of the spread in the electron energy, amounting to 50 eV, several resonances can be simultaneously excited. At the much higher energy of 6.67 keV, chosen to be just above the threshold for electron-impact excitation of helium-like iron, no dielectronic excitations are possible. The lines observed are exclusively caused by direct excitation.

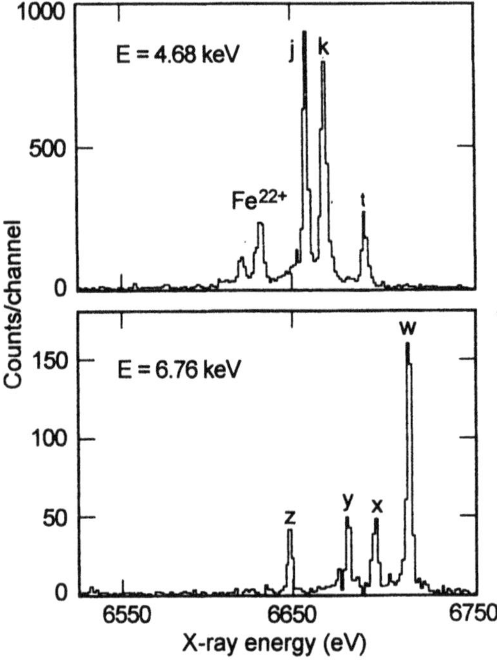

Fig. 6.3. K x-ray spectra of iron measured at the EBIT at two different electron-beam energies. At 4.68 keV all lines are induced by KLL dielectronic recombination. With 6.76 keV, the electron energy is 60 eV above the threshold for electron-impact excitation of a K-shell electron in helium-like iron. This energy region is free of resonances and the lines are solely induced by direct excitation. The lines are labeled in Gabriel's notation, with the letters w, x, y, and z referring to helium-like levels $1s2p\,{}^1P_1$, $1s2p\,{}^3P_2$, $1s2p\,{}^3P_1$ and $1s2s\,{}^3S_1$, respectively [6.59] © 1992 APS

Figures 6.4,5 show examples for DR measurements at storage rings. There, the width of the observed lines is limited by the availability of intense cold electron targets. Particularly interesting is the study of very heavy few-electron ions. With the DR measurements of such systems relativistic effects on the transition probabilities and the corresponding wavefunctions are investigated. At the ESR storage ring at GSI Darmstadt, dielectronic recombination involving $\Delta n = 0$ transitions on lithium-like Au^{76+} have been measured [6.63,64] using the electron cooler as an electron target. Figure 6.4 shows the measured rate coefficient for an energy range covering the processes

$$\mathrm{Au}^{76+}(1s^2 2s) + e \;\rightarrow\; \mathrm{Au}^{75+}(1s^2 2p_{1/2,3/2} n\ell_j)^{**} \;\rightarrow\; \mathrm{Au}^{75+} + \hbar\omega\,.$$

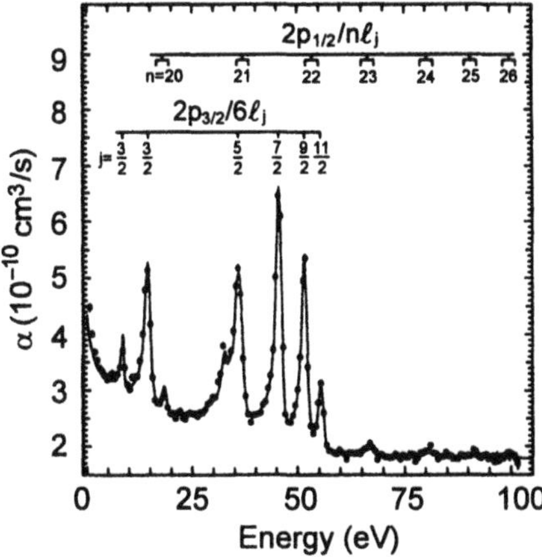

Fig. 6.4. Rate coefficient for recombination of lithium-like $Au^{76+}(1s^2 2s)$ measured [6.63,64] at the ESR of GSI. The *solid line* represents a calculation taking into account $\Delta n = 0$ dielectronic resonances plus a smooth background from radiative recombination. [6.64] © 1995 Elsevier

The spectrum is dominated by the $1s^2 2p_{3/2} 6\ell_j$ resonances. A remarkably good agreement with a fully relativistic calculation [6.64] is obtained including the DR processes plus the smooth background from radiative recombination.

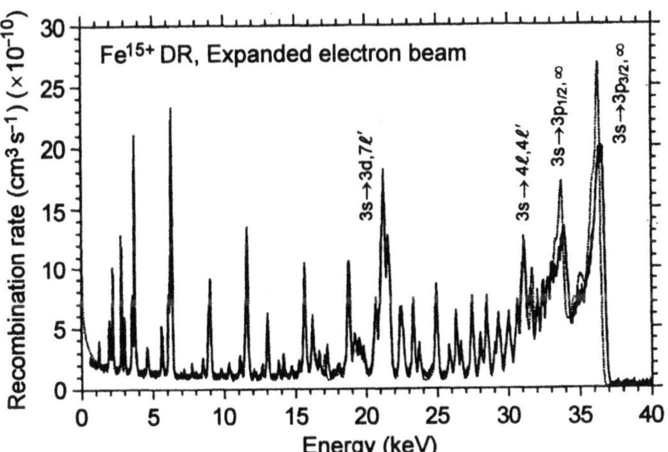

Fig. 6.5. High-precision DR measurement with an expanded electron beam at the TSR in Heidelberg. The lines are due to $\Delta n = 0$ dielectronic recombination of Fe^{15+}. The *dotted curve* represents a theoretical calculation convoluted with a velocity distribution given by a transverse temperature of 15 meV and a longitudinal temperature of 0.15 meV. [6.57] © 1995 Elsevier

Figure 6.5 is a demonstration [6.57] of the high resolution obtained with an adiabatically expanded electron beam (Sect. 2.3.2). The $\Delta n = 0$ resonances were obtained with sodium-like iron ions stored and cooled in the Heidelberg Storage ring TSR. The data could be well fitted with a theoretical calculation including a convolution with a very narrow velocity distribution corresponding to transverse and longitudinal electron-beam temperatures of $T_\perp = 15$ meV and $T_\parallel = 0.15$ meV, respectively, (2.34).

6.1.5 Interference Between Dielectronic and Radiative Recombination

In the examples of DR presented above, radiative recombination (RR) appeared as a background process. Both processes, however, can have identical initial and final states and thus are not distinguishable and should reveal interference patterns. The interference should be largest when the probabilities for both processes are nearly equal. This condition is most nearly fulfilled for the heaviest ions. A large interference has been observed [6.65] in the KLL dielectronic resonances of highly charged uranium in the SuperEBIT. There the excitation functions have been measured by ramping the electron beam through the resonances monitoring the stabilizing $K\alpha$ radiation.

Figure 6.6 shows such an excitation function for the $K\alpha_2$ radiation due to an $L_{12} \to K$ transition of several charge states of uranium. The peaks originate from photons emitted through stabilizing of states produced by DR and the continuum results from RR photons.

The line shape of the K-$L_{12}L_3$ resonances in Fig. 6.6 shows a typical asymmetry with a long high-energy tail. As shown in the figure the data can be fitted with a Fano [6.73,74] profile given by

$$\sigma(E_\mathrm{e}) = \sigma_\mathrm{a}\frac{(q_F+\epsilon)^2}{1+\epsilon^2} + \sigma_b\,, \qquad \epsilon = \frac{E_\mathrm{e}-E_\mathrm{r}}{\Gamma/2}\,, \tag{6.15}$$

where σ_a and σ_b are the interfering and noninterfering parts of the continuum cross section, respectively, and q_F is the Fano lineshape factor. Small values of q_F mean large interference. In the fitting procedure, relative resonance strengths within each ionization stage and resonance widths are all fixed to theoretical predictions [6.75]. For the strong peaks due to B-like and Be-like ions the values deduced from the fit [6.65] are $q_F = 4.8$ and 2.9, respectively. The peaks according to the KL_3L_3 resonance also measured in that experiment do not coincide with any radiative recombination photons. They reveal a symmetric lineshape.

Radiative interference effects in KLL resonances due to overlapping *intermediate* states and coupling of the different decay channels have been analyzed theoretically [6.76,77] in the framework of QED. For the doubly excited states $2s^2_{1/2}, J=0$ and $2p^2_{1/2}, J=0$ formed in helium-like U^{90+} and Pb^{80+}, interference contributions to the DR rate were found to reach about 25%. The problems of interference between DR and RR processes are also considered in [6.39,78–80].

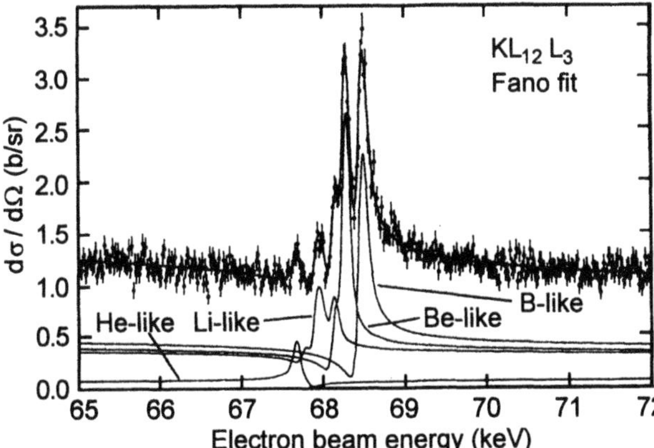

Fig. 6.6. $KL_{12}L_3$ resonances of U^{87+} up to U^{90+} observed [6.65] in the EBIT of LLNL. The data are fit with Fano profiles. Relative resonance strengths within each ionization stage and resonance widths are all fixed to theoretical predictions [6.75] © 1995 APS

6.1.6 Binding Energies

The high resolution and precision demonstrated with very cold electron targets suggests [6.54,64] to employ the method for an accurate determination of single resonances of few electron configurations. Such measurements can provide a test of quantum-electrodynamic (QED) and electron-correlation contributions to electronic binding energies.

Candidates for such measurements are mainly lithium-like and helium-like ions. For instance, transitions of the kind $1s^2 2s_{1/2} \to 1s^2 2p_{1/2} n\ell$, $1s^2 2p_{3/2} n\ell$ can be easily accessed because of the small excitation energy. For high $n\ell$, differences in the binding energies are expected to converge to the respective energy differences of the core configurations.

Adiabatically expanded electron beams have been shown [6.54,56,57] to have temperatures near $T_\perp = 10$ meV and $T_\parallel = 0.1$ meV. The contribution to the DR linewidth in the center-of-mass system is (nonrelativistically) $\Delta E_\mathrm{e} = T_\perp + T_\parallel + 2(E_{cm} T_\parallel)^{1/2}$. In experiments on lithium-like neon and argon [6.54] linewidths of 10 meV have already been observed and a total uncertainty of ± 20 meV has been reached.

Determination of the location E_r of a resonance on an absolute scale with a high accuracy is not a trivial task because it entails the solution of a number of issues related to storage-ring and electron-cooling techniques (see also Sect. 2.3.2). The problems encountered are:

- space-charge uncertainties
- drag-force corrections
- electron and ion-beam energy calibration
- monitoring of particle trajectories.

With a dedicated electron target [6.81], in addition to the electron cooler, the problems will relax considerably because a fast switching of the cooler voltage between cooling and measurement conditions can be avoided.

6.2 Radiative Electron Capture

6.2.1 Comparison with Radiative Recombination

In fast collisions of highly charged heavy ions with light target atoms the dominant electron transfer process is the radiative capture (REC) of an electron initially bound to the target nucleus:

$$X^{q+} + Y \rightarrow X^{(q-1)+}(n\ell) + Y^{+} + \hbar\omega \,. \tag{6.16}$$

The capture of a *free* electron (Radiative Recombination) by a bare heavy ion can only proceed via emission of a photon because of momentum and energy conservation. This argument, though less strict, can still be applied for the capture of a weakly bound electron. By photon emission, the momentum matching between initial and final electronic states is much relieved explaining the dominance of REC over nonradiative electron capture at high relative velocity. In a nonrelativistic approximation, the Coulomb charge transfer may be shown [6.82] to fall off with the projectile velocity v as $\propto v^{-12}$ as compared to $\propto v^{-5}$ for the REC [6.83]. Experimentally, the REC has been identified [6.84–86] through accelerator-based experiments in the early 1970s. Many experimental and theoretical studies of the process followed.

Figure 6.7 illustrates the REC process as an electron transition from a bound state of the target into a bound state of the projectile. In general, this is a complicated problem to solve theoretically because it involves two-center scattering dynamics with corresponding three-body wavefunctions. For a general review on high-energy collisions see for instance Refs. [6.87–89]. The problem simplifies when heavy projectiles and light targets are considered where the effects of the initial binding are small. In the *impulse approximation* the initial state is simply treated as a continuum state of the projectile. The target only serves to provide an initial momentum distribution along with some binding energy of the electron. In this case one can completely resort to the problem of radiative recombination. The cross section for REC is then given [6.90,86] as a convolution of the RR cross section with the momentum distribution of the target electron. In the inertial system of the projectile the double-differential cross section may be written as

$$\frac{\partial^2 \sigma_{\mathrm{REC}}}{\partial \Omega_x' \partial E_x'} = \sum_{n\ell} g_{n\ell} \int \mathrm{d}^3 \boldsymbol{p}_{\mathrm{e}}\, |\Phi_{n\ell}(\boldsymbol{p}_{\mathrm{e}})|^2 \, \frac{\partial \sigma_{\mathrm{RR}}(\boldsymbol{p}_{\mathrm{e}}')}{\partial \Omega_x'} \, \delta[E_x' - (E_i' - E_f')]. \tag{6.17}$$

where the summation runs over all target subshells $n\ell$ each carrying $g_{n\ell}$ equivalent electrons. In (6.17) all primed quantities refer to the rest frame of the projectile. The integration is performed over the target electron momentum $\boldsymbol{p}_{\mathrm{e}}$, whereas the $\Phi_{n\ell}(\boldsymbol{p}_{\mathrm{e}})$ denote the target wavefunctions in momentum space and $\partial\sigma_{\mathrm{RR}}(\boldsymbol{p}_{\mathrm{e}}')/\partial\Omega_x'$ is the differential RR cross section taken at the momentum $\boldsymbol{p}_{\mathrm{e}}'$ that has been transformed into the projectile frame. The delta

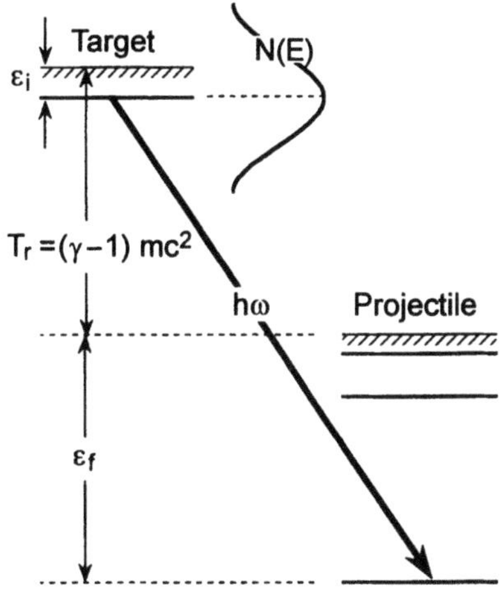

Fig. 6.7. Radiative electron capture viewed from the inertial system of the heavy projectile. An electron transition from a bound target state into a projectile state occurs. The initial state is characterized by its electronic binding energy ϵ_i and electron-energy distribution $N(E)$ and the final state by the binding energy ϵ_f, respectively. The relative motion of the target is qualified by the kinetic energy $T_\mathrm{r} = (\gamma - 1)m_\mathrm{e}c_o^2$ of an electron moving with the same velocity as the target atom.

function ensures energy conservation where, in the projectile frame, the x-ray energy E'_x equals the difference between initial E'_i and final E'_f electron energy.

The *Lorentz transformations* [6.91,92] relevant for the present situation are

$$\begin{aligned} E'_i &= \gamma\,(E_i - \beta c_o p_z) = \gamma m_\mathrm{e} c_o^2 - \gamma|\epsilon_i| - \gamma\beta c_o p_z\,, \\ E'_f &= m_\mathrm{e} c_o^2 - |\epsilon'_f|\,, \end{aligned} \tag{6.18}$$

where ϵ_i and ϵ'_f denote the electronic binding energies in their respective rest frames and β is the collision velocity divided by the velocity of light c_o, whereas γ denotes the Lorentz parameter $\gamma = (1 - \beta^2)^{-1/2}$. At this point, we also give the transformation of the x-ray energy E'_x into the laboratory frame

$$E_x = \frac{E'_x}{\gamma\,(1 - \beta\cos\vartheta)} \tag{6.19}$$

depending on the emission angle ϑ relative to the ion-beam direction. The angular and solid-angle transformations are

$$\cos\vartheta' = \frac{\cos\vartheta - \beta}{1 - \beta\cos\vartheta}\,, \tag{6.20}$$

$$\frac{\partial\Omega'_x}{\partial\Omega_x} = \frac{1}{\gamma^2\,(1 - \beta\cos\vartheta)^2}\,. \tag{6.21}$$

The transformations (6.18) lead to

$$E'_x = E'_i - E'_f = (\gamma - 1)m_\mathrm{e}c_o^2 + |\epsilon'_f| - \gamma|\epsilon_i| - \gamma\beta c_o p_z \tag{6.22}$$

and consequently to

$$\frac{\partial^2 \sigma_{\rm REC}}{\partial \Omega'_x \partial E'_x} = \sum_{n\ell} g_{n\ell} \int \mathrm{d}^3 \boldsymbol{p}_{\rm e} \, |\Phi_{n\ell}(\boldsymbol{p}_{\rm e})|^2$$
$$\times \frac{\partial \sigma_{\rm RR}(\boldsymbol{p}'_{\rm e})}{\partial \Omega'_x} \delta \left[E'_x - (T_{\rm r} + |\epsilon'_f| - \gamma |\epsilon_i| - \gamma \beta c_o p_z) \right], \tag{6.23}$$

where we used $T_{\rm r} = (\gamma - 1) m_{\rm e} c_o^2$ for the kinetic energy of an electron traveling with the projectile velocity βc_o. The vector

$$\boldsymbol{p}'_{\rm e} = \boldsymbol{p}'_{\perp} + \left(\frac{p'_{\parallel}}{\gamma} + \frac{\beta |\epsilon_i|}{c_o} \right) \hat{\boldsymbol{e}}_{\parallel} \tag{6.24}$$

appearing in the argument of $\sigma_{\rm RR}$ defines the axis relative to which the solid angle element $\partial \Omega'_x$, into which the photon is emitted, has to be taken. For small perpendicular momenta $\boldsymbol{p}'_{\perp}$ the direction of $\boldsymbol{p}'_{\rm e}$ nearly coincides with the beam direction represented by the unit vector $\hat{\boldsymbol{e}}_{\parallel}$. The transverse momenta can be taken into account in a numerical integration of (6.23). For the integration over the $\boldsymbol{p}_{\perp}$ components an analytical expression has been derived by *Ichihara* et al. [6.93].

An approximate solution of (6.23) can be obtained by taking the slowly varying recombination cross section outside the integral. Then the REC cross section reduces to

$$\frac{\partial^2 \sigma_{\rm REC}}{\partial \Omega'_x \partial E'_x} \simeq \frac{1}{\gamma \beta c_o} \mathcal{J}(p_z) \left(\frac{\partial \sigma_{\rm RR}}{\partial \Omega'_x} \right)_{p'_z = \gamma p_z - \gamma \beta m_e c_o} \tag{6.25}$$

with the Compton profile

$$\mathcal{J}(p_z) = 2\pi \sum_{n\ell} g_{n\ell} \int_{p_z}^{\infty} |\Phi_{n\ell}(p_{\rm e})|^2 p_{\rm e} \mathrm{d}p_{\rm e} \tag{6.26}$$

giving the probability of finding a momentum component of a target electron along the z direction.

Within the formalism given above, through the exact (6.23) and approximate (6.25) formulae, one is left with the problem of choosing appropriate target wavefunctions and RR cross sections. This will be outlined in the three subsections to follow when we consider the REC lineshape and the total and angular-differential cross sections.

6.2.2 Line Shape

The main influence of the target binding is to spread the REC photons into a broad spectral distribution as determined by the term $\gamma \beta c_o p_z$ in the delta function of (6.23) and by the averaging over the random distribution of the electron momenta described by the spherically averaged wavefunctions $\Phi_{n\ell}(\boldsymbol{p}_{\rm e})$. For light target atoms it is generally sufficient to use non-relativistic wavefunctions. Accurate wavefunctions can be obtained with the Roothaan-Hartree-Fock expansion technique [6.94,95] in which single-electron orbitals are expanded in the basis of Slater-type orbitals of the form

$$\begin{aligned}\chi_{n\ell}(\boldsymbol{r}) &= N_n\, Y_{\ell m}(\theta,\phi)\, r^{n-1}\, \mathrm{e}^{-\zeta r}, \\ N_n &= (2n!)^{-1/2}(2\zeta)^{n+1/2},\end{aligned} \tag{6.27}$$

where $Y_{\ell m}(\theta, \phi)$ denote the normalized spherical harmonics. For practical computational purposes it is very efficient to use tabulated wavefunctions as for example, those of *Clementi* and *Roetti* [6.96] who present their calculations as *double-zeta* functions. A double-zeta function is a very accurate approximation of a Roothaan-Hartree-Fock function in which each single-electron orbital is described by *two* Slater functions. By Fourier transformation, the spherically averaged target wavefunctions can be obtained as

$$\Phi_{n\ell}(p_\mathrm{e}) = \sum_{k=1}^{2}\sum_{n_i l_i} C_{n_i l_i k}\, \tilde{\chi}_{n_i l_i}(\zeta_{n_i l_i k}, p_\mathrm{e}) \tag{6.28}$$

where the coefficients $C_{n_i l_i k}$ and $\zeta_{n_i l_i k}$ can be taken from the tabulation [6.96] and the functions $\tilde{\chi}_{n_i l_i}(\zeta_{n_i l_i k}, p)$ are the Slater-type radial functions [6.97,98] in momentum representation. For convenience, the innermost orbitals are listed

$$\begin{aligned}\chi_{1s}(p_\mathrm{e}) &= \frac{2^{3/2}}{\pi}\frac{\zeta^{5/2}}{(\zeta^2+p_e^2)^2}, \\ \chi_{2s}(p_\mathrm{e}) &= \frac{2^{3/2}\zeta^{5/2}}{5^{1/2}\pi}\frac{3\zeta^2-p_e^2}{(\zeta^2+p_e^2)^3}, \\ \chi_{2p}(p_\mathrm{e}) &= \frac{2^{7/2}\zeta^{7/2}}{3^{1/2}\pi}\frac{p_\mathrm{e}}{(\zeta^2+p_e^2)^3}, \\ \chi_{3s}(p_\mathrm{e}) &= \frac{2^4\zeta^{9/2}}{5^{1/2}\pi}\frac{\zeta^2-p_e^2}{(\zeta^2+p_e^2)^4}, \\ \chi_{3p}(p_\mathrm{e}) &= \frac{2^4\zeta^{7/2}}{5^{1/2}3\pi}\frac{p_\mathrm{e}\,(p_e^2-5\zeta^2)}{(\zeta^2+p_e^2)^4},\end{aligned} \tag{6.29}$$

given in atomic units, where one atomic unit of momentum equals $m_\mathrm{e}e^2/\hbar \simeq$ 3729 eV$/c_o$.

For the accurate analysis of REC line shapes the kind of approximation used for the RR cross section is not very critical because only the variation over a small energy interval is required. The nonrelativistic Stobbe cross section for radiative recombination [6.99] can be used for this purpose up to a specific beam energy of several hundred MeV/u (compare also the discussion in the next section). For capture of an electron into the $1s$ state of the projectile the Stobbe cross section per $1s$ vacancy is given by [6.99]:

$$\begin{aligned}\frac{\partial\sigma_{\mathrm{RR},1s}^{\mathrm{Stobbe}}}{\partial\Omega_x'} &= \frac{2^4\pi\hbar e^2}{m_\mathrm{e}^2c_o^3}\left(\frac{\eta^3}{1+\eta^2}\right)^2\frac{\exp\left(-4\eta\arctan(1/\eta)\right)}{1-\exp\left(-2\pi\eta\right)} \\ &\quad\times\frac{\sin^2\vartheta'}{(1-\beta\cos\vartheta')^4}, \qquad \text{with } \eta \equiv \frac{\alpha Z}{\beta},\end{aligned} \tag{6.30}$$

where η is the Sommerfeld parameter and α denotes the fine-structure constant. The prime on the solid angle Ω'_x and on the photon-emission angle ϑ' is used to indicate the ions inertial system.

An example of an REC line profile calculated according to (6.23) is displayed in Fig. 6.8 for the case of bare Dy^{66+} ions colliding with argon target atoms at a specific energy of 292 MeV/u. For this calculation the electron and photon momenta were correctly treated in three dimensions [compare the discussion following (6.24)]. In Fig. 6.8 the individual contributions from the argon $n\ell$ subshells are displayed together with their sum. In Fig. 6.9 the theoretical results are compared with a measurement [6.100] conducted at the ESR storage ring at GSI.

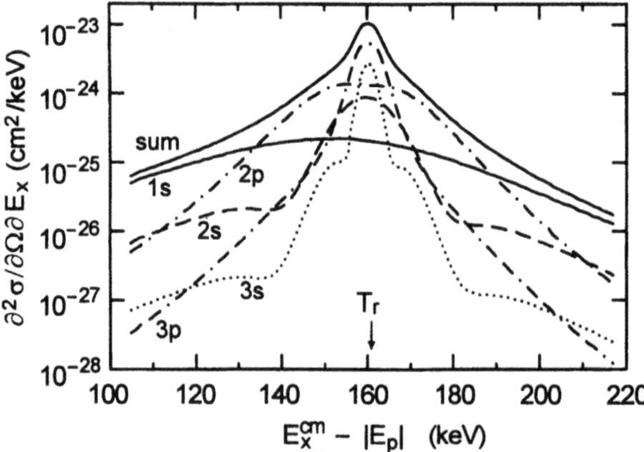

Fig. 6.8. Calculated x-ray line profile for radiative electron capture into the 1s state by bare Dy^{66+} ions. The individual contributions from the initial subshells of an argon target are shown as indicated [6.100] © 1993 IOP

The approximation (6.25) brings about a direct link between REC and Compton scattering [6.101] mediated by the Compton profile (6.26). Despite this appealing idea, it was pointed out [6.100] that it is not appropriate for

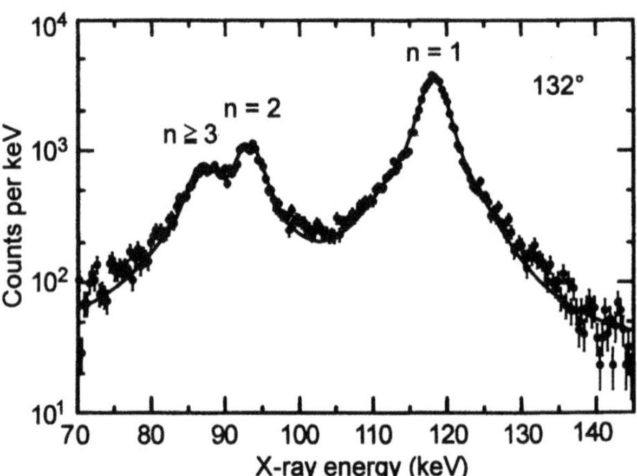

Fig. 6.9. Experimental REC spectrum for $Dy^{66+} \to Ar$ collisions at 292 MeV/u [6.100] © 1993 IOP

accurate analysis of line shapes and line positions. Because of the decrease of the recombination cross section with increasing x-ray energy, the REC line has a pronounced asymmetry and is shifted to lower energy with respect to the position (6.22) with $p_z = 0$. In Fig. 6.9 a nearly perfect agreement between experiment and a detailed calculation is revealed. There, the theoretical line profile is fixed *a priori* including also the experimental detection efficiency, whereas the absolute height and position is obtained from experiment.

6.2.3 Total Cross Section

Differential cross sections for radiative electron capture, which are relativistically exact, have been calculated [6.93] starting from a relativistic treatment [6.102] of photoionization or radiative recombination. The total cross section is obtained by integrating the differential cross section (6.23) over the solid angle and summing over the final states. For single-electron targets total cross sections were tabulated by *Ichihara* et al. [6.103]. In the nonrelativistic dipole approximation the integrated Stobbe cross section for radiative recombination can be obtained by summing over the final projectile states. Because at high velocity the cross section into various main shells falls off with $\sim n^{-3}$ only the innermost projectile shells contribute appreciably.

In Fig. 6.10 the total REC cross section [6.104] is plotted as a function of a scaled kinetic energy which is the inverse of the squared Sommerfeld parameter η

$$1/\eta^2 \simeq 40.31 \, \frac{E_{\mathrm{kin}}(\mathrm{MeV/u})}{Z^2} \, .$$

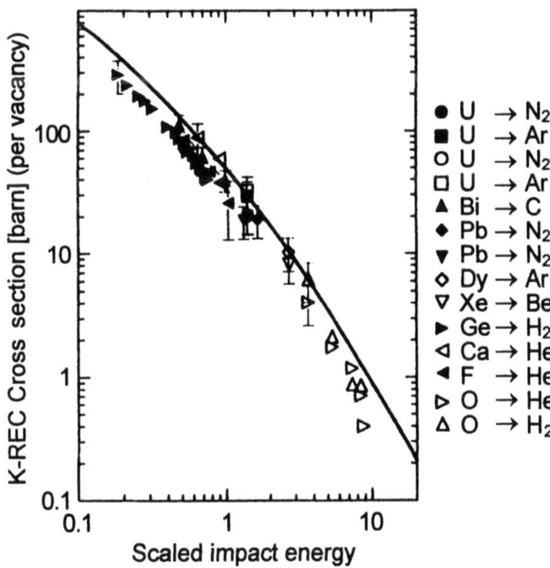

Fig. 6.10. Total cross sections for REC into the 1s state of heavy projectiles as a function of the scaled impact energy (40 $\times$ E [MeV/u] / Z^2). The solid line represents the nonrelativistic dipole approximation according to (6.30) [6.104] © 1995 APS

Such a scaling is suggested by the form of the Stobbe formula where the collision energy only enters through the Sommerfeld parameter. The measured data [6.100,104–109] follow a universal curve surprisingly well and are in good overall agreement with the theoretical curve lying only slightly above the measured data points. An explanation for this behavior in a regime where the nonrelativistic approximation is expected to break down is the mutual cancellation of relativistic effects. This has been analyzed in some detail theoretically [6.110] and it was found that the nonrelativistic dipole approximation shows remarkable agreement with full relativistic multipole results for s subshells even at high energies (cf. Sect. 5.1.2). This is not so for other $\ell \neq 0$ states and also not for angular distributions. Because capture into the K shell is the dominant contribution, the net relativistic effects from other shells are expected to be too small to be detected in the *total* cross section.

We should note, that part of the experimental data were obtained with solid targets. Some differences in the REC cross section between gas and solid targets have been noted by *Tribedi* et al. [6.111] who also investigated an ion-solid-state effect employing channeled ions [6.112,113].

6.2.4 Angular Distribution

For REC into the projectile K shell, the photon angular distribution may be obtained from formula (6.30) as

$$\frac{\partial \sigma_{\mathrm{RR},1s}}{\partial \Omega'_x} \propto \frac{\sin^2\vartheta'}{(1-\beta\cos\vartheta')^4}\,, \tag{6.31}$$

where the effect of retardation is already included in lowest order giving rise to the term $(1-\beta\cos\vartheta')^4$. The same leading term is obtained in the relativistic *Sauter* formula [6.114] which originates from the lowest order expansion in αZ of the relativistic matrix element

$$\begin{aligned}\frac{\partial \sigma^{\mathrm{Sauter}}_{\mathrm{RR},1s}}{\partial \Omega'_x} &= \alpha Z^5 \frac{\beta}{\gamma^3}\left(\frac{\hbar}{m_\mathrm{e}^2 c_o}\right)^2 \left(\frac{m_\mathrm{e}^2 c_o^2}{E'_x}\right)^3 \\ &\times \frac{\sin^2\vartheta'}{(1-\beta\cos\vartheta')^4}\left[1+\frac{\gamma(\gamma-1)(\gamma-2)}{2}(1-\beta\cos\vartheta')\right].\end{aligned} \tag{6.32}$$

Only at high values of the Lorentz factor γ, the second term in the square brackets gains importance and the two angular distributions become different. The distribution (6.31) can be transformed into the laboratory frame using the Lorentz transformation (6.20,21). The result is a simple

$$\frac{\partial \sigma_{\mathrm{RR},1s}}{\partial \Omega_x} \propto \sin^2\vartheta \tag{6.33}$$

distribution. This perfect cancellation of retardation and Lorentz transformation was first noted by *Spindler* et al. [6.115,116]. The cancellation occurs only if the velocities responsible for retardation and for Lorentz transformation are the same. For radiative-recombination studies at an electron cooler, for instance, the low collision energies cause only negligible retardation whereas the high laboratory velocities require enormous Doppler transformations.

Experimentally the cancellation has been confirmed by *Anholt* et al. [6.109] observing the angular distribution for bare Xe^{54+} ions traversing a thin Be foil at a specific collision energy of 197 MeV/u. The experimental results are shown in Fig. 6.11 together with a fully relativistic calculation of *Ichihara* et al. [6.93]. Besides a normalization factor of 0.8 a nearly perfect agreement can be stated. For this example the detailed theoretical calculation still agrees, rather well, with the $\sin^2\vartheta$ dependence.

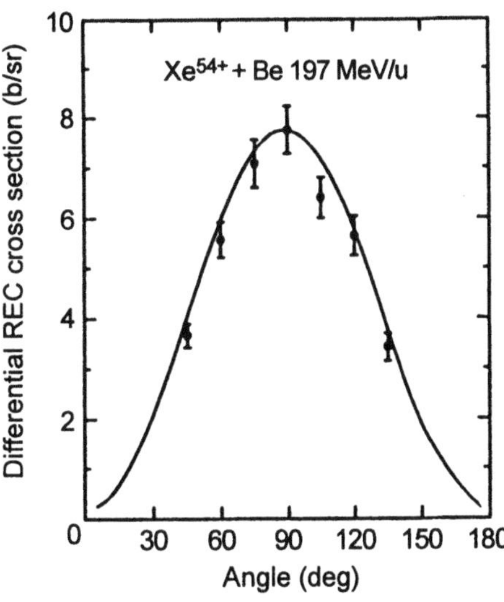

Fig. 6.11. Angular distribution of x rays following K REC for 197-MeV/u Xe^{54+} on Be. Measured values are from *Anholt* et al. [6.109]. The *solid line* represents the fully relativistic calculation of *Ichihara* et al. [6.93] containing a normalization factor of 0.8 which is within the systematic uncertainty of the experiment. [6.93] © 1994 APS

The same theory predicts [6.93] noticeable deviations from the symmetric distribution for very heavy ions at high collision energies as shown in Fig. 6.12 for 295-MeV/u U^{92+} on nitrogen. The asymmetry is attributed to spin-flip transitions which mainly contribute to the forward emission. Without electron spin, the differential cross section for REC into an $\ell = 0$ state vanishes in the forward and in the backward direction because the initial and final states have the same angular-momentum projection $m_\ell = 0$, so that the longitudinal emission of a photon with angular momentum ± 1 is forbidden. Relativistically, REC into s states becomes possible also with photon emission in the forward and backward direction through magnetic interactions leading to spin flip. The spin-flip contributions to the angular-differential cross section are very sensitive to the details of the wave function. Therefore, measurements of angular distributions near the forward and backward direction can stringently test the theory.

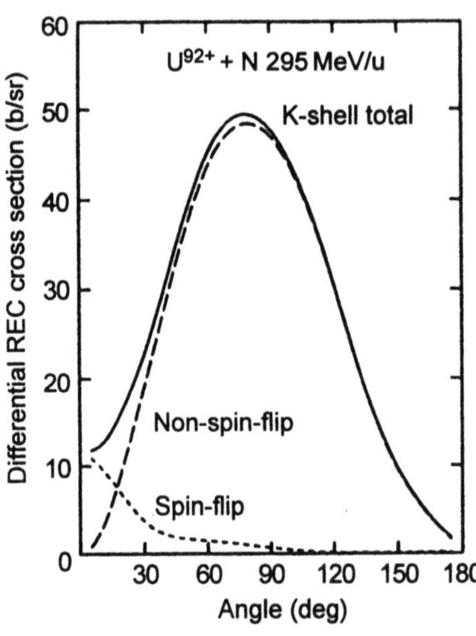

Fig. 6.12. Calculated angular-differential K-REC cross section for 295-MeV/u U^{92+} on N as a function of the laboratory angle. Spin-flip contributions mainly contributing in the forward direction cause a noticable asymmetry. [6.93] © 1994 APS

The situation for REC into $\ell > 0$ states is different from the case $\ell = 0$, as forward and backward emission can be accommodated without spin flip because of the coupling between the electron spin and the orbital angular momentum. A profound discussion of angular distributions into various subshells has been presented by *Eichler* et al. [6.117]. Already the *nonrelativistic* angular distribution is different from (6.33). For capture into $2p$ states, for instance, a

$$\frac{\partial \sigma_{\mathrm{RR},2p}}{\partial \Omega_x} \quad \propto \quad 1 + \mathrm{const} \times \sin^2\vartheta \tag{6.34}$$

distribution is obtained [6.115] which is still symmetric around 90°.

Appreciable deviations from symmetry around 90° have been observed by *Stöhlker* et al. [6.118] for the REC into the L-shell of He-like U^{90+} projectiles. The results of this work are displayed in Fig. 6.13a,b. In this measurement, using 89 MeV/u $U^{90+} \to$ C collisions, the L-REC radiation is spectroscopically resolved into the two subshell contributions $L_{1,2}$ and L_3. Besides an experimental normalization factor of 0.65, the experimental angular distributions for both peaks agree very well with the fully relativistic calculation [6.118,93]. Figure 6.13a,b reveals a forward shift of the $2s_{1/2}$ distribution and a slight backward shift of the $2p_{1/2}$ and $2p_{3/2}$ distributions, an effect difficult to detect without spectral separation. The form of the $2s_{1/2}$ contribution, shown as dashed curve, is similar to the one depicted in Fig. 6.12 for the $1s_{1/2}$ REC. Its forward shift is still noticeable in the measured sum of both $j = 1/2$ contributions.

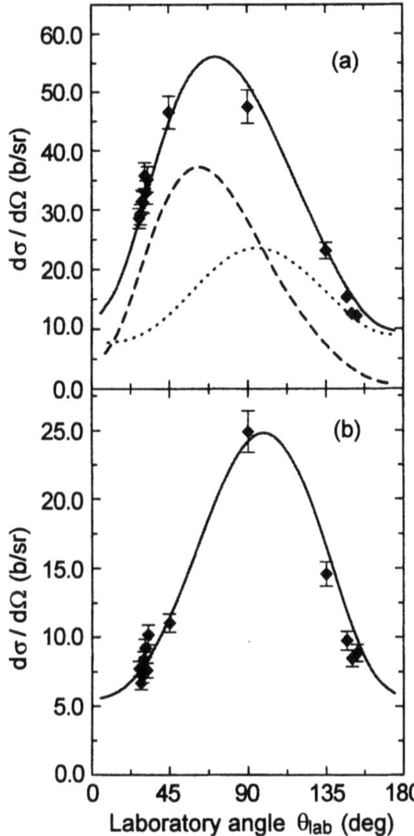

Fig. 6.13a,b. Angular distribution of L-REC radiation for 89 MeV/u $U^{90+} \rightarrow C$ collisions. Contributions from capture into the $2s_{1/2}, 2p_{1/2}$ and into $2p_{3/2}$ states are spectroscopically resolved. The *solid lines* represent the results of detailed relativistic calculations. Individual contributions calculated for $2s_{1/2}$ (*dashed line*) and for $2p_{1/2}$ (*dotted line*) are also shown. [6.118] © 1994 APS

6.3 Resonant Transfer and Excitation

In fast ion-atom collisions, a resonant electron capture can occur similar to the dielectronic capture of a free electron introduced in Sect. 6.1. The process, referred to as *Resonant Transfer and Excitation* (RTE), may be formally expressed (cf. (6.1)) as

$$X^{q+} + Y \rightarrow X^{(q-1)+^{**}} + Y^+ \rightarrow \left\{ \begin{array}{l} X^{(q-1)+^*} + \hbar\omega \\ \text{or} \\ X^{q+} + e \end{array} \right. + Y^+ . \qquad (6.35)$$

It involves the capture of a target electron simultaneously with the excitation of the projectile in a single collision, leading to the formation of the intermediate doubly excited state $X^{(q-1)+^{**}}$ of the projectile. As in DR, this doubly excited state may decay either via x-ray or Auger-electron emission. Since its first experimental identification [6.119,120] a large number of experiments followed [6.121] mainly confining to ions of low nuclear charge. Interest in the study of RTE has largely been motivated through the analogy with DR which now is well established [6.120,122–125]. Simultaneous capture and excitation can be either correlated as in the RTE (6.35) or uncorrelated in which case it is referred to as *Nonresonant Transfer and Excitation* (NTE). Both effects have been identified [6.126,127] and can be distinguished by their different dependence on the projectile velocity. Although interference between NTE and RTE is possible one has tried, in experiments, to reduce NTE in favor of a clear RTE signature.

Experimentally, RTE can be observed as a structure on the energy dependence of the total electron capture cross section [6.128,129] or through x-ray [6.120,130] or Auger [6.131] emission. The cross section for simultaneous x-ray production and charge exchange in a single collision can be obtained by the coincidence between the detection of a photon and the detection of a specific charge state emerging from the interaction region.

Considering the initial distribution of the electron momentum in the target atom, there is a close correspondence between RTE and REC, and virtually the same kind of approximations as discussed in Sect. 6.2 are possible. Specifically, the RTE cross section can be obtained by convolution of the calculated DR cross section with the momentum distribution of the target electrons in analogy to the integrations (6.23) and (6.25). It turned out that the resulting broadening of the RTE structures is the main limitation of the method. For projectiles of low and intermediate atomic number Z, the width of the Compton profile allows, at most, to resolve the main resonances corresponding to different principal quantum numbers. Only at very high Z, the size of the fine-structure splitting in the L shell becomes comparable to the width of the Compton profile.

Figure 6.14a,b shows the results of an RTE experiment [6.129] for $U^{90+} \to H_2$ collisions. The total cross section for single-electron capture as a function of the collision energy shows a structure from which the smooth nonresonant charge-exchange background has been subtracted. The experimental RTE cross section thus obtained is shown in Fig. 6.14b. The three peaks arise, with increasing energy, from the Li-like intermediate states

$$
\begin{aligned}
&- \quad 1s_{1/2}2s_{1/2}^2,\ 1s_{1/2}2s_{1/2}2p_{1/2},\ 1s_{1/2}2p_{1/2}^2 \\
&- \quad 1s_{1/2}2s_{1/2}2p_{3/2},\ 1s_{1/2}2p_{1/2}2p_{3/2} \qquad\qquad (6.36) \\
&- \quad 1s_{1/2}2p_{3/2}^2\ .
\end{aligned}
$$

The measurement agrees well with relativistic calculations [6.132–134] for the resonances shown as the solid curve.

The dominant stabilization of the doubly excited states (6.36) occurs through $n = 2 \to 1$ x-ray transitions energetically grouped into the two $K\alpha_1$ and $K\alpha_2$ lines of Li-like U^{89+}. The $K\alpha_1$ line originates from $2p_{3/2} \to 1s_{1/2}$ transitions whereas the $K\alpha_2$ line, $2p_{1/2} \to 1s_{1/2}$, is blended from $2s_{1/2} \to 1s_{1/2}$ transitions. Through an x-ray/particle coincidence measurement [6.135,136] the RTE for $U^{90+} \to C$ was measured making use of the selectivity provided by the $K\alpha$ lines. Because of the closed K shell, the re-

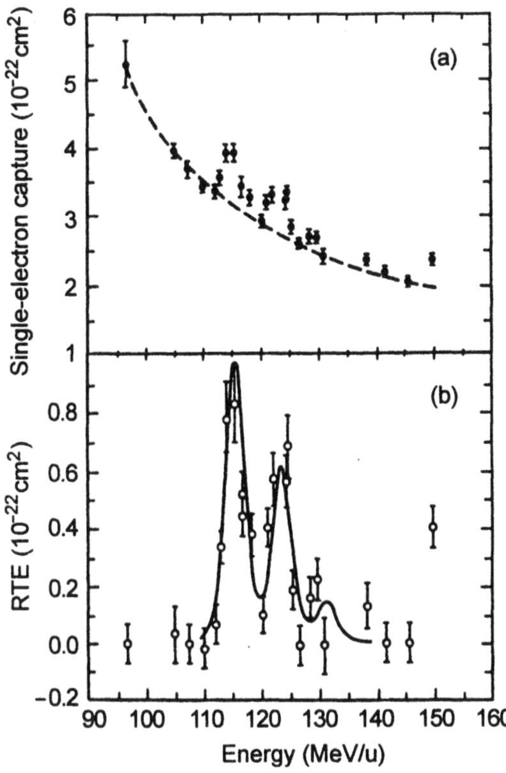

Fig. 6.14a,b. Electron-capture cross section as a function of the collision energy for $U^{90+} \rightarrow H_2$. (a) measured data points and empirical fit to electron-capture background (*dashed curve*). (b) *Open circles*, experimentally determined RTE cross section (total single-electron--capture cross section minus empirical fit). *Solid curve*, theoretical calculations [6.132] of RTE cross sections. [6.129] © 1990 APS

quirement of a $K\alpha$ emission in addition to the single-electron capture is an unambiguous signature for the RTE reducing the background considerably. Figure 6.15 shows the x-ray spectra observed at 30° coincident with

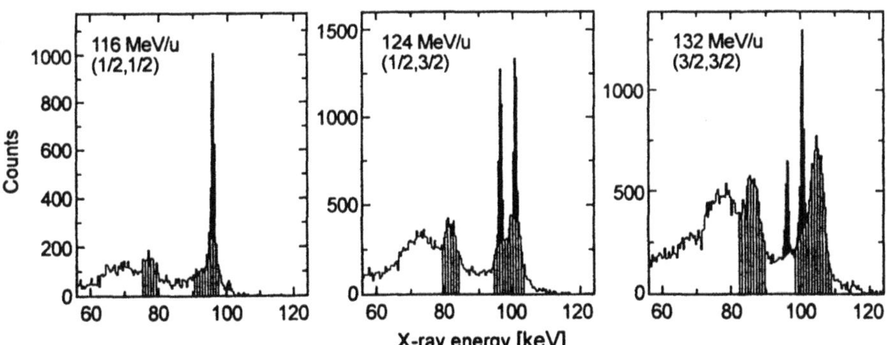

Fig. 6.15. X-ray spectra coincident with one-electron capture for $U^{90+} \rightarrow C$ collisions transformed into the projectile system. The collision energies are chosen to sample the KLL resonances. The hatched areas indicate the L- and M-REC contributions. [6.135] © 1995 Elsevier

one-electron capture. The spectra are transformed into the projectile frame showing the $K\alpha_{1,2}$ lines at 96 and 100 keV. The broad features, originating from REC into the L and M shell, shift with impact energy. When the collision energy is scanned over the KLL resonances, the $K\alpha_1$ and $K\alpha_2$ lines receive intensity which changes with the respective RTE subshell population. The corresponding differential cross section for $K\alpha_1$ and $K\alpha_2$ emission and the summed RTE cross section is shown in Fig. 6.16. There, the observed line width is somewhat larger than for the $U^{90+} \to H_2$ case of Fig. 6.14a,b, which can be attributed to the broader Compton profile for the solid carbon target. Discriminating between the two $K\alpha_{1,2}$ lines, however, more than compensates for the increased linewidth. From the measured differential cross section for $K\alpha_{1,2}$ x-ray production, subshell-differential RTE cross sections were derived [6.135] which were found to agree well with relativistic calculations including the Breit interaction.

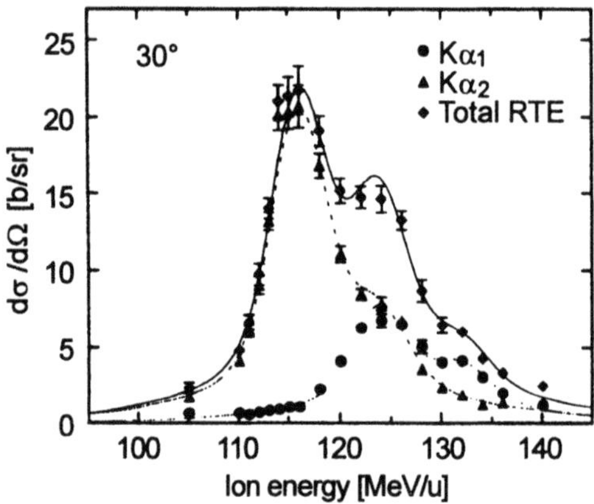

Fig. 6.16. Differential $K\alpha_1$, $K\alpha_2$ and summed RTE cross sections for $U^{90+} \to C$ collisions. The *lines* represent a fit to the experimental data points [6.135] © 1995 Elsevier

6.4 Three-Body Recombination

A common loss mechanism of highly charged ions colliding with charged particles is the following recombination with electrons

$$X^{q+} + e + e \to X^{(q-1)+} + e\,. \tag{6.37}$$

This process is inverse to electron-impact ionization and is called *three-body* or *ternary* recombination. Another ternary process, three-body ion recombination, is principally also possible

$$X^{q+} + e + X^{q'+} \to X^{(q-1)+} + X^{q'+}\,. \tag{6.38}$$

The presence of a third particle, an electron in the process (6.37) and an ion in process (6.38), is necessary to fulfill the momentum and energy conservation. The density of recombining ions per unit of time can be expressed as

$$dn_q/dt = -\alpha_{\mathrm{TR}}\, n_q n_{\mathrm{e}}, \tag{6.39}$$

where n_q and n_{e} are the ion and electron densities. The coefficient α_{TR} is called the *recombination* rate coefficient. In the literature, one can also find another definition of the recombination coefficient, namely, the product $n_e\alpha_{\mathrm{TR}}$ is called the TR rate, e.g., [6.137].

Under equilibrium conditions, the rates of recombination and ionization have to be equal. For (6.37) the equality reads

$$\alpha_{TR}\, n_q n_{\mathrm{e}} = \langle\sigma_{\mathrm{i}} v\rangle\, n_{q-1} n_{\mathrm{e}}\,, \tag{6.40}$$

where

$$\langle\sigma_{\mathrm{i}} v\rangle = \int \sigma_i(v) v f(\boldsymbol{v}, T)\, \mathrm{d}^3\boldsymbol{v} \tag{6.41}$$

and $f(\boldsymbol{v}, T)$ is the Maxwell velocity distribution function of the interacting particles with a temperature T. Using the Saha equation (2.8), which bounds the quantities n_q and n_{q+1}, one arrives at the detailed-balance equation for the rates in question

$$\alpha_{\mathrm{TR}} = \frac{n_{q-1}}{n_q}\langle\sigma_{\mathrm{i}} v\rangle = \left(\frac{2\pi\hbar^2}{m_{\mathrm{e}} T_{\mathrm{e}}}\right)^{3/2} \frac{g_{q-1}}{2g_q}\, n_{\mathrm{e}}\langle\sigma_{\mathrm{i}} v\rangle\, \exp(I_{q-1}/T_{\mathrm{e}}), \tag{6.42}$$

where I_{q-1} is the binding energy of the captured electron. Therefore, knowing, from theory or experiment, the ionization rate coefficient $\langle\sigma_{\mathrm{i}} v\rangle$ and its dependence on the electron temperature T_{e} one can estimate the respective rate coefficient α_{TR} for three-body recombination. For example, the use of the the semi-empirical Lotz formula [6.138] or the classical Thomson formula in eq. (6.42) lead to identical results, e.g., [6.139]

$$\alpha_{\mathrm{TR}} \simeq \frac{16\pi^2 e^4 \hbar^3}{5 m_{\mathrm{e}}^2} N \frac{g_{q-1}}{2g_q} \frac{n_{\mathrm{e}}}{I_{q-1} T_{\mathrm{e}}^2} \tag{6.43}$$

where N is the number of equivalent electrons in the shell which the electron is captured to.

In the case of electron recombination of a bare nucleus to a state with principal quantum number n, one has:

$$g_{q-1} = 2n^2, \quad g_q = 1, \quad N = 1, \quad I_{q-1} = \frac{Z^2\mathrm{Ry}}{n^2}\,, \tag{6.44}$$

where Z is the nuclear charge of the resulting H-like ion and

$$\alpha_{\mathrm{TR}}(n) = \frac{32\pi^2\hbar^5 n_{\mathrm{e}} n^4}{5 Z^2 m_{\mathrm{e}}^3 T_{\mathrm{e}}^2}\,. \tag{6.45}$$

The total TR rate coefficient is defined by the sum

$$\alpha_{\mathrm{TR}}(T_{\mathrm{e}}) = \sum_{n=1}^{n=n_{\max}} \alpha_{\mathrm{TR}}(n), \quad n_{\max}^2 = \frac{Z^2\mathrm{Ry}}{T_{\mathrm{e}}}\,. \tag{6.46}$$

The maximum principal quantum number n_{max} is limited by the electron temperature T_e because ions being in higher excited states n, such that $I_n \ll T_e$, will be reionized and therefore will not contribute to the net recombination. The total recombination rate is then equal to

$$\alpha_{TR}(T_e) = 5.8 \times 10^{-27}\ \mathrm{cm^3 s^{-1}}\ Z^3 n_e[\mathrm{cm^{-3}}]\, T_e[\mathrm{eV}]^{-9/2}. \tag{6.47}$$

It follows that the three-body recombination process is important at low temperatures and high densities. The total rate coefficient (6.47) obtained by the simple consideration shown above, reveals the same dependence on density, temperature and nuclear charge as more elaborate treatments [6.140,141]. Only the numerical factor of 5.8 in (6.47) is about a factor of two larger.

Considering the recombination (6.38) involving two ions plus one electron, a similar treatment is possible as with the ion-electron-electron collision. However, the use of simple formulae describing the heavy-ion impact ionization will be inaccurate. Furthermore, one has to clarify under which conditions Saha equilibrium can be assumed.

There is an important interplay of the individual collisional and radiative processes responsible for the stabilization of a final charge state. A net lowering of the charge state can be achieved by reactions which have been grouped in three categories. The first is *collisional recombination* which is a three-body recombination populating high n states followed by collisionally induced transitions to more tightly bound states. Secondly, *radiative recombination* effectively contributes to the population of low lying bound states. The third mechanism is a three-body recombination followed by a stabilizing decay that is dominated by radiative transitions. The net effect of these processes is known as *collisional-radiative recombination* and the theory for it was first laid out for hydrogenic plasmas by *Bates* and coworkers [6.142–144] followed by a large number of works from which we mention only a few [6.140,145–148]. The subject was extended to highly charged ions [6.141,149] in order to estimate recombination losses in an electron cooler for heavy ions.

According to the treatment sketched, the total rate coefficient for collisional-radiative recombination in a plasma with electrons and completely stripped ions of charge number Z can be represented in the form

$$\alpha_{tot} = \alpha_{TR} + \alpha_{RR} + \alpha_{rad}, \tag{6.48}$$

where α_{TR} is the three-body recombination rate given by (6.47), α_{RR} is the radiative-recombination rate coefficient and α_{rad} is the rate coefficient for the processes involving radiative stabilization.

Figure 6.17 shows the three contributions to the total rate coefficient for collisional-radiative recombination of bare uranium ions with electrons of density $n_e = 10^8$ cm^{-3} as a function of the electron temperature T_e.

The influence of external magnetic and electric fields on the collisional recombination processes has also been treated [6.149–151]. While the effects of an electric field is expected to be small, in a strongly magnetized plasma

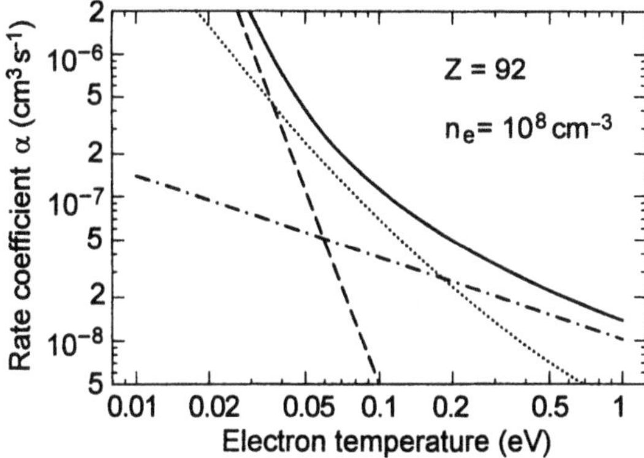

Fig. 6.17. Total rate coefficient for recombination of bare uranium ions with free electrons at a density $n_e = 10^8\,\mathrm{cm}^{-3}$ as a function of the electron temperature T_e. The individual contributions according to (6.48) are given. *Dashed* the collisional contribution α_{TR}; *dotted* α_{rad} involving radiative stabilization; *dash-dotted* α_{RR} due to radiative recombination; solid: sum α_{tot} of the previous three. From [6.141] © 1989 Gordon and Breach

the calculated three-body recombination rate is reduced by about a factor of 10 as compared to a field-free plasma.

Despite these thorough considerations, three-body recombination can not directly cause the enormously enhanced recombination rates observed in merged-beam experiments with highly charged ions (cf. Sect. 5.1.3). It has been concluded [6.152] that for a more complete understanding of recombination processes at very low energies it is necessary to treat the collisional-recombination complex in a time-dependent model accounting for a temperature asymmetry in the merged-beams experiments and including the influence of electric and magnetic fields.

Appendices

A.1 Numerical Data for Electronic Binding Energies

Table A.1. Ionization energies of the ground state and transition energies (in cm^{-1}) between $n = 1$ and $n = 2$ states in H-like ions, from [A.1]

Z	$1s_{1/2}$	$2p_{1/2} - 1s_{1/2}$	$2s_{1/2} - 2p_{1/2}$	$2p_{3/2} - 2p_{1/2}$
1	-1096787711×10^{-4}	8225891870×10^{-5}	352868×10^{-6}	36588804×10^{-8}
2	-4389088766×10^{-4}	3291792922×10^{-4}	46840×10^{-5}	58571673×10^{-7}
3	-987661008×10^{-3}	740734337×10^{-3}	20922×10^{-4}	2965848×10^{-5}
4	-1756018793×10^{-3}	1316979229×10^{-3}	59972×10^{-4}	937585×10^{-4}
5	-274410790×10^{-2}	205799808×10^{-2}	13495×10^{-3}	2289724×10^{-4}
6	-395206167×10^{-2}	296387822×10^{-2}	26084×10^{-3}	474970×10^{-3}
7	-53800898×10^{-1}	40347618×10^{-1}	4541×10^{-2}	880323×10^{-3}
8	-70283947×10^{-1}	52707829×10^{-1}	7325×10^{-2}	1502540×10^{-3}
9	-88972425×10^{-1}	66721213×10^{-1}	11151×10^{-2}	240814×10^{-2}
10	-109868772×10^{-1}	82389376×10^{-1}	1621×10^{-1}	367269×10^{-2}
11	-13297680×10^{0}	9971493×10^{0}	2271×10^{-1}	538093×10^{-2}
12	-15829950×10^{0}	11869985×10^{0}	3088×10^{-1}	762680×10^{-2}
13	-18584143×10^{0}	13934725×10^{0}	409×10^{-1}	1051360×10^{-1}
14	-21560631×10^{0}	16165960×10^{0}	5306×10^{-1}	141540×10^{-1}
15	-24759942×10^{0}	18564051×10^{0}	6756×10^{-1}	186703×10^{-1}
16	-28182526×10^{0}	21129300×10^{0}	8486×10^{-1}	241944×10^{-1}
17	-31828982×10^{0}	23862117×10^{0}	10456×10^{-1}	308680×10^{-1}
18	-35699895×10^{0}	26762896×10^{0}	12757×10^{-1}	388428×10^{-1}
19	-397957842×10^{0}	29831985×10^{0}	1539×10^{0}	482805×10^{-1}
20	-44117409×10^{0}	33069905×10^{0}	1838×10^{0}	593533×10^{-1}
21	-4866551×10^{1}	3647716×10^{1}	2177×10^{0}	722436×10^{-1}
22	-5344074×10^{1}	4005419×10^{1}	2556×10^{0}	871447×10^{-1}
23	-5844392×10^{1}	4380155×10^{1}	2980×10^{0}	1042604×10^{-1}
24	-6367585×10^{1}	4771979×10^{1}	3450×10^{0}	123806×10^{0}
25	-6913743×10^{1}	5180952×10^{1}	3972×10^{0}	146008×10^{0}
26	-7482955×10^{1}	5607135×10^{1}	4546×10^{0}	171104×10^{0}
27	-8075321×10^{1}	6050595×10^{1}	5176×10^{0}	199344×10^{0}
28	-8690935×10^{1}	6511396×10^{1}	5864×10^{0}	230989×10^{0}
29	-9329909×10^{1}	6989614×10^{1}	6619×10^{0}	266315×10^{0}
30	-9992345×10^{1}	7485319×10^{1}	7438×10^{0}	305608×10^{0}

Table A.1. Continued

Z	$1s_{1/2}$	$2p_{1/2}-1s_{1/2}$	$2s_{1/2}-2p_{1/2}$	$2p_{3/2}-2p_{1/2}$
31	-10678368×10^1	7998597×10^1	8325×10^0	349168×10^0
32	-11388088×10^1	8529521×10^1	9288×10^0	397308×10^0
33	-12121631×10^1	9078178×10^1	10323×10^0	450355×10^0
34	-12879128×10^1	9644656×10^1	1144×10^1	508650×10^0
35	-13660714×10^1	10229049×10^1	1264×10^1	572549×10^0
36	-14466528×10^1	10831447×10^1	1393×10^1	642423×10^0
37	-1529672×10^2	1145196×10^2	1530×10^1	718657×10^0
38	-1615144×10^2	1209068×10^2	1677×10^1	801654×10^0
39	-1703084×10^2	1274772×10^2	1834×10^1	891830×10^0
40	-1793509×10^2	1342318×10^2	2001×10^1	989622×10^0
41	-1886434×10^2	1411719×10^2	2180×10^1	1095482×10^0
42	-1981878×10^2	1482985×10^2	2372×10^1	120988×10^1
43	-2079861×10^2	1556131×10^2	2572×10^1	133330×10^1
44	-2180399×10^2	1631168×10^2	2788×10^1	146626×10^1
45	-2283514×10^2	1708109×10^2	3015×10^1	160929×10^1
46	-2389224×10^2	1786970×10^2	3255×10^1	176292×10^1
47	-2497552×10^2	1867763×10^2	3509×10^1	192774×10^1
48	-2608517×10^2	1950502×10^2	3782×10^1	210433×10^1
49	-2722146×10^2	2035206×10^2	4065×10^1	229331×10^1
50	-2838458×10^2	2121888×10^2	4367×10^1	249532×10^1
51	-2957478×10^2	2210563×10^2	4687×10^1	271103×10^1
52	-3079228×10^2	2301247×10^2	5031×10^1	294111×10^1
53	-3203748×10^2	2393917×10^2	5376×10^1	318630×10^1
54	-3331046×10^2	2488732×10^2	5757×10^1	344733×10^1
55	-3461166×10^2	2585568×10^2	6145×10^1	372499×10^1
56	-3594125×10^2	2684486×10^2	6561×10^1	402006×10^1
57	-3729959×10^2	2785511×10^2	6997×10^1	433339×10^1
58	-3688696×10^2	2888662×10^2	7458×10^1	466585×10^1
59	-4010370×10^2	2993963×10^2	7942×10^1	501833×10^1
60	-4155012×10^2	3101433×10^2	8454×10^1	539179×10^1
61	-4302653×10^2	3211093×10^2	9000×10^1	578719×10^1
62	-4453328×10^2	3322964×10^2	9584×10^1	620555×10^1
63	-4607090×10^2	3437090×10^2	1018×10^2	664793×10^1
64	-4763956×10^2	3553475×10^2	1082×10^2	711542×10^1
65	-4923986×10^2	3672163×10^2	1149×10^2	760919×10^1
66	-5087220×10^2	3793183×10^2	1218×10^2	813042×10^1
67	-5253640×10^2	3916504×10^2	1298×10^2	868032×10^1
68	-5423421×10^2	4042283×10^2	1370×10^2	926026×10^1
69	-5596467×10^2	4170412×10^2	1456×10^2	987151×10^1
70	-5772896×10^2	4300991×10^2	1546×10^2	1051551×10^1
71	-595276×10^3	443406×10^3	1639×10^2	1119375×10^1
72	-613609×10^3	456962×10^3	1739×10^2	1190772×10^1
73	-632296×10^3	470775×10^3	1843×10^2	126591×10^2
74	-651338×10^3	484842×10^3	1959×10^2	134494×10^2
75	-670749×10^3	499175×10^3	2074×10^2	142805×10^2
76	-690527×10^3	513772×10^3	2200×10^2	152542×10^2
77	-710682×10^3	528639×10^3	2335×10^2	160724×10^2

Table A.1. Continued

Z	$1s_{1/2}$	$2p_{1/2} - 1s_{1/2}$	$2s_{1/2} - 2p_{1/2}$	$2p_{3/2} - 2p_{1/2}$
78	-731220×10^3	543781×10^3	2477×10^2	170371×10^2
79	-752147×10^3	559201×10^3	2629×10^2	180504×10^2
80	-773467×10^3	574902×10^3	2795×10^2	191144×10^2
81	-795195×10^3	590894×10^3	2967×10^2	202316×10^2
82	-817333×10^3	607178×10^3	3154×10^2	2140417×10^2
83	-839888×10^3	623758×10^3	3356×10^2	2263475×10^2
84	-862874×10^3	640645×10^3	3567×10^2	2392604×10^1
85	-886290×10^3	657835×10^3	3805×10^2	2528066×10^1
86	-910146×10^3	675335×10^3	4067×10^2	2670159×10^1
87	-934466×10^3	693166×10^3	4333×10^2	2819211×10^1
88	-959246×10^3	711319×10^3	4627×10^2	2975529×10^1
89	-984504×10^3	729811×10^3	4938×10^2	3139464×10^1
90	-1010237×10^3	748633×10^3	5290×10^2	3311354×10^1
91	-1036480×10^3	767816×10^3	5648×10^2	349163×10^2
92	-1063210×10^3	787333×10^3	6073×10^2	368061×10^2
93	-1090481×10^3	807234×10^3	6498×10^2	387882×10^2
94	-1118259×10^3	827480×10^3	7005×10^2	408659×10^2
95	-1146608×10^3	848130×10^3	7512×10^2	430454×10^2
96	-1175498×10^3	869149×10^3	8100×10^2	453305×10^2
97	-1204979×10^3	890580×10^3	8716×10^2	477275×10^2
98	-1235031×10^3	912400×10^3	9424×10^2	502412×10^2
99	-1265693×10^3	934637×10^3	1020×10^3	528781×10^2
100	-1296975×10^3	957299×10^3	1105×10^3	55645×10^3
101	-1328909×10^3	980407×10^3	1197×10^3	58548×10^3
102	-1361516×10^3	1003976×10^3	1296×10^3	61595×10^3
103	-1394776×10^3	1027979×10^3	1411×10^3	64794×10^3
104	-1428797×10^3	1052513×10^3	1527×10^3	68154×10^3
105	-1463444×10^3	1077437×10^3	1674×10^3	71680×10^3
106	-149889×10^4	110291×10^4	1827×10^3	75387×10^3
107	-153512×10^4	112891×10^4	1995×10^6	79283×10^3
108	-157211×10^4	115540×10^4	2188×10^3	83378×10^3
109	-160993×10^4	118244×10^4	2403×10^3	87685×10^3
110	-164860×10^4	121003×10^4	2645×10^3	92216×10^3

Table A.2. Calculated wavelengths for the Lyman series of H-like ions with nuclear charge Z, from [A.2]

Transition	J_1 - J_0	Wavelength [nm]
$Z = 1$		
$2p - 1s$	1/2 - 1/2	$1.2156736449413(35)\times10^2$
$2p - 1s$	3/2 - 1/2	$1.2156682376464(35)\times10^2$
$3p - 1s$	1/2 - 1/2	$1.0257229656831(24)\times10^2$
$3p - 1s$	3/2 - 1/2	$1.0257218250807(24)\times10^2$
$4p - 1s$	1/2 - 1/2	$9.725370271934(22)\times10^1$
$4p - 1s$	3/2 - 1/2	$9.725365946098(22)\times10^1$

Table A.2. Continued

Transition	J_1 - J_0	Wavelength [nm]
$Z = 1$		
$5p - 1s$	1/2 - 1/2	$9.497431209963(21) \times 10^1$
$5p - 1s$	3/2 - 1/2	$9.497429097741(21) \times 10^1$
$6p - 1s$	1/2 - 1/2	$9.378034804914(20) \times 10^1$
$6p - 1s$	3/2 - 1/2	$9.378033613105(20) \times 10^1$
$7p - 1s$	1/2 - 1/2	$9.307482491503(20) \times 10^1$
$7p - 1s$	3/2 - 1/2	$9.307481752226(20) \times 10^1$
$8p - 1s$	1/2 - 1/2	$9.262256702800(20) \times 10^1$
$8p - 1s$	3/2 - 1/2	$9.262256212344(20) \times 10^1$
$9p - 1s$	1/2 - 1/2	$9.231503195238(19) \times 10^1$
$9p - 1s$	3/2 - 1/2	$9.231502853059(19) \times 10^1$
$10p - 1s$	1/2 - 1/2	$9.209630396344(19) \times 10^1$
$10p - 1s$	3/2 - 1/2	$9.209630148077(19) \times 10^1$
$11p - 1s$	1/2 - 1/2	$9.193513607335(19) \times 10^1$
$11p - 1s$	3/2 - 1/2	$9.193513421461(19) \times 10^1$
$12p - 1s$	1/2 - 1/2	$9.181293191406(19) \times 10^1$
$12p - 1s$	3/2 - 1/2	$9.181293048615(19) \times 10^1$
$13p - 1s$	1/2 - 1/2	$9.171805295155(19) \times 10^1$
$13p - 1s$	3/2 - 1/2	$9.171805183079(19) \times 10^1$
$14p - 1s$	1/2 - 1/2	$9.164290884280(19) \times 10^1$
$14p - 1s$	3/2 - 1/2	$9.164290794693(19) \times 10^1$
$15p - 1s$	1/2 - 1/2	$9.158237610932(19) \times 10^1$
$15p - 1s$	3/2 - 1/2	$9.158237538190(19) \times 10^1$
$16p - 1s$	1/2 - 1/2	$9.153289399123(19) \times 10^1$
$16p - 1s$	3/2 - 1/2	$9.153289339250(19) \times 10^1$
$17p - 1s$	1/2 - 1/2	$9.149192493670(19) \times 10^1$
$17p - 1s$	3/2 - 1/2	$9.149192443799(19) \times 10^1$
$18p - 1s$	1/2 - 1/2	$9.145762071267(19) \times 10^1$
$18p - 1s$	3/2 - 1/2	$9.145762029286(19) \times 10^1$
$19p - 1s$	1/2 - 1/2	$9.142860914660(19) \times 10^1$
$19p - 1s$	3/2 - 1/2	$9.142860878987(19) \times 10^1$
$20p - 1s$	1/2 - 1/2	$9.140385410855(19) \times 10^1$
$20p - 1s$	3/2 - 1/2	$9.140385380286(19) \times 10^1$
$Z = 10$		
$2p - 1s$	1/2 - 1/2	$1.2137488222(57) \times 10^0$
$2p - 1s$	3/2 - 1/2	$1.2132080091(57) \times 10^0$
$3p - 1s$	1/2 - 1/2	$1.0239629762(40) \times 10^0$
$3p - 1s$	3/2 - 1/2	$1.0238488797(40) \times 10^0$
$4p - 1s$	1/2 - 1/2	$9.708461597(36) \times 10^{-1}$
$4p - 1s$	3/2 - 1/2	$9.708028916(36) \times 10^{-1}$
$5p - 1s$	1/2 - 1/2	$9.480882936(34) \times 10^{-1}$
$5p - 1s$	3/2 - 1/2	$9.480671688(34) \times 10^{-1}$
$6p - 1s$	1/2 - 1/2	$9.361699085(33) \times 10^{-1}$
$6p - 1s$	3/2 - 1/2	$9.361579901(33) \times 10^{-1}$
$7p - 1s$	1/2 - 1/2	$9.291282836(33) \times 10^{-1}$
$7p - 1s$	3/2 - 1/2	$9.291208912(33) \times 10^{-1}$

Table A.2. Continued

Transition	J_1 - J_0	Wavelength [nm]
$Z = 10$		
$8p - 1s$	1/2 - 1/2	$9.246149456(33)\times10^{-1}$
$8p - 1s$	3/2 - 1/2	$9.246100416(33)\times10^{-1}$
$9p - 1s$	1/2 - 1/2	$9.215461597(32)\times10^{-1}$
$9p - 1s$	3/2 - 1/2	$9.215427385(32)\times10^{-1}$
$10p - 1s$	1/2 - 1/2	$9.193637119(32)\times10^{-1}$
$10p - 1s$	3/2 - 1/2	$9.193612297(32)\times10^{-1}$
$11p - 1s$	1/2 - 1/2	$9.177556932(32)\times10^{-1}$
$11p - 1s$	3/2 - 1/2	$9.177538349(32)\times10^{-1}$
$12p - 1s$	1/2 - 1/2	$9.165364907(32)\times10^{-1}$
$12p - 1s$	3/2 - 1/2	$9.165350632(32)\times10^{-1}$
$13p - 1s$	1/2 - 1/2	$9.155899478(32)\times10^{-1}$
$13p - 1s$	3/2 - 1/2	$9.155888274(32)\times10^{-1}$
$14p - 1s$	1/2 - 1/2	$9.148403152(32)\times10^{-1}$
$14p - 1s$	3/2 - 1/2	$9.148394196(32)\times10^{-1}$
$15p - 1s$	1/2 - 1/2	$9.142364650(32)\times10^{-1}$
$15p - 1s$	3/2 - 1/2	$9.142357379(32)\times10^{-1}$
$16p - 1s$	1/2 - 1/2	$9.137428661(32)\times10^{-1}$
$16p - 1s$	3/2 - 1/2	$9.137422676(32)\times10^{-1}$
$17p - 1s$	1/2 - 1/2	$9.133341984(32)\times10^{-1}$
$17p - 1s$	3/2 - 1/2	$9.133336999(32)\times10^{-1}$
$18p - 1s$	1/2 - 1/2	$9.129920206(32)\times10^{-1}$
$18p - 1s$	3/2 - 1/2	$9.129916010(32)\times10^{-1}$
$19p - 1s$	1/2 - 1/2	$9.127026423(32)\times10^{-1}$
$19p - 1s$	3/2 - 1/2	$9.127022857(32)\times10^{-1}$
$20p - 1s$	1/2 - 1/2	$9.124557257(32)\times10^{-1}$
$20p - 1s$	3/2 - 1/2	$9.124554202(32)\times10^{-1}$
$Z = 26$		
$2p - 1s$	1/2 - 1/2	$1.78344909(55)\times10^{-1}$
$2p - 1s$	3/2 - 1/2	$1.77802344(55)\times10^{-1}$
$3p - 1s$	1/2 - 1/2	$1.50350123(39)\times10^{-1}$
$3p - 1s$	3/2 - 1/2	$1.50235531(39)\times10^{-1}$
$4p - 1s$	1/2 - 1/2	$1.42534361(35)\times10^{-1}$
$4p - 1s$	3/2 - 1/2	$1.42490927(35)\times10^{-1}$
$5p - 1s$	1/2 - 1/2	$1.39191188(33)\times10^{-1}$
$5p - 1s$	3/2 - 1/2	$1.39169995(33)\times10^{-1}$
$6p - 1s$	1/2 - 1/2	$1.37442394(33)\times10^{-1}$
$6p - 1s$	3/2 - 1/2	$1.37430444(33)\times10^{-1}$
$7p - 1s$	1/2 - 1/2	$1.36410069(32)\times10^{-1}$
$7p - 1s$	3/2 - 1/2	$1.36402660(32)\times10^{-1}$
$8p - 1s$	1/2 - 1/2	$1.35748845(32)\times10^{-1}$
$8p - 1s$	3/2 - 1/2	$1.35743932(32)\times10^{-1}$
$9p - 1s$	1/2 - 1/2	$1.35299496(31)\times10^{-1}$
$9p - 1s$	3/2 - 1/2	$1.35296069(31)\times10^{-1}$
$10p - 1s$	1/2 - 1/2	$1.34980069(31)\times10^{-1}$
$10p - 1s$	3/2 - 1/2	$1.34977583(31)\times10^{-1}$

Table A.2. Continued

Transition	J_1 - J_0	Wavelength [nm]
$Z = 26$		
$11p - 1s$	1/2 - 1/2	$1.34744802(31)\times10^{-1}$
$11p - 1s$	3/2 - 1/2	$1.34742942(31)\times10^{-1}$
$12p - 1s$	1/2 - 1/2	$1.34566477(31)\times10^{-1}$
$12p - 1s$	3/2 - 1/2	$1.34565048(31)\times10^{-1}$
$13p - 1s$	1/2 - 1/2	$1.34428069(31)\times10^{-1}$
$13p - 1s$	3/2 - 1/2	$1.34426948(31)\times10^{-1}$
$14p - 1s$	1/2 - 1/2	$1.34318479(31)\times10^{-1}$
$14p - 1s$	3/2 - 1/2	$1.34317583(31)\times10^{-1}$
$15p - 1s$	1/2 - 1/2	$1.34230218(31)\times10^{-1}$
$15p - 1s$	3/2 - 1/2	$1.34229491(31)\times10^{-1}$
$16p - 1s$	1/2 - 1/2	$1.34158085(31)\times10^{-1}$
$16p - 1s$	3/2 - 1/2	$1.34157486(31)\times10^{-1}$
$17p - 1s$	1/2 - 1/2	$1.34098373(31)\times10^{-1}$
$17p - 1s$	3/2 - 1/2	$1.34097874(31)\times10^{-1}$
$18p - 1s$	1/2 - 1/2	$1.34048383(31)\times10^{-1}$
$18p - 1s$	3/2 - 1/2	$1.34047963(31)\times10^{-1}$
$19p - 1s$	1/2 - 1/2	$1.34006111(31)\times10^{-1}$
$19p - 1s$	3/2 - 1/2	$1.34005755(31)\times10^{-1}$
$20p - 1s$	1/2 - 1/2	$1.33970047(31)\times10^{-1}$
$20p - 1s$	3/2 - 1/2	$1.33969741(31)\times10^{-1}$
$Z = 28$		
$2p - 1s$	1/2 - 1/2	$1.53577731(66)\times10^{-1}$
$2p - 1s$	3/2 - 1/2	$1.53034852(65)\times10^{-1}$
$3p - 1s$	1/2 - 1/2	$1.29453337(47)\times10^{-1}$
$3p - 1s$	3/2 - 1/2	$1.29338654(46)\times10^{-1}$
$4p - 1s$	1/2 - 1/2	$1.22721283(42)\times10^{-1}$
$4p - 1s$	3/2 - 1/2	$1.22677818(42)\times10^{-1}$
$5p - 1s$	1/2 - 1/2	$1.19842558(40)\times10^{-1}$
$5p - 1s$	3/2 - 1/2	$1.19821352(40)\times10^{-1}$
$6p - 1s$	1/2 - 1/2	$1.18337047(39)\times10^{-1}$
$6p - 1s$	3/2 - 1/2	$1.18325090(39)\times10^{-1}$
$7p - 1s$	1/2 - 1/2	$1.17448480(38)\times10^{-1}$
$7p - 1s$	3/2 - 1/2	$1.17441068(38)\times10^{-1}$
$8p - 1s$	1/2 - 1/2	$1.16879409(38)\times10^{-1}$
$8p - 1s$	3/2 - 1/2	$1.16874494(38)\times10^{-1}$
$9p - 1s$	1/2 - 1/2	$1.16492723(38)\times10^{-1}$
$9p - 1s$	3/2 - 1/2	$1.16489295(38)\times10^{-1}$
$10p - 1s$	1/2 - 1/2	$1.16217864(37)\times10^{-1}$
$10p - 1s$	3/2 - 1/2	$1.16215378(37)\times10^{-1}$
$11p - 1s$	1/2 - 1/2	$1.16015437(37)\times10^{-1}$
$11p - 1s$	3/2 - 1/2	$1.16013577(37)\times10^{-1}$
$12p - 1s$	1/2 - 1/2	$1.15862013(37)\times10^{-1}$
$12p - 1s$	3/2 - 1/2	$1.15860584(37)\times10^{-1}$
$13p - 1s$	1/2 - 1/2	$1.15742937(37)\times10^{-1}$
$13p - 1s$	3/2 - 1/2	$1.15741816(37)\times10^{-1}$

Table A.2. Continued

Transition	J_1 - J_0	Wavelength [nm]
$Z = 28$		
$14p - 1s$	1/2 - 1/2	$1.15648659(37)\times10^{-1}$
$14p - 1s$	3/2 - 1/2	$1.15647762(37)\times10^{-1}$
$15p - 1s$	1/2 - 1/2	$1.15572732(37)\times10^{-1}$
$15p - 1s$	3/2 - 1/2	$1.15572005(37)\times10^{-1}$
$16p - 1s$	1/2 - 1/2	$1.15510682(37)\times10^{-1}$
$16p - 1s$	3/2 - 1/2	$1.15510083(37)\times10^{-1}$
$17p - 1s$	1/2 - 1/2	$1.15459317(37)\times10^{-1}$
$17p - 1s$	3/2 - 1/2	$1.15458819(37)\times10^{-1}$
$18p - 1s$	1/2 - 1/2	$1.15416317(37)\times10^{-1}$
$18p - 1s$	3/2 - 1/2	$1.15415897(37)\times10^{-1}$
$19p - 1s$	1/2 - 1/2	$1.15379957(37)\times10^{-1}$
$19p - 1s$	3/2 - 1/2	$1.15379601(37)\times10^{-1}$
$20p - 1s$	1/2 - 1/2	$1.15348937(37)\times10^{-1}$
$20p - 1s$	3/2 - 1/2	$1.15348631(37)\times10^{-1}$
$Z = 42$		
$2p - 1s$	1/2 - 1/2	$6.743336(16)\times10^{-2}$
$2p - 1s$	3/2 - 1/2	$6.688767(16)\times10^{-2}$
$3p - 1s$	1/2 - 1/2	$5.677094(11)\times10^{-2}$
$3p - 1s$	3/2 - 1/2	$5.665544(11)\times10^{-2}$
$4p - 1s$	1/2 - 1/2	$5.380841(10)\times10^{-2}$
$4p - 1s$	3/2 - 1/2	$5.376467(10)\times10^{-2}$
$5p - 1s$	1/2 - 1/2	$5.254524(10)\times10^{-2}$
$5p - 1s$	3/2 - 1/2	$5.252392(10)\times10^{-2}$
$6p - 1s$	1/2 - 1/2	$5.1885982(99)\times10^{-2}$
$6p - 1s$	3/2 - 1/2	$5.1873974(99)\times10^{-2}$
$7p - 1s$	1/2 - 1/2	$5.1497478(98)\times10^{-2}$
$7p - 1s$	3/2 - 1/2	$5.1490040(98)\times10^{-2}$
$8p - 1s$	1/2 - 1/2	$5.1248961(97)\times10^{-2}$
$8p - 1s$	3/2 - 1/2	$5.1244032(97)\times10^{-2}$
$9p - 1s$	1/2 - 1/2	$5.1080251(96)\times10^{-2}$
$9p - 1s$	3/2 - 1/2	$5.1076816(96)\times10^{-2}$
$10p - 1s$	1/2 - 1/2	$5.0960424(96)\times10^{-2}$
$10p - 1s$	3/2 - 1/2	$5.0957934(96)\times10^{-2}$
$11p - 1s$	1/2 - 1/2	$5.0872231(95)\times10^{-2}$
$11p - 1s$	3/2 - 1/2	$5.0870367(95)\times10^{-2}$
$12p - 1s$	1/2 - 1/2	$5.0805423(95)\times10^{-2}$
$12p - 1s$	3/2 - 1/2	$5.0803992(95)\times10^{-2}$
$13p - 1s$	1/2 - 1/2	$5.0753595(95)\times10^{-2}$
$13p - 1s$	3/2 - 1/2	$5.0752473(95)\times10^{-2}$
$14p - 1s$	1/2 - 1/2	$5.0712577(95)\times10^{-2}$
$14p - 1s$	3/2 - 1/2	$5.0711681(95)\times10^{-2}$
$15p - 1s$	1/2 - 1/2	$5.0679555(95)\times10^{-2}$
$15p - 1s$	3/2 - 1/2	$5.0678828(95)\times10^{-2}$
$16p - 1s$	1/2 - 1/2	$5.0652577(94)\times10^{-2}$
$16p - 1s$	3/2 - 1/2	$5.0651978(94)\times10^{-2}$

Table A.2. Continued

Transition	J_1 - J_0	Wavelength [nm]
$Z = 42$		
$17p - 1s$	1/2 - 1/2	$5.0630250(94)\times10^{-2}$
$17p - 1s$	3/2 - 1/2	$5.0629751(94)\times10^{-2}$
$18p - 1s$	1/2 - 1/2	$5.0611564(94)\times10^{-2}$
$18p - 1s$	3/2 - 1/2	$5.0611144(94)\times10^{-2}$
$19p - 1s$	1/2 - 1/2	$5.0595766(94)\times10^{-2}$
$19p - 1s$	3/2 - 1/2	$5.0595410(94)\times10^{-2}$
$20p - 1s$	1/2 - 1/2	$5.0582292(94)\times10^{-2}$
$20p - 1s$	3/2 - 1/2	$5.0581986(94)\times10^{-2}$
$Z = 56$		
$2p - 1s$	1/2 - 1/2	$3.725423(31)\times10^{-2}$
$2p - 1s$	3/2 - 1/2	$3.670460(30)\times10^{-2}$
$3p - 1s$	1/2 - 1/2	$3.130721(22)\times10^{-2}$
$3p - 1s$	3/2 - 1/2	$3.119055(22)\times10^{-2}$
$4p - 1s$	1/2 - 1/2	$2.966542(19)\times10^{-2}$
$4p - 1s$	3/2 - 1/2	$2.962131(19)\times10^{-2}$
$5p - 1s$	1/2 - 1/2	$2.896838(19)\times10^{-2}$
$5p - 1s$	3/2 - 1/2	$2.894692(19)\times10^{-2}$
$6p - 1s$	1/2 - 1/2	$2.860570(18)\times10^{-2}$
$6p - 1s$	3/2 - 1/2	$2.859362(18)\times10^{-2}$
$7p - 1s$	1/2 - 1/2	$2.839245(18)\times10^{-2}$
$7p - 1s$	3/2 - 1/2	$2.838497(18)\times10^{-2}$
$8p - 1s$	1/2 - 1/2	$2.825627(18)\times10^{-2}$
$8p - 1s$	3/2 - 1/2	$2.825132(18)\times10^{-2}$
$9p - 1s$	1/2 - 1/2	$2.816395(18)\times10^{-2}$
$9p - 1s$	3/2 - 1/2	$2.816051(18)\times10^{-2}$
$10p - 1s$	1/2 - 1/2	$2.809846(17)\times10^{-2}$
$10p - 1s$	3/2 - 1/2	$2.809596(17)\times10^{-2}$
$11p - 1s$	1/2 - 1/2	$2.805030(17)\times10^{-2}$
$11p - 1s$	3/2 - 1/2	$2.804843(17)\times10^{-2}$
$12p - 1s$	1/2 - 1/2	$2.801385(17)\times10^{-2}$
$12p - 1s$	3/2 - 1/2	$2.801241(17)\times10^{-2}$
$13p - 1s$	1/2 - 1/2	$2.798559(17)\times10^{-2}$
$13p - 1s$	3/2 - 1/2	$2.798446(17)\times10^{-2}$
$14p - 1s$	1/2 - 1/2	$2.796324(17)\times10^{-2}$
$14p - 1s$	3/2 - 1/2	$2.796234(17)\times10^{-2}$
$15p - 1s$	1/2 - 1/2	$2.794525(17)\times10^{-2}$
$15p - 1s$	3/2 - 1/2	$2.794452(17)\times10^{-2}$
$16p - 1s$	1/2 - 1/2	$2.793056(17)\times10^{-2}$
$16p - 1s$	3/2 - 1/2	$2.792996(17)\times10^{-2}$
$17p - 1s$	1/2 - 1/2	$2.791841(17)\times10^{-2}$
$17p - 1s$	3/2 - 1/2	$2.791791(17)\times10^{-2}$
$18p - 1s$	1/2 - 1/2	$2.790825(17)\times10^{-2}$
$18p - 1s$	3/2 - 1/2	$2.790783(17)\times10^{-2}$
$19p - 1s$	1/2 - 1/2	$2.789965(17)\times10^{-2}$
$19p - 1s$	3/2 - 1/2	$2.789930(17)\times10^{-2}$

Table A.2. Continued

Transition	J_1 - J_0	Wavelength [nm]
$Z = 56$		
$20p - 1s$	1/2 - 1/2	$2.789233(17)\times10^{-2}$
$20p - 1s$	3/2 - 1/2	$2.789202(17)\times10^{-2}$
$Z = 80$		
$2p - 1s$	1/2 - 1/2	$1.740070(64)\times10^{-2}$
$2p - 1s$	3/2 - 1/2	$1.684073(60)\times10^{-2}$
$3p - 1s$	1/2 - 1/2	$1.455168(45)\times10^{-2}$
$3p - 1s$	3/2 - 1/2	$1.443210(44)\times10^{-2}$
$4p - 1s$	1/2 - 1/2	$1.377882(40)\times10^{-2}$
$4p - 1s$	3/2 - 1/2	$1.373376(40)\times10^{-2}$
$5p - 1s$	1/2 - 1/2	$1.345446(38)\times10^{-2}$
$5p - 1s$	3/2 - 1/2	$1.343263(38)\times10^{-2}$
$6p - 1s$	1/2 - 1/2	$1.328706(37)\times10^{-2}$
$6p - 1s$	3/2 - 1/2	$1.327482(37)\times10^{-2}$
$7p - 1s$	1/2 - 1/2	$1.318923(37)\times10^{-2}$
$7p - 1s$	3/2 - 1/2	$1.318168(37)\times10^{-2}$
$8p - 1s$	1/2 - 1/2	$1.312704(36)\times10^{-2}$
$8p - 1s$	3/2 - 1/2	$1.312206(36)\times10^{-2}$
$9p - 1s$	1/2 - 1/2	$1.308504(36)\times10^{-2}$
$9p - 1s$	3/2 - 1/2	$1.308157(36)\times10^{-2}$
$10p - 1s$	1/2 - 1/2	$1.305533(36)\times10^{-2}$
$10p - 1s$	3/2 - 1/2	$1.305282(36)\times10^{-2}$
$11p - 1s$	1/2 - 1/2	$1.303354(36)\times10^{-2}$
$11p - 1s$	3/2 - 1/2	$1.303166(36)\times10^{-2}$
$12p - 1s$	1/2 - 1/2	$1.301708(36)\times10^{-2}$
$12p - 1s$	3/2 - 1/2	$1.301564(36)\times10^{-2}$
$13p - 1s$	1/2 - 1/2	$1.300434(36)\times10^{-2}$
$13p - 1s$	3/2 - 1/2	$1.300321(36)\times10^{-2}$
$14p - 1s$	1/2 - 1/2	$1.299428(36)\times10^{-2}$
$14p - 1s$	3/2 - 1/2	$1.299338(36)\times10^{-2}$
$15p - 1s$	1/2 - 1/2	$1.298619(36)\times10^{-2}$
$15p - 1s$	3/2 - 1/2	$1.298546(36)\times10^{-2}$
$16p - 1s$	1/2 - 1/2	$1.297960(36)\times10^{-2}$
$16p - 1s$	3/2 - 1/2	$1.297900(35)\times10^{-2}$
$17p - 1s$	1/2 - 1/2	$1.297415(35)\times10^{-2}$
$17p - 1s$	3/2 - 1/2	$1.297365(35)\times10^{-2}$
$18p - 1s$	1/2 - 1/2	$1.296959(35)\times10^{-2}$
$18p - 1s$	3/2 - 1/2	$1.296917(35)\times10^{-2}$
$19p - 1s$	1/2 - 1/2	$1.296575(35)\times10^{-2}$
$19p - 1s$	3/2 - 1/2	$1.296539(35)\times10^{-2}$
$20p - 1s$	1/2 - 1/2	$1.296247(35)\times10^{-2}$
$20p - 1s$	3/2 - 1/2	$1.296217(35)\times10^{-2}$
$Z = 92$		
$2p - 1s$	1/2 - 1/2	$1.271243(83)\times10^{-2}$
$2p - 1s$	3/2 - 1/2	$1.214419(76)\times10^{-2}$

Table A.2. Continued

Transition	J_1 - J_0	Wavelength [nm]
$Z = 92$		
$3p - 1s$	1/2 - 1/2	$1.059081(58) \times 10^{-2}$
$3p - 1s$	3/2 - 1/2	$1.046907(56) \times 10^{-2}$
$4p - 1s$	1/2 - 1/2	$1.002293(52) \times 10^{-2}$
$4p - 1s$	3/2 - 1/2	$9.97720(51) \times 10^{-3}$
$5p - 1s$	1/2 - 1/2	$9.78667(49) \times 10^{-3}$
$5p - 1s$	3/2 - 1/2	$9.76458(49) \times 10^{-3}$
$6p - 1s$	1/2 - 1/2	$9.66548(48) \times 10^{-3}$
$6p - 1s$	3/2 - 1/2	$9.65313(48) \times 10^{-3}$
$7p - 1s$	1/2 - 1/2	$9.59496(47) \times 10^{-3}$
$7p - 1s$	3/2 - 1/2	$9.58736(47) \times 10^{-3}$
$8p - 1s$	1/2 - 1/2	$9.55029(47) \times 10^{-3}$
$8p - 1s$	3/2 - 1/2	$9.54528(47) \times 10^{-3}$
$9p - 1s$	1/2 - 1/2	$9.52020(46) \times 10^{-3}$
$9p - 1s$	3/2 - 1/2	$9.51672(46) \times 10^{-3}$
$10p - 1s$	1/2 - 1/2	$9.49896(46) \times 10^{-3}$
$10p - 1s$	3/2 - 1/2	$9.49645(46) \times 10^{-3}$
$11p - 1s$	1/2 - 1/2	$9.48341(46) \times 10^{-3}$
$11p - 1s$	3/2 - 1/2	$9.48153(46) \times 10^{-3}$
$12p - 1s$	1/2 - 1/2	$9.47168(46) \times 10^{-3}$
$12p - 1s$	3/2 - 1/2	$9.47024(46) \times 10^{-3}$
$13p - 1s$	1/2 - 1/2	$9.46262(46) \times 10^{-3}$
$13p - 1s$	3/2 - 1/2	$9.46149(46) \times 10^{-3}$
$14p - 1s$	1/2 - 1/2	$9.45546(46) \times 10^{-3}$
$14p - 1s$	3/2 - 1/2	$9.45457(46) \times 10^{-3}$
$15p - 1s$	1/2 - 1/2	$9.44972(46) \times 10^{-3}$
$15p - 1s$	3/2 - 1/2	$9.44900(46) \times 10^{-3}$
$16p - 1s$	1/2 - 1/2	$9.44504(46) \times 10^{-3}$
$16p - 1s$	3/2 - 1/2	$9.44445(46) \times 10^{-3}$
$17p - 1s$	1/2 - 1/2	$9.44118(46) \times 10^{-3}$
$17p - 1s$	3/2 - 1/2	$9.44068(46) \times 10^{-3}$
$18p - 1s$	1/2 - 1/2	$9.43795(46) \times 10^{-3}$
$18p - 1s$	3/2 - 1/2	$9.43754(46) \times 10^{-3}$
$19p - 1s$	1/2 - 1/2	$9.43523(46) \times 10^{-3}$
$19p - 1s$	3/2 - 1/2	$9.43487(46) \times 10^{-3}$
$20p - 1s$	1/2 - 1/2	$9.43291(46) \times 10^{-3}$
$20p - 1s$	3/2 - 1/2	$9.43261(46) \times 10^{-3}$

Table A.3. Calculated wavelengths for Paschen, Brackett and Pfund series in H-like ions with nuclear charge Z, from [A.2]

Transition	J_1 - J_0	Wavelength [nm]
$Z = 1$		
$4f-3d$	5/2 - 3/2	1875.61308963(27)
$4f-3d$	5/2 - 5/2	1875.62580217(27)
$4f-3d$	7/2 - 3/2	1875.61040811(27)
$4f-3d$	7/2 - 5/2	1875.62312062(27)
$5f-3d$	5/2 - 3/2	1282.15971017(12)
$5f-3d$	5/2 - 5/2	1282.16565076(12)
$5f-3d$	7/2 - 3/2	1282.15906860(12)
$5f-3d$	7/2 - 5/2	1282.16500918(12)
$6f-3d$	5/2 - 3/2	1094.110205334(91)
$6f-3d$	5/2 - 5/2	1094.114531139(91)
$6f-3d$	7/2 - 3/2	1094.109934974(91)
$6f-3d$	7/2 - 5/2	1094.114260776(91)
$5g-4f$	7/2 - 5/2	4052.2673264(12)
$5g-4f$	7/2 - 7/2	4052.2798432(12)
$5g-4f$	9/2 - 5/2	4052.2634813(12)
$5g-4f$	9/2 - 7/2	4052.2759980(12)
$6g-4f$	7/2 - 5/2	2625.87157273(52)
$6g-4f$	7/2 - 7/2	2625.87682858(52)
$6g-4f$	9/2 - 5/2	2625.87063836(52)
$6g-4f$	9/2 - 7/2	2625.87589420(52)
$7g-4f$	7/2 - 5/2	2166.12355936(36)
$7g-4f$	7/2 - 7/2	2166.12713589(36)
$7g-4f$	9/2 - 5/2	2166.12315895(36)
$7g-4f$	9/2 - 7/2	2166.12673548(36)
$6h-5g$	9/2 - 7/2	7459.8671613(42)
$6h-5g$	9/2 - 9/2	7459.8801923(42)
$6h-5g$	11/2 - 7/2	7459.8621340(42)
$6h-5g$	11/2 - 9/2	7459.8751649(42)
$7h-5g$	9/2 - 7/2	4653.7853576(16)
$7h-5g$	9/2 - 9/2	4653.7904290(16)
$7h-5g$	11/2 - 7/2	4653.7841255(16)
$7h-5g$	11/2 - 9/2	4653.7891969(16)
$8h-5g$	9/2 - 7/2	3740.5627113(10)
$8h-5g$	9/2 - 9/2	3740.5659876(10)
$8h-5g$	11/2 - 7/2	3740.5621780(10)
$8h-5g$	11/2 - 9/2	3740.5654544(10)
$Z = 10$		
$4f-3d$	5/2 - 3/2	18.73223097(26)
$4f-3d$	5/2 - 5/2	18.74493784(26)
$4f-3d$	7/2 - 3/2	18.72955415(26)
$4f-3d$	7/2 - 5/2	18.74225739(26)
$5f-3d$	5/2 - 3/2	12.80755854(12)
$5f-3d$	5/2 - 5/2	12.81349734(12)
$5f-3d$	7/2 - 3/2	12.80691775(12)
$5f-3d$	7/2 - 5/2	12.81285596(12)
$6f-3d$	5/2 - 3/2	10.929702175(91)
$6f-3d$	5/2 - 5/2	10.934026849(91)

Table A.3. Continued

Transition	J_1 - J_0	Wavelength [nm]
$Z = 10$		
$6f - 3d$	7/2 - 3/2	10.929432108(91)
$6f - 3d$	7/2 - 5/2	10.933756569(91)
$5g - 4f$	7/2 - 5/2	40.4876845(12)
$5g - 4f$	7/2 - 7/2	40.5001953(12)
$5g - 4f$	9/2 - 5/2	40.4838435(12)
$5g - 4f$	9/2 - 7/2	40.4963518(12)
$6g - 4f$	7/2 - 5/2	26.23836306(52)
$6g - 4f$	7/2 - 7/2	26.24361673(52)
$6g - 4f$	9/2 - 5/2	26.23742943(52)
$6g - 4f$	9/2 - 7/2	26.24268272(52)
$7g - 4f$	7/2 - 5/2	21.64498533(35)
$7g - 4f$	7/2 - 7/2	21.64856044(35)
$7g - 4f$	9/2 - 5/2	21.64458521(35)
$7g - 4f$	9/2 - 7/2	21.64816018(35)
$6h - 5g$	9/2 - 7/2	7.45455408(42)
$6h - 5g$	9/2 - 9/2	7.45585654(42)
$6h - 5g$	11/2 - 7/2	7.45405176(42)
$6h - 5g$	11/2 - 9/2	7.45535404(42)
$7h - 5g$	9/2 - 7/2	4.65071691(16)
$7h - 5g$	9/2 - 9/2	4.65122382(16)
$7h - 5g$	11/2 - 7/2	4.65059377(16)
$7h - 5g$	11/2 - 9/2	4.65110066(16)
$8h - 5g$	9/2 - 7/2	3.73814981(10)
$8h - 5g$	9/2 - 9/2	3.73847730(10)
$8h - 5g$	11/2 - 7/2	3.73809652(10)
$8h - 5g$	11/2 - 9/2	3.73842400(10)
$Z = 26$		
$4f - 3d$	5/2 - 3/2	2.75875916(26)
$4f - 3d$	5/2 - 5/2	2.77147093(26)
$4f - 3d$	7/2 - 3/2	2.75610164(26)
$4f - 3d$	7/2 - 5/2	2.76878888(26)
$5f - 3d$	5/2 - 3/2	1.88818858(12)
$5f - 3d$	5/2 - 5/2	1.89413476(12)
$5f - 3d$	7/2 - 3/2	1.88755051(12)
$5f - 3d$	7/2 - 5/2	1.89349267(12)
$6f - 3d$	5/2 - 3/2	1.611839612(91)
$6f - 3d$	5/2 - 5/2	1.616170639(91)
$6f - 3d$	7/2 - 3/2	1.611570453(91)
$6f - 3d$	7/2 - 5/2	1.615900033(91)
$5g - 4f$	7/2 - 5/2	5.9771075(12)
$5g - 4f$	7/2 - 7/2	5.9896203(12)
$5g - 4f$	9/2 - 5/2	5.9732787(12)
$5g - 4f$	9/2 - 7/2	5.9857755(12)
$6g - 4f$	7/2 - 5/2	3.87550722(52)
$6g - 4f$	7/2 - 7/2	3.88076390(52)
$6g - 4f$	9/2 - 5/2	3.87457510(52)
$6g - 4f$	9/2 - 7/2	3.87982925(52)
$7g - 4f$	7/2 - 5/2	3.19751167(35)

Table A.3. Continued

Transition	J_1 - J_0	Wavelength [nm]
$Z = 26$		
$7g - 4f$	7/2 - 7/2	3.20108914(35)
$7g - 4f$	9/2 - 5/2	3.19711201(35)
$7g - 4f$	9/2 - 7/2	3.20068858(35)
$6h - 5g$	9/2 - 7/2	1.10147091(42)
$6h - 5g$	9/2 - 9/2	1.10277351(42)
$6h - 5g$	11/2 - 7/2	1.10096948(42)
$6h - 5g$	11/2 - 9/2	1.10227089(42)
$7h - 5g$	9/2 - 7/2	0.68739292(16)
$7h - 5g$	9/2 - 9/2	0.68790001(16)
$7h - 5g$	11/2 - 7/2	0.68726989(16)
$7h - 5g$	11/2 - 9/2	0.68777679(16)
$8h - 5g$	9/2 - 7/2	0.55255817(10)
$8h - 5g$	9/2 - 9/2	0.55288579(10)
$8h - 5g$	11/2 - 7/2	0.55250490(10)
$8h - 5g$	11/2 - 9/2	0.55283246(10)
$Z = 28$		
$4f - 3d$	5/2 - 3/2	2.37674658(26)
$4f - 3d$	5/2 - 5/2	2.38945931(26)
$4f - 3d$	7/2 - 3/2	2.37409267(26)
$4f - 3d$	7/2 - 5/2	2.38677696(26)
$5f - 3d$	5/2 - 3/2	1.62704634(12)
$5f - 3d$	5/2 - 5/2	1.63299393(12)
$5f - 3d$	7/2 - 3/2	1.62640878(12)
$5f - 3d$	7/2 - 5/2	1.63235170(12)
$6f - 3d$	5/2 - 3/2	1.388998199(91)
$6f - 3d$	5/2 - 5/2	1.393330437(91)
$6f - 3d$	7/2 - 3/2	1.388729211(91)
$6f - 3d$	7/2 - 5/2	1.393059768(91)
$5g - 4f$	7/2 - 5/2	5.1517737(12)
$5g - 4f$	7/2 - 7/2	5.1642870(12)
$5g - 4f$	9/2 - 5/2	5.1479472(12)
$5g - 4f$	9/2 - 7/2	5.1604419(12)
$6g - 4f$	7/2 - 5/2	3.34069056(52)
$6g - 4f$	7/2 - 7/2	3.34594783(52)
$6g - 4f$	9/2 - 5/2	3.33975872(52)
$6g - 4f$	9/2 - 7/2	3.34501306(52)
$7g - 4f$	7/2 - 5/2	2.75633295(35)
$7g - 4f$	7/2 - 7/2	2.75991088(35)
$7g - 4f$	9/2 - 5/2	2.75593337(35)
$7g - 4f$	9/2 - 7/2	2.75951026(35)
$6h - 5g$	9/2 - 7/2	0.94953453(42)
$6h - 5g$	9/2 - 9/2	0.95083716(42)
$6h - 5g$	11/2 - 7/2	0.94903326(42)
$6h - 5g$	11/2 - 9/2	0.95033452(42)
$7h - 5g$	9/2 - 7/2	0.59260852(16)
$7h - 5g$	9/2 - 9/2	0.59311564(16)
$7h - 5g$	11/2 - 7/2	0.59248550(16)
$7h - 5g$	11/2 - 9/2	0.59299241(16)

Table A.3. Continued

Transition	J_1 - J_0	Wavelength [nm]
$Z = 28$		
$8h - 5g$	9/2 - 7/2	0.47637353(10)
$8h - 5g$	9/2 - 9/2	0.47670117(10)
$8h - 5g$	11/2 - 7/2	0.47632027(10)
$8h - 5g$	11/2 - 9/2	0.47664784(10)
$Z = 42$		
$4f - 3d$	5/2 - 3/2	1.04834116(26)
$4f - 3d$	5/2 - 5/2	1.06106274(26)
$4f - 3d$	7/2 - 3/2	1.04571993(26)
$4f - 3d$	7/2 - 5/2	1.05837758(26)
$5f - 3d$	5/2 - 3/2	0.71894918(12)
$5f - 3d$	5/2 - 5/2	0.72490964(12)
$5f - 3d$	7/2 - 3/2	0.71831625(12)
$5f - 3d$	7/2 - 5/2	0.72426617(12)
$6f - 3d$	5/2 - 3/2	0.614088309(90)
$6f - 3d$	5/2 - 5/2	0.618431608(91)
$6f - 3d$	7/2 - 3/2	0.613820871(90)
$6f - 3d$	7/2 - 5/2	0.618160374(91)
$5g - 4f$	7/2 - 5/2	2.2817806(12)
$5g - 4f$	7/2 - 7/2	2.2942979(12)
$5g - 4f$	9/2 - 5/2	2.2779749(12)
$5g - 4f$	9/2 - 7/2	2.2904504(12)
$6g - 4f$	7/2 - 5/2	1.48093188(52)
$6g - 4f$	7/2 - 7/2	1.48619446(52)
$6g - 4f$	9/2 - 5/2	1.48000260(52)
$6g - 4f$	9/2 - 7/2	1.48525856(52)
$7g - 4f$	7/2 - 5/2	1.22218864(35)
$7g - 4f$	7/2 - 7/2	1.22577071(35)
$7g - 4f$	9/2 - 5/2	1.22178982(35)
$7g - 4f$	9/2 - 7/2	1.22536956(35)
$6h - 5g$	9/2 - 7/2	0.42119551(42)
$6h - 5g$	9/2 - 9/2	0.42249843(42)
$6h - 5g$	11/2 - 7/2	0.42069575(42)
$6h - 5g$	11/2 - 9/2	0.42199557(42)
$7h - 5g$	9/2 - 7/2	0.26300796(16)
$7h - 5g$	9/2 - 9/2	0.26351540(16)
$7h - 5g$	11/2 - 7/2	0.26288511(16)
$7h - 5g$	11/2 - 9/2	0.26339207(16)
$8h - 5g$	9/2 - 7/2	0.21145128(10)
$8h - 5g$	9/2 - 9/2	0.21177915(10)
$8h - 5g$	11/2 - 7/2	0.21139806(10)
$8h - 5g$	11/2 - 9/2	0.21172576(10)
$Z = 80$		
$4f - 3d$	5/2 - 3/2	0.27848469(24)
$4f - 3d$	5/2 - 5/2	0.29125085(26)
$4f - 3d$	7/2 - 3/2	0.27601646(23)
$4f - 3d$	7/2 - 5/2	0.28855223(26)
$5f - 3d$	5/2 - 3/2	0.19264401(11)

Table A.3. Continued

Transition	J_1 - J_0	Wavelength [nm]
$Z = 80$		
$5f - 3d$	5/2 - 5/2	0.19866787(12)
$5f - 3d$	7/2 - 3/2	0.19203331(11)
$5f - 3d$	7/2 - 5/2	0.19801845(12)
$6f - 3d$	5/2 - 3/2	0.164973021(85)
$6f - 3d$	5/2 - 5/2	0.169370902(90)
$6f - 3d$	7/2 - 3/2	0.164713095(85)
$6f - 3d$	7/2 - 5/2	0.169096945(89)
$5g - 4f$	7/2 - 5/2	0.6186060(12)
$5g - 4f$	7/2 - 7/2	0.6311429(12)
$5g - 4f$	9/2 - 5/2	0.6148982(11)
$5g - 4f$	9/2 - 7/2	0.6272838(12)
$6g - 4f$	7/2 - 5/2	0.40317974(51)
$6g - 4f$	7/2 - 7/2	0.40846792(52)
$6g - 4f$	9/2 - 5/2	0.40226264(50)
$6g - 4f$	9/2 - 7/2	0.40752663(52)
$7g - 4f$	7/2 - 5/2	0.33313407(34)
$7g - 4f$	7/2 - 7/2	0.33673619(35)
$7g - 4f$	9/2 - 5/2	0.33273894(34)
$7g - 4f$	9/2 - 7/2	0.33633247(35)
$6h - 5g$	9/2 - 7/2	0.11502325(41)
$6h - 5g$	9/2 - 9/2	0.11632753(42)
$6h - 5g$	11/2 - 7/2	0.11453059(41)
$6h - 5g$	11/2 - 9/2	0.11582365(42)
$7h - 5g$	9/2 - 7/2	0.07200386(16)
$7h - 5g$	9/2 - 9/2	0.07251280(16)
$7h - 5g$	11/2 - 7/2	0.07188178(16)
$7h - 5g$	11/2 - 9/2	0.07238899(16)
$8h - 5g$	9/2 - 7/2	0.05792841(10)
$8h - 5g$	9/2 - 9/2	0.05825737(10)
$8h - 5g$	11/2 - 7/2	0.05787539(10)
$8h - 5g$	11/2 - 9/2	0.05820375(10)
$Z = 92$		
$4f - 3d$	5/2 - 3/2	0.20703015(23)
$4f - 3d$	5/2 - 5/2	0.21981762(26)
$4f - 3d$	7/2 - 3/2	0.20462926(23)
$4f - 3d$	7/2 - 5/2	0.21711291(25)
$5f - 3d$	5/2 - 3/2	0.14378168(11)
$5f - 3d$	5/2 - 5/2	0.14983516(12)
$5f - 3d$	7/2 - 3/2	0.14318108(11)
$5f - 3d$	7/2 - 5/2	0.14918304(12)
$6f - 3d$	5/2 - 3/2	0.123276435(83)
$6f - 3d$	5/2 - 5/2	0.127699861(89)
$6f - 3d$	7/2 - 3/2	0.123019957(83)
$6f - 3d$	7/2 - 5/2	0.127424667(89)
$5g - 4f$	7/2 - 5/2	0.4642750(11)
$5g - 4f$	7/2 - 7/2	0.4768209(12)
$5g - 4f$	9/2 - 5/2	0.4606105(11)
$5g - 4f$	9/2 - 7/2	0.4729565(12)

Table A.3. Continued

Transition	J_1 - J_0	Wavelength [nm]
$Z = 92$		
$6g - 4f$	7/2 - 5/2	0.30316651(50)
$6g - 4f$	7/2 - 7/2	0.30846633(52)
$6g - 4f$	9/2 - 5/2	0.30225489(50)
$6g - 4f$	9/2 - 7/2	0.30752261(52)
$7g - 4f$	7/2 - 5/2	0.25063167(34)
$7g - 4f$	7/2 - 7/2	0.25424291(35)
$7g - 4f$	9/2 - 5/2	0.25023821(34)
$7g - 4f$	9/2 - 7/2	0.25383804(35)
$6h - 5g$	9/2 - 7/2	0.08661374(41)
$6h - 5g$	9/2 - 9/2	0.08791864(42)
$6h - 5g$	11/2 - 7/2	0.08612422(40)
$6h - 5g$	11/2 - 9/2	0.08741430(42)
$7h - 5g$	9/2 - 7/2	0.05428043(16)
$7h - 5g$	9/2 - 9/2	0.05479006(16)
$7h - 5g$	11/2 - 7/2	0.05415870(16)
$7h - 5g$	11/2 - 9/2	0.05466603(16)
$8h - 5g$	9/2 - 7/2	0.04368292(10)
$8h - 5g$	9/2 - 9/2	0.04401237(10)
$8h - 5g$	11/2 - 7/2	0.04362999(10)
$8h - 5g$	11/2 - 9/2	0.04395864(10)

Table A.4. Atomic energy levels in a.u. (1 a.u. = 2 Ry = 27.2114 eV) for $n = 1$ and $n = 2$ singlet and triplet states in He-like ions, from [A.3]

Z	$1\,^1S_0$	$2\,^1S_0$	$2\,^1P_1$	$2\,^3S_1$	$2\,^3P_0$	$2\,^3P_1$	$2\,^3P_2$
3	2.779961	0.540908	-	0.6107986	0.5277241	0.5277479	0.5277383
4	5.655943	1.185020	-	1.2974536	1.1750420	1.1750946	1.1750270
5	9.532325	2.078995	-	2.2346975	2.0734244	2.0734983	2.0732582
6	14.409738	3.223173	-	3.4225604	3.2225342	3.2225911	3.2219723
7	20.288915	4.617846	4.460516	4.8612263	4.6223725	4.6223327	4.6210068
8	27.170717	6.263310	6.077303	6.5509613	6.2731130	6.2728452	6.2703318
9	35.056151	8.159902	7.944584	8.4920953	8.1750086	8.1743193	8.1699548
10	43.946361	10.307993	10.062534	10.685017	10.328414	10.327044	10.319945
11	53.842700	12.708008	12.431376	13.130165	12.733687	12.731304	12.720336
12	64.746633	15.360422	15.051280	15.828034	15.391312	15.387516	15.371240
13	76.659815	18.265756	17.922486	18.779174	18.301761	18.296079	18.272720
14	89.584050	21.424583	21.045164	21.984185	21.465625	21.457531	21.424918
15	103.521316	24.837528	24.419541	25.443719	24.883485	24.872403	24.827916
16	118.473770	28.505270	28.045781	29.158486	28.556044	28.541370	28.481877
17	134.443729	32.428542	31.924093	33.129248	32.483990	32.465106	32.386907
18	151.433691	36.608128	36.054640	37.356821	36.668101	36.644409	36.543159
19	169.446344	41.044873	40.437518	41.842085	41.109281	41.080218	40.950863
20	188.484525	45.739674	45.072904	46.585964	45.808350	45.773422	45.610121
21	208.551266	50.693484	49.960959	51.589448	50.766235	50.725046	50.521094
22	229.649770	55.907329	55.101732	56.853580	55.984001	55.936255	55.684033
23	251.783499	61.382247	60.495525	62.379507	61.462771	61.407918	61.099123
24	274.955995	67.119401	66.141860	68.168286	67.203442	67.142247	66.766592
26	324.432617	79.385200	78.194158	80.539636	79.475864	79.401745	78.859506
28	378.112524	92.715019	91.259614	93.978220	92.811520	92.726141	91.964718
30	436.03120	107.12010	105.33957	108.49559	107.22161	107.12768	106.08426

Table A.4. Continued

Z	$1\,^1S_0$	$2\,^1S_0$	$2\,^1P_1$	$2\,^3S_1$	$2\,^3P_0$	$2\,^3P_1$	$2\,^3P_2$
32	498.22835	122.61289	120.43566	124.10447	122.71846	122.61957	121.22040
34	564.74687	139.20685	136.54984	140.81872	139.31554	139.21593	137.37562
36	635.63339	156.91668	153.68449	158.65330	157.02749	156.93192	154.55253
38	710.93857	175.75828	171.84225	177.62453	175.87011	175.78380	172.75389
40	790.71669	195.74883	191.02594	197.74990	195.86056	195.78908	191.98266
42	875.02561	216.90675	211.23875	219.04827	217.01728	216.96658	212.24196
44	963.93030	239.25228	232.48394	241.54027	239.36028	239.33664	233.53514
46	1057.49844	262.80688	254.76501	265.24780	262.91108	262.92109	255.86575
47	1106.05415	275.04479	266.29516	277.56472	275.14655	275.17600	267.42122
48	1155.80244	287.59364	278.08570	290.19440	287.69270	287.74342	279.23740
50	1258.92259	313.63767	302.44983	316.40570	313.73012	313.82890	303.65408
52	1366.93888	340.96521	327.86154	343.90838	341.05000	341.20466	329.11982
54	1479.94933	369.60598	354.32503	372.73282	369.68122	369.90002	355.63899
56	1598.04815	399.59056	381.84486	402.91012	399.65455	399.94626	383.21597
58	1721.33798	430.95196	410.42559	434.47392	431.00333	431.37724	411.85556
60	1849.93167	463.72611	440.07215	467.46073	463.76323	464.22927	441.56265
62	1983.94179	497.95003	470.78968	501.90862	497.97269	498.54136	472.34450
64	2123.50855	533.66699	502.58336	537.86132	533.67325	534.35574	504.19993
66	2268.77497	570.92210	535.45877	575.36492	570.90957	571.71785	537.14109
68	2419.87609	609.76122	569.42164	614.46612	609.72981	610.67661	571.17160
70	2576.96162	650.23442	604.47796	655.21600	650.18591	651.28489	606.29747
72	2740.22939	692.40143	640.63398	697.67541	692.33437	693.60008	642.52501
74	2909.84877	736.32000	677.89610	741.90323	736.23610	737.68419	679.86075
76	3086.04309	782.05910	716.27108	787.96979	781.95744	783.60474	718.31140
78	3269.00407	829.68551	755.76588	835.94321	829.57017	831.43468	757.88408
80	3458.96537	879.27588	796.38774	885.90167	879.15261	881.25373	798.58608
82	3656.2039	930.9176	838.1441	937.9599	930.7906	933.1492	840.4250
84	3860.9789	984.6988	881.0431	992.1313	984.5774	987.2162	883.4085
86	4073.5338	1040.7093	925.0922	1048.5837	1040.6146	1043.5582	927.5451
88	4294.3025	1099.0756	970.3004	1107.4216	1099.0172	1102.2920	972.8430
90	4523.5687	1159.9045	1016.6762	1168.7536	1159.9080	1163.5426	1019.3111
92	4761.7311	1223.3279	1064.2287	1232.7146	1223.4249	1227.4511	1066.9583
94	5009.2119	1289.4886	1112.9674	1299.4507	1289.7206	1294.1727	1115.7940
96	5266.5222	1358.5532	1162.9017	1369.1323	1358.9663	1363.8820	1165.8282
98	5533.9779	1430.6851	1214.0419	1441.9265	1431.3506	1436.7719	1217.0707
100	5812.5127	1506.0685	1266.3983	1518.0218	1507.0883	1513.0592	1269.5324

A.2 Numerical Data for Electromagnetic Decay

Table A.5. Wavelengths λ (in Å), transition probabilities A (in s^{-1}) and oscillator strengths f for E1 transitions in He-like ions, from [A.4]

Z	Transition	Multiplet	λ(Å)	$J_i - J_k$	$A_{ki}(s^{-1})$	f_{ik}
2	$1s^2 - 1s2p$	^1S–^3P	591.33	0–1	1.787×10^2	2.810×10^{-8}
		^1S–^1P	584.25	0–1	1.799×10^9	2.761×10^{-1}
	$1s2s - 1s2p$	^3S–^3P	10832.	1–2	1.022×10^7	2.995×10^{-1}
			10832.	1–1	1.022×10^7	1.797×10^{-1}
			10831.	1–0	1.022×10^7	5.990×10^{-2}
		^3S–^1P	8864.8	1–1	1.552	1.828×10^{-8}
		^1S–^1P	20584.	0–1	1.976×10^6	1.255×10^{-1}

Table A.5. Continued

Z	Transition	Multiplet	λ(Å)	J_i - J_k	A_{ki} (s^{-1})	f_{ik}
3	$1s^2-1s2p$	^1S–^3P	202.30	0–1	1.790×10^4	3.295×10^{-7}
		^1S–^1P	199.26	0–1	2.556×10^{10}	4.565×10^{-1}
	$1s2s-1s2p$	^3S–^3P	5485.6	1–2	2.276×10^7	1.712×10^{-1}
			5486.2	1–1	2.276×10^7	1.027×10^{-1}
			5484.7	1–0	2.278×10^7	3.424×10^{-2}
		^3S–^1P	3879.7	1–1	4.190×10^1	9.456×10^{-8}
		^1S–^1P	9583.4	0–1	5.148×10^6	7.088×10^{-2}
4	$1s^2$-$1s2p$	^1S-^3P	101.68	0-1	4.002×10^5	1.861×10^{-6}
		^1S-^1P	100.25	0-1	1.220×10^{11}	5.513×10^{-1}
	$1s2s$-$1s2p$	^3S-^3P	3721.7	1-2	$3.427.\times10^7$	1.186×10^{-1}
			3723.7	1-1	3.421×10^7	7.112×10^{-2}
			3722.1	1-0	3.426×10^7	2.372×10^{-2}
		^3S-^1P	2441.7	1-1	3.853×10^2	3.444×10^{-7}
		^1S-^1P	6143.3	0-1	8.757×10^6	4.955×10^{-2}
5	$1s^2$-$1s2p$	^1S-^3P	61.086	0-1	4.226×10^6	7.093×10^{-6}
		^1S-^1P	60.311	0-1	3.719×10^{11}	6.085×10^{-1}
	$1s2s$-$1s2p$	^3S-^3P	2822.3	1-2	4.557×10^7	9.070×10^{-2}
			2826.5	1-1	4.536×10^7	5.433×10^{-2}
			2825.2	1-0	4.542×10^7	1.812×10^{-2}
		^3S-^1P	1771.9	1-1	2.061×10^3	9.701×10^{-7}
		^1S-^1P	4491.8	0-1	1.262×10^7	3.818×10^{-2}
6	$1s^2$-$1s2p$	^1S-^3P	40.728	0-1	2.830×10^7	2.111×10^{-5}
		^1S-^1P	40.266	0-1	8.864×10^{11}	6.463×10^{-1}
	$1s2s$-$1s2p$	^3S-^3P	2271.5	1-2	5.702×10^7	7.352×10^{-2}
			2278.5	1-1	5.648×10^7	4.396×10^{-2}
			2277.9	1-0	5.652×10^7	1.466×10^{-2}
		^3S-^1P	1386.7	1-1	7.952×10^3	2.293×10^{-6}
		^1S-^1P	3527.4	0-1	1.668×10^7	3.111×10^{-2}
7	$1s^2$-$1s2p$	^1S-^3P	29.083	0-1	1.394×10^8	5.303×10^{-5}
		^1S-^1P	28.786	0-1	1.806×10^{12}	6.731×10^{-1}
	$1s2s$-$1s2p$	^3S-^3P	1896.7	1-2	6.891×10^7	6.195×10^{-2}
			1907.3	1-1	6.774×10^7	3.694×10^{-2}
			1907.6	1-0	6.770×10^7	1.231×10^{-2}
		^3S-^1P	1137.1	1-1	2.465×10^4	4.778×10^{-6}
		^1S-^1P	2896.0	0-1	2.092×10^7	2.630×10^{-2}
8	$1s^2$-$1s2p$	^1S-^3P	21.803	0-1	5.499×10^8	1.176×10^{-4}
		^1S-^1P	21.601	0-1	3.302×10^{12}	6.929×10^{-1}
	$1s2s$-$1s2p$	^3S-^3P	1623.6	1-2	8.149×10^7	5.368×10^{-2}
			1638.3	1-1	7.925×10^7	3.189×10^{-2}
			1639.9	1-0	7.902×10^7	1.062×10^{-2}
		^3S-^1P	961.95	1-1	6.533×10^4	9.063×10^{-6}
		^1S-^1P	2449.6	0-1	2.537×10^7	2.283×10^{-2}

Table A.5. Continued

Z	Transition	Multiplet	λ(Å)	J_i - J_k	A_{ki} (s^{-1})	f_{ik}
9	$1s^2$-$1s2p$	^1S-^3P	16.950	0-1	1.833×10^{9}	2.369×10^{-4}
		^1S-^1P	16.806	0-1	5.574×10^{12}	7.080×10^{-1}
	$1s2s$-$1s2p$	^3S-^3P	1414.4	1-2	9.506×10^{7}	4.752×10^{-2}
			1433.8	1-1	9.113×10^{7}	2.809×10^{-2}
			1436.9	1-0	9.054×10^{7}	9.342×10^{-3}
		^3S-^1P	832.19	1-1	1.539×10^{5}	1.598×10^{-5}
		^1S-^1P	2116.1	0-1	3.011×10^{7}	2.021×10^{-2}
10	$1s^2$-$1s2p$	^1S-^3P	13.553	0-1	5.356×10^{9}	4.425×10^{-4}
		^1S-^1P	13.447	0-1	8.851×10^{12}	7.198×10^{-1}
	$1s2s$-$1s2p$	^3S-^3P	1248.1	1-2	1.099×10^{8}	4.278×10^{-2}
			1272.8	1-1	1.034×10^{8}	2.513×10^{-2}
			1277.7	1-0	1.023×10^{8}	8.344×10^{-3}
		^3S-^1P	731.96	1-1	3.307×10^{5}	2.656×10^{-5}
		^1S-^1P	1856.3	0-1	3.521×10^{7}	1.819×10^{-2}
11	$1s^2$-$1s2p$	^1S-^3P	11.083	0-1	1.405×10^{10}	7.764×10^{-4}
		^1S-^1P	11.003	0-1	1.339×10^{13}	7.291×10^{-1}
	$1s2s$-$1s2p$	^3S-^3P	1111.8	1-2	1.265×10^{8}	3.908×10^{-2}
			1142.3	1-1	1.163×10^{8}	2.276×10^{-2}
			1149.2	1-0	1.143×10^{8}	7.544×10^{-3}
		^3S-^1P	652.03	1-1	6.600×10^{5}	4.207×10^{-5}
		^1S-^1P	1647.1	0-1	4.077×10^{7}	1.658×10^{-2}
12	$1s^2$-$1s2p$	^1S-^3P	9.2310	0-1	3.375×10^{10}	1.293×10^{-3}
		^1S-^1P	9.1685	0-1	1.948×10^{13}	7.364×10^{-1}
	$1s2s$-$1s2p$	^3S-^3P	997.46	1-2	1.453×10^{8}	3.612×10^{-2}
			1034.3	1-1	1.299×10^{8}	2.083×10^{-2}
			1043.3	1-0	1.266×10^{8}	6.889×10^{-3}
		^3S-^1P	586.59	1-1	1.240×10^{6}	6.399×10^{-5}
		^1S-^1P	1473.9	0-1	4.696×10^{7}	1.529×10^{-2}
13	$1s^2$-$1s2p$	^1S-^3P	7.8068	0-1	7.519×10^{10}	2.061×10^{-3}
		^1S-^1P	7.7571	0-1	2.742×10^{13}	7.420×10^{-1}
	$1s2s$-$1s2p$	^3S-^3P	899.65	1-2	1.669×10^{8}	3.375×10^{-2}
			943.16	1-1	1.441×10^{8}	1.992×10^{-2}
			954.38	1-0	1.393×10^{8}	6.343×10^{-3}
		^3S-^1P	531.85	1-1	2.217×10^{6}	9.400×10^{-5}
		^1S-^1P	1327.3	0-1	5.393×10^{7}	1.424×10^{-2}
14	$1s^2$-$1s2p$	^1S-^3P	6.6881	0-1	1.570×10^{11}	3.159×10^{-3}
		^1S-^1P	6.6478	0-1	3.755×10^{13}	7.463×10^{-1}
	$1s2s$-$1s2p$	^3S-^3P	814.70	1-2	1.919×10^{8}	3.183×10^{-2}
			865.15	1-1	1.592×10^{8}	1.786×10^{-2}
			878.65	1-0	1.524×10^{8}	5.880×10^{-3}
		^3S-^1P	485.22	1-1	3.796×10^{6}	1.340×10^{-4}
		^1S-^1P	1200.9	0-1	6.191×10^{7}	1.339×10^{-2}

Table A.5. Continued

Z	Transition	Multiplet	λ(Å)	J_i - J_k	A_{ki} (s^{-1})	f_{ik}
15	$1s^2$-$1s2p$	^1S-^3P	5.7933	0-1	3.099×10^{11}	4.678×10^{-3}
		^1S-^1P	5.7601	0-1	5.021×10^{13}	7.492×10^{-1}
	$1s2s$-$1s2p$	^3S-^3P	739.90	1-2	2.214×10^{8}	3.028×10^{-2}
			797.52	1-1	1.751×10^{8}	1.670×10^{-2}
			813.29	1-0	1.659×10^{8}	5.485×10^{-3}
		^3S-^1P	444.88	1-1	6.265×10^{6}	1.859×10^{-4}
		^1S-^1P	1090.1	0-1	7.118×10^{7}	1.268×10^{-2}
16	$1s^2$-$1s2p$	^1S-^3P	5.0664	0-1	5.818×10^{11}	6.717×10^{-3}
		^1S-^1P	5.0386	0-1	6.576×10^{13}	7.509×10^{-1}
	$1s2s$-$1s2p$	^3S-^3P	673.41	1-2	2.562×10^{8}	2.903×10^{-2}
			738.33	1-1	1.920×10^{8}	1.569×10^{-2}
			756.31	1-0	1.799×10^{8}	5.142×10^{-3}
		^3S-^1P	409.48	1-1	1.001×10^{7}	2.517×10^{-4}
		^1S-^1P	991.61	0-1	8.208×10^{7}	1.210×10^{-2}
17	$1s^2$-$1s2p$	^1S-^3P	4.4679	0-1	1.044×10^{12}	9.374×10^{-3}
		^1S-^1P	4.4444	0-1	8.458×10^{13}	7.514×10^{-1}
	$1s2s$-$1s2p$	^3S-^3P	613.78	1-2	2.978×10^{8}	2.803×10^{-2}
			686.05	1-1	2.097×10^{8}	1.480×10^{-2}
			706.13	1-0	1.944×10^{8}	4.843×10^{-3}
		^3S-^1P	378.07	1-1	1.556×10^{7}	3.335×10^{-4}
		^1S-^1P	903.23	0-1	9.505×10^{7}	1.163×10^{-2}
18	$1s^2$-$1s2p$	^1S-^3P	3.9693	0-1	1.799×10^{12}	1.274×10^{-2}
		^1S-^1P	3.9490	0-1	1.070×10^{14}	7.506×10^{-1}
	$1s2s$-$1s2p$	^3S-^3P	559.98	1-2	3.477×10^{8}	2.724×10^{-2}
			639.56	1-1	2.284×10^{8}	1.401×10^{-2}
			661.57	1-0	2.094×10^{8}	4.580×10^{-3}
		^3S-^1P	349.90	1-1	2.358×10^{7}	4.329×10^{-4}
		^1S-^1P	823.20	0-1	1.106×10^{8}	1.124×10^{-2}
19	$1s^2$-$1s2p$	^1S-^3P	3.5495	0-1	2.984×10^{12}	1.691×10^{-2}
		^1S-^1P	3.5318	0-1	1.334×10^{14}	7.486×10^{-1}
	$1s2s$-$1s2p$	^3S-^3P	511.25	1-2	4.079×10^{8}	2.664×10^{-2}
			598.05	1-1	2.479×10^{8}	1.329×10^{-2}
			621.77	1-0	2.250×10^{8}	4.346×10^{-3}
		^3S-^1P	324.39	1-1	3.494×10^{7}	5.512×10^{-4}
		^1S-^1P	750.19	0-1	1.296×10^{8}	1.093×10^{-2}
20	$1s^2$-$1s2p$	^1S-^3P	3.1927	0-1	4.782×10^{12}	2.192×10^{-2}
		^1S-^1P	3.1771	0-1	1.642×10^{14}	7.453×10^{-1}
	$1s2s$-$1s2p$	^3S-^3P	466.91	1-2	4.809×10^{8}	2.619×10^{-2}
			560.75	1-1	2.681×10^{8}	1.264×10^{-2}
			585.94	1-0	2.412×10^{8}	4.138×10^{-3}
		^3S-^1P	301.13	1-1	5.073×10^{7}	6.896×10^{-4}
		^1S-^1P	683.34	0-1	1.528×10^{8}	1.069×10^{-2}

Table A.5. Continued

Z	Transition	Multiplet	λ(Å)	J_i - J_k	A_{ki} (s^{-1})	f_{ik}
21	$1s^2$-$1s2p$	^1S-^3P	2.8869	0-1	7.421×10^{12}	2.782×10^{-2}
		^1S-^1P	2.8730	0-1	1.996×10^{14}	7.409×10^{-1}
	$1s2s$-$1s2p$	^3S-^3P	426.48	1-2	5.697×10^{8}	2.589×10^{-2}
			527.11	1-1	2.890×10^{8}	1.204×10^{-2}
			553.48	1-0	2.581×10^{8}	3.952×10^{-3}
		^3S-^1P	279.79	1-1	7.228×10^{7}	8.483×10^{-4}
		^1S-^1P	622.00	0-1	1.812×10^{8}	1.051×10^{-2}
22	$1s^2$-$1s2p$	^1S-^3P	2.6229	0-1	1.117×10^{13}	3.458×10^{-2}
		^1S-^1P	2.6104	0-1	2.399×10^{14}	7.352×10^{-1}
	$1s2s$-$1s2p$	^3S-^3P	389.58	1-2	6.782×10^{8}	2.572×10^{-2}
			496.70	1-1	3.103×10^{8}	1.148×10^{-2}
			523.97	1-0	2.758×10^{8}	3.783×10^{-3}
		^3S-^1P	260.09	1-1	1.013×10^{8}	1.027×10^{-3}
		^1S-^1P	565.58	0-1	2.165×10^{8}	1.038×10^{-2}
23	$1s^2$-$1s2p$	^1S-^3P	2.3933	0-1	1.636×10^{13}	4.215×10^{-2}
		^1S-^1P	2.3819	0-1	2.855×10^{14}	7.285×10^{-1}
	$1s2s$-$1s2p$	^3S-^3P	355.86	1-2	8.109×10^{8}	2.566×10^{-2}
			468.96	1-1	3.323×10^{8}	1.096×10^{-2}
			497.02	1-0	2.941×10^{8}	3.631×10^{-3}
		^3S-^1P	241.85	1-1	1.397×10^{8}	1.225×10^{-3}
		^1S-^1P	513.84	0-1	2.602×10^{8}	1.030×10^{-2}
24	$1s^2$-$1s2p$	^1S-^3P	2.1925	0-1	2.334×10^{13}	5.046×10^{-2}
		^1S-^1P	2.1820	0-1	3.366×10^{14}	7.208×10^{-1}
	$1s2s$-$1s2p$	^3S-^3P	325.06	1-2	9.735×10^{8}	2.570×10^{-2}
			444.07	1-1	3.538×10^{8}	1.046×10^{-2}
			472.24	1-0	3.134×10^{8}	3.492×10^{-3}
		^3S-^1P	224.85	1-1	1.901×10^{8}	1.441×10^{-3}
		^1S-^1P	466.10	0-1	3.151×10^{8}	1.026×10^{-2}
25	$1s^2$-$1s2p$	^1S-^3P	2.0158	0-1	3.249×10^{13}	5.937×10^{-2}
		^1S-^1P	2.0061	0-1	3.935×10^{14}	7.123×10^{-1}
	$1s2s$-$1s2p$	^3S-^3P	296.91	1-2	1.173×10^{9}	2.584×10^{-2}
			421.24	1-1	3.756×10^{8}	9.993×10^{-3}
			449.42	1-0	3.335×10^{8}	3.366×10^{-3}
		^3S-^1P	209.02	1-1	2.554×10^{8}	1.673×10^{-3}
		^1S-^1P	422.45	0-1	3.837×10^{8}	1.027×10^{-2}
26	$1s^2$-$1s2p$	^1S-^3P	1.8595	0-1	4.421×10^{13}	6.876×10^{-2}
		^1S-^1P	1.8504	0-1	4.566×10^{14}	7.031×10^{-1}
	$1s2s$-$1s2p$	^3S-^3P	271.19	1-2	1.419×10^{9}	2.607×10^{-2}
			400.42	1-1	3.973×10^{8}	9.551×10^{-3}
			428.32	1-0	3.545×10^{8}	3.250×10^{-3}
		^3S-^1P	194.26	1-1	3.392×10^{8}	1.919×10^{-3}
		^1S-^1P	382.55	0-1	4.697×10^{8}	1.030×10^{-2}

Table A.5. Continued

Z	Transition	Multiplet	λ(Å)	J_i - J_k	A_{ki} (s^{-1})	f_{ik}
27	$1s^2$-$1s2p$	^1S-^3P	1.7206	0-1	5.893×10^{13}	7.846×10^{-2}
		^1S-^1P	1.7120	0-1	5.260×10^{14}	6.934×10^{-1}
	$1s2s$-$1s2p$	^3S-^3P	247.71	1-2	1.721×10^{9}	2.638×10^{-2}
			381.36	1-1	4.189×10^{8}	9.134×10^{-3}
			408.74	1-0	3.765×10^{8}	3.144×10^{-3}
		^3S-^1P	180.47	1-1	4.459×10^{8}	2.177×10^{-3}
		^1S-^1P	346.15	0-1	5.777×10^{8}	1.038×10^{-2}
28	$1s^2$-$1s2p$	^1S-^3P	1.5966	0-1	7.705×10^{13}	8.834×10^{-2}
		^1S-^1P	1.5884	0-1	6.022×10^{14}	6.833×10^{-1}
	$1s2s$-$1s2p$	^3S-^3P	226.29	1-2	2.092×10^{9}	2.677×10^{-2}
			363.90	1-1	4.402×10^{8}	8.739×10^{-3}
			390.53	1-0	3.996×10^{8}	3.046×10^{-3}
		^3S-^1P	167.60	1-1	5.806×10^{8}	2.445×10^{-3}
		^1S-^1P	313.06	0-1	7.134×10^{8}	1.048×10^{-2}
29	$1s^2$-$1s2p$	^1S-^3P	1.4854	0-1	9.899×10^{13}	9.823×10^{-2}
		^1S-^1P	1.4776	0-1	6.855×10^{14}	6.731×10^{-1}
	$1s2s$-$1s2p$	^3S-^3P	206.76	1-2	2.549×10^{9}	2.723×10^{-2}
			347.86	1-1	4.611×10^{8}	8.365×10^{-3}
			373.54	1-0	4.239×10^{8}	2.956×10^{-3}
		^3S-^1P	155.58	1-1	7.498×10^{8}	2.721×10^{-3}
		^1S-^1P	283.05	0-1	8.839×10^{8}	1.062×10^{-2}
30	$1s^2$-$1s2p$	^1S-^3P	1.3853	0-1	1.251×10^{14}	1.080×10^{-1}
		^1S-^1P	1.3778	0-1	7.763×10^{14}	6.629×10^{-1}
	$1s2s$-$1s2p$	^3S-^3P	188.96	1-2	3.112×10^{9}	2.776×10^{-2}
			333.09	1-1	4.818×10^{8}	8.013×10^{-3}
			357.65	1-0	4.493×10^{8}	2.872×10^{-3}
		^3S-^1P	144.37	1-1	9.610×10^{8}	3.003×10^{-3}
		^1S-^1P	255.90	0-1	1.098×10^{9}	1.078×10^{-2}
31	$1s^2$-$1s2p$	^1S-^3P	1.2949	0-1	1.559×10^{14}	1.176×10^{-1}
		^1S-^1P	1.2877	0-1	8.752×10^{14}	6.527×10^{-1}
	$1s2s$-$1s2p$	^3S-^3P	172.74	1-2	3.804×10^{9}	2.836×10^{-2}
			319.46	1-1	5.021×10^{8}	7.682×10^{-3}
			342.74	1-0	4.760×10^{8}	2.794×10^{-3}
		^3S-^1P	133.92	1-1	1.223×10^{9}	3.290×10^{-3}
		^1S-^1P	231.37	0-1	1.367×10^{9}	1.097×10^{-2}
32	$1s^2$-$1s2p$	^1S-^3P	1.2131	0-1	1.916×10^{14}	1.268×10^{-1}
		^1S-^1P	1.2060	0-1	9.824×10^{14}	6.427×10^{-1}
	$1s2s$-$1s2p$	^3S-^3P	157.98	1-2	4.654×10^{9}	2.902×10^{-2}
			306.84	1-1	5.221×10^{8}	7.370×10^{-3}
			328.74	1-0	$5.040\times{}^{1}0^{8}$	2.722×10^{-3}
		^3S-^1P	124.19	1-1	1.549×10^{9}	3.581×10^{-3}
		^1S-^1P	209.27	0-1	1.705×10^{9}	1.119×10^{-2}

Table A.5. Continued

Z	Transition	Multiplet	λ(Å)	J_i - J_k	A_{ki} (s^{-1})	f_{ik}
33	$1s^2$-$1s2p$	^1S-^3P	1.1386	0-1	2.327×10^{14}	1.357×10^{-1}
		^1S-^1P	1.1318	0-1	1.099×10^{15}	6.329×10^{-1}
	$1s2s$-$1s2p$	^3S-^3P	144.55	1-2	5.697×10^{9}	2.974×10^{-2}
			295.15	1-1	5.419×10^{8}	7.077×10^{-3}
			315.55	1-0	5.335×10^{8}	2.655×10^{-3}
		^3S-^1P	115.14	1-1	1.950×10^{9}	3.875×10^{-3}
		^1S-^1P	189.38	0-1	2.128×10^{9}	1.144×10^{-2}
34	$1s^2$-$1s2p$	^1S-^3P	1.0707	0-1	2.795×10^{14}	1.441×10^{-1}
		^1S-^1P	1.0641	0-1	1.224×10^{15}	6.235×10^{-1}
	$1s2s$-$1s2p$	^3S-^3P	132.33	1-2	6.977×10^{9}	3.053×10^{-2}
			284.28	1-1	5.614×10^{8}	6.802×10^{-3}
			303.11	1-0	5.646×10^{8}	2.592×10^{-3}
		^3S-^1P	106.73	1-1	2.443×10^{9}	4.173×10^{-3}
		^1S-^1P	171.48	0-1	2.656×10^{9}	1.171×10^{-2}
35	$1s^2$-$1s2p$	^1S-^3P	1.0087	0-1	3.324×10^{14}	1.521×10^{-1}
		^1S-^1P	1.0022	0-1	1.360×10^{15}	6.143×10^{-1}
	$1s2s$-$1s2p$	^3S-^3P	121.22	1-2	8.543×10^{9}	3.137×10^{-2}
			274.15	1-1	5.807×10^{8}	6.543×10^{-3}
			291.37	1-0	5.972×10^{8}	2.534×10^{-3}
		^3S-^1P	98.933	1-1	3.048×10^{9}	4.473×10^{-3}
		^1S-^1P	155.41	0-1	3.316×10^{9}	1.201×10^{-2}
36	$1s^2$-$1s2p$	^1S-^3P	.95181	0-1	3.917×10^{14}	1.596×10^{-1}
		^1S-^1P	.94540	0-1	1.506×10^{15}	6.056×10^{-1}
	$1s2s$-$1s2p$	^3S-^3P	111.11	1-2	1.046×10^{10}	3.226×10^{-2}
			264.69	1-1	5.999×10^{8}	6.301×10^{-3}
			280.25	1-0	6.316×10^{8}	2.479×10^{-3}
		^3S-^1P	91.699	1-1	3.789×10^{9}	4.776×10^{-3}
		^1S-^1P	140.97	0-1	4.137×10^{9}	1.233×10^{-2}
37	$1s^2$-$1s2p$	^1S-^3P	.89955	0-1	4.579×10^{14}	1.666×10^{-1}
		^1S-^1P	.89323	0-1	1.664×10^{15}	5.972×10^{-1}
	$1s2s$-$1s2p$	^3S-^3P	101.91	1-2	1.280×10^{10}	3.321×10^{-2}
			255.84	1-1	6.190×10^{8}	6.074×10^{-3}
			269.71	1-0	6.678×10^{8}	2.427×10^{-3}
		^3S-^1P	84.998	1-1	4.692×10^{9}	5.082×10^{-3}
		^1S-^1P	128.00	0-1	5.158×10^{9}	1.267×10^{-2}
38	$1s^2$-$1s2p$	^1S-^3P	.85141	0-1	5.313×10^{14}	1.732×10^{-1}
		^1S-^1P	.84518	0-1	1.834×10^{15}	5.892×10^{-1}
	$1s2s$-$1s2p$	^3S-^3P	93.547	1-2	1.565×10^{10}	3.422×10^{-2}
			247.53	1-1	6.380×10^{8}	5.861×10^{-3}
			259.71	1-0	7.059×10^{8}	2.379×10^{-3}
		^3S-^1P	78.798	1-1	5.791×10^{9}	5.391×10^{-3}
		^1S-^1P	116.35	0-1	6.422×10^{9}	1.303×10^{-2}

Table A.5. Continued

Z	Transition	Multiplet	λ(Å)	J_i - J_k	A_{ki} (s^{-1})	f_{ik}
39	$1s^2$-$1s2p$	^1S-^3P	.80697	0-1	6.122×10^{14}	1.793×10^{-1}
		^1S-^1P	.80082	0-1	2.016×10^{15}	5.815×10^{-1}
	$1s2s$-$1s2p$	^3S-^3P	85.934	1-2	1.912×10^{10}	3.527×10^{-2}
			239.73	1-1	6.570×10^{8}	5.660×10^{-3}
			250.20	1-0	7.460×10^{8}	2.334×10^{-3}
		^3S-^1P	73.063	1-1	7.126×10^{9}	5.703×10^{-3}
		^1S-^1P	105.88	0-1	7.985×10^{9}	1.342×10^{-2}
40	$1s^2$-$1s2p$	^1S-^3P	.76586	0-1	7.011×10^{14}	1.850×10^{-1}
		^1S-^1P	.75978	0-1	2.212×10^{15}	5.742×10^{-1}
	$1s2s$-$1s2p$	^3S-^3P	79.004	1-2	2.333×10^{10}	3.638×10^{-2}
			232.37	1-1	6.760×10^{8}	5.472×10^{-3}
			241.16	1-0	7.884×10^{8}	2.291×10^{-3}
		^3S-^1P	67.763	1-1	8.743×10^{9}	6.019×10^{-3}
		^1S-^1P	96.473	0-1	9.912×10^{9}	1.383×10^{-2}
41	$1s^2$-$1s2p$	^1S-^3P	.72777	0-1	7.983×10^{14}	1.902×10^{-1}
		^1S-^1P	.71174	0-1	2.421×10^{15}	5.673×10^{-1}
	$1s2s$-$1s2p$	^3S-^3P	72.694	1-2	2.843×10^{10}	3.754×10^{-2}
			225.45	1-1	6.948×10^{8}	5.295×10^{-3}
			232.57	1-0	8.328×10^{8}	2.251×10^{-3}
		^3S-^1P	62.867	1-1	1.070×10^{10}	6.338×10^{-3}
		^1S-^1P	88.011	0-1	1.228×10^{10}	1.426×10^{-2}
42	$1s^2$-$1s2p$	^1S-^3P	.69239	0-1	9.043×10^{14}	1.950×10^{-1}
		^1S-^1P	.68642	0-1	2.645×10^{15}	5.606×10^{-1}
	$1s2s$-$1s2p$	^3S-^3P	66.943	1-2	3.460×10^{10}	3.875×10^{-2}
			218.88	1-1	7.140×10^{8}	5.128×10^{-3}
			224.34	1-0	8.799×10^{8}	2.213×10^{-3}
		^3S-^1P	58.343	1-1	1.306×10^{10}	6.662×10^{-3}
		^1S-^1P	80.387	0-1	1.519×10^{10}	1.471×10^{-2}
43	$1s^2$-$1s2p$	^1S-^3P	.65948	0-1	1.020×10^{15}	1.994×10^{-1}
		^1S-^1P	.65355	0-1	2.885×10^{15}	5.542×10^{-1}
	$1s2s$-$1s2p$	^3S-^3P	61.701	1-2	4.205×10^{10}	4.000×10^{-2}
			212.67	1-1	7.331×10^{8}	4.971×10^{-3}
			216.51	1-0	9.295×10^{8}	2.177×10^{-3}
		^3S-^1P	54.167	1-1	1.589×10^{10}	6.991×10^{-3}
		^1S-^1P	73.517	0-1	1.874×10^{10}	1.519×10^{-2}
44	$1s^2$-$1s2p$	^1S-^3P	.62881	0-1	1.144×10^{15}	2.035×10^{-1}
		^1S-^1P	.62292	0-1	3.141×10^{15}	5.481×10^{-1}
	$1s2s$-$1s2p$	^3S-^3P	56.918	1-2	5.104×10^{10}	4.131×10^{-2}
			206.76	1-1	7.525×10^{8}	4.823×10^{-3}
			209.01	1-0	9.819×10^{8}	2.144×10^{-3}
		^3S-^1P	50.311	1-1	1.930×10^{10}	7.325×10^{-3}
		^1S-^1P	67.318	0-1	2.308×10^{10}	1.568×10^{-2}

Table A.5. Continued

Z	Transition	Multiplet	λ(Å)	J_i - J_k	A_{ki} (s^{-1})	f_{ik}
45	$1s^2$-$1s2p$	^1S-^3P	.60019	0-1	1.279×10^{15}	2.072×10^{-1}
		^1S-^1P	.59433	0-1	3.413×10^{15}	5.423×10^{-1}
	$1s2s$-$1s2p$	^3S-^3P	52.552	1-2	6.183×10^{10}	4.267×10^{-2}
			201.16	1-1	7.719×10^{8}	4.683×10^{-3}
			201.84	1-0	1.037×10^{9}	2.112×10^{-3}
		^3S-^1P	46.751	1-1	2.339×10^{10}	7.664×10^{-3}
		^1S-^1P	61.720	0-1	2.835×10^{10}	1.619×10^{-2}
46	$1s^2$-$1s2p$	^1S-^3P	.57343	0-1	1.425×10^{15}	2.107×10^{-1}
		^1S-^1P	.56760	0-1	3.704×10^{15}	5.367×10^{-1}
	$1s2s$-$1s2p$	^3S-^3P	48.564	1-2	7.479×10^{10}	4.407×10^{-2}
			195.83	1-1	7.915×10^{8}	4.550×10^{-3}
			194.99	1-0	1.095×10^{9}	2.081×10^{-3}
		^3S-^1P	43.465	1-1	2.827×10^{10}	8.008×10^{-3}
		^1S-^1P	56.658	0-1	3.475×10^{2}	1.673×10^{-2}
47	$1s^2$-$1s2p$	^1S-^3P	.54838	0-1	1.581×10^{15}	2.138×10^{-1}
		^1S-^1P	.54258	0-1	4.012×10^{15}	5.312×10^{-1}
	$1s2s$-$1s2p$	^3S-^3P	44.919	1-2	9.030×10^{10}	4.552×10^{-2}
			190.74	1-1	8.112×10^{8}	4.425×10^{-3}
			188.42	1-0	1.157×10^{9}	2.053×10^{-3}
		^3S-^1P	40.430	1-1	3.411×10^{10}	8.359×10^{-3}
		^1S-^1P	52.075	0-1	4.250×10^{10}	1.728×10^{-2}
48	$1s^2$-$1s2p$	^1S-^3P	.52489	0-1	1.749×10^{15}	2.167×10^{-1}
		^1S-^1P	.51911	0-1	4.340×10^{15}	5.260×10^{-1}
	$1s2s$-$1s2p$	^3S-^3P	41.584	1-2	1.088×10^{11}	4.702×10^{-2}
			185.90	1-1	8.310×10^{8}	4.305×10^{-3}
			182.13	1-0	1.222×10^{9}	2.025×10^{-3}
		^3S-^1P	37.629	1-1	4.106×10^{10}	8.715×10^{-3}
		^1S-^1P	47.921	0-1	5.185×10^{10}	1.785×10^{-2}
49	$1s^2$-$1s2p$	^1S-^3P	.50284	0-1	1.928×10^{15}	2.193×10^{-1}
		^1S-^1P	.49708	0-1	4.688×10^{15}	5.210×10^{-1}
	$1s2s$-$1s2p$	^3S-^3P	38.530	1-2	1.309×10^{11}	4.857×10^{-2}
			181.27	1-1	8.509×10^{8}	4.192×10^{-3}
			176.10	1-0	1.290×10^{9}	2.000×10^{-3}
		^3S-^1P	35.040	1-1	4.932×10^{10}	9.078×10^{-3}
		^1S-^1P	44.152	0-1	6.311×10^{10}	1.844×10^{-2}
50	$1s^2$-$1s2p$	^1S-^3P	.48210	0-1	2.120×10^{15}	2.216×10^{-1}
		^1S-^1P	.47637	0-1	5.057×10^{15}	5.161×10^{-1}
	$1s2s$-$1s2p$	^3S-^3P	35.731	1-2	1.573×10^{11}	5.017×10^{-2}
			176.82	1-1	8.713×10^{8}	4.084×10^{-3}
			170.29	1-0	1.363×10^{9}	1.975×10^{-3}
		^3S-^1P	32.648	1-1	5.911×10^{10}	9.446×10^{-3}
		^1S-^1P	40.726	0-1	7.774×10^{10}	1.906×10^{-2}

Table A.5. Continued

Z	Transition	Multiplet	λ(Å)	J_i - J_k	A_{ki} (s^{-1})	f_{ik}
51	$1s^2$-$1s2p$	^1S-^3P	.46259	0-1	2.325×10^{15}	2.238×10^{-1}
		^1S-^1P	.45687	0-1	5.447×10^{15}	5.113×10^{-1}
	$1s2s$-$1s2p$	^3S-^3P	33.166	1-2	1.885×10^{11}	5.181×10^{-2}
			172.59	1-1	8.914×10^{8}	3.981×10^{-3}
			164.75	1-0	1.439×10^{9}	1.952×10^{-3}
		^3S-^1P	30.438	1-1	7.072×10^{10}	9.823×10^{-3}
		^1S-^1P	37.610	0-1	9.284×10^{10}	1.969×10^{-2}
52	$1s^2$-$1s2p$	^1S-^3P	.44420	0-1	2.544×10^{15}	2.258×10^{-1}
		^1S-^1P	.43850	0-1	5.859×10^{15}	5.067×10^{-1}
	$1s2s$-$1s2p$	^3S-^3P	30.810	1-2	2.256×10^{11}	5.350×10^{-2}
			168.52	1-1	9.120×10^{8}	3.883×10^{-3}
			159.40	1-0	1.519×10^{9}	1.929×10^{-3}
		^3S-^1P	28.394	1-1	8.444×10^{10}	1.021×10^{-2}
		^1S-^1P	34.771	0-1	1.122×10^{11}	2.034×10^{-2}
53	$1s^2$-$1s2p$	^1S-^3P	.42685	0-1	2.776×10^{15}	2.275×10^{-1}
		^1S-^1P	.42116	0-1	6.295×10^{15}	5.022×10^{-1}
	$1s2s$-$1s2p$	^3S-^3P	28.645	1-2	2.694×10^{11}	5.524×10^{-2}
			164.61	1-1	9.329×10^{8}	3.789×10^{-3}
			154.26	1-0	1.605×10^{9}	1.908×10^{-3}
		^3S-^1P	26.503	1-1	1.006×10^{11}	1.060×10^{-2}
		^1S-^1P	32.182	0-1	1.353×10^{11}	2.101×10^{-2}
54	$1s^2$-$1s2p$	^1S-^3P	.41046	0-1	3.024×10^{15}	2.291×10^{-1}
		^1S-^1P	.40478	0-1	6.754×10^{15}	4.977×10^{-1}
	$1s2s$-$1s2p$	^3S-^3P	26.655	1-2	3.212×10^{11}	5.702×10^{-2}
			160.84	1-1	9.540×10^{8}	3.700×10^{-3}
			149.31	1-0	1.695×10^{9}	1.888×10^{-3}
		^3S-^1P	24.752	1-1	1.197×10^{11}	1.100×10^{-2}
		^1S-^1P	29.817	0-1	1.628×10^{11}	2.170×10^{-2}
55	$1s^2$-$1s2p$	^1S-^3P	.39497	0-1	3.286×10^{15}	2.305×10^{-1}
		^1S-^1P	.38930	0-1	7.239×10^{15}	4.934×10^{-1}
	$1s2s$-$1s2p$	^3S-^3P	24.823	1-2	3.823×10^{11}	5.886×10^{-2}
			157.22	1-1	9.753×10^{8}	3.614×10^{-3}
			144.54	1-0	1.790×10^{9}	1.869×10^{-3}
		^3S-^1P	23.131	1-1	1.421×10^{11}	1.140×10^{-2}
		^1S-^1P	27.655	0-1	1.954×10^{11}	2.241×10^{-2}
56	$1s^2$-$1s2p$	$1S$-$3P$	.38030	0-1	3.564×10^{15}	2.318×10^{-1}
		^1S-^1P	.37464	0-1	7.749×10^{15}	4.891×10^{-1}
	$1s2s$-$1s2p$	^3S-^3P	23.135	1-2	4.542×10^{11}	6.074×10^{-2}
			153.73	1-1	9.969×10^{8}	3.532×10^{-3}
			139.96	1-0	1.891×10^{9}	1.851×10^{-3}
		^3S-^1P	21.630	1-1	1.685×10^{11}	1.182×10^{-2}
		^1S-^1P	25.676	0-1	2.341×10^{11}	2.314×10^{-2}

Table A.5. Continued

Z	Transition	Multiplet	λ(Å)	J_i - J_k	A_{ki} (s^{-1})	f_{ik}
57	$1s^2$-$1s2p$	^1S-^3P	.36640	0-1	3.858×10^{15}	2.330×10^{-1}
		^1S-^1P	.36074	0-1	8.285×10^{15}	4.849×10^{-1}
	$1s2s$-$1s2p$	^3S-^3P	21.580	1-2	5.386×10^{11}	6.267×10^{-2}
			150.37	1-1	1.019×10^{9}	3.453×10^{-3}
			135.53	1-0	1.997×10^{9}	1.834×10^{-3}
		^3S-^1P	20.238	1-1	1.993×10^{11}	1.224×10^{-2}
		^1S-^1P	23.862	0-1	2.798×10^{11}	2.388×10^{-2}
58	$1s^2$-$1s2p$	^1S-^3P	.35322	0-1	4.170×10^{15}	2.340×10^{-1}
		^1S-^1P	.34757	0-1	8.849×10^{15}	4.808×10^{-1}
	$1s2s$-$1s2p$	^3S-^3P	20.144	1-2	6.376×10^{11}	6.465×10^{-2}
			147.14	1-1	$1.041\times{}^{1}0^{9}$	3.377×10^{-3}
			131.28	1-0	2.109×10^{9}	1.817×10^{-3}
		^3S-^1P	18.947	1-1	2.354×10^{11}	1.267×10^{-2}
		^1S-^1P	22.197	0-1	3.337×10^{11}	2.465×10^{-2}
59	$1s^2$-$1s2p$	^1S-^3P	.34070	0-1	4.498×10^{15}	2.348×10^{-1}
		^1S-^1P	.33506	0-1	9.441×10^{15}	4.767×10^{-1}
	$1s2s$-$1s2p$	^3S-^3P	18.819	1-2	7.535×10^{11}	6.667×10^{-2}
			144.00	1-1	1.063×10^{9}	3.304×10^{-3}
			127.18	1-0	2.228×10^{9}	1.801×10^{-3}
		^3S-^1P	17.748	1-1	2.775×10^{11}	1.311×10^{-2}
		^1S-^1P	20.668	0-1	3.972×10^{11}	2.544×10^{-2}
60	$1s^2$-$1s2p$	^1S-^3P	.32881	0-1	4.845×10^{15}	2.356×10^{-1}
		^1S-^1P	.32318	0-1	1.006×10^{16}	4.727×10^{-1}
	$1s2s$-$1s2p$	^3S-^3P	17.593	1-2	8.889×10^{11}	6.875×10^{-2}
			141.00	1-1	1.085×10^{9}	3.234×10^{-3}
			123.23	1-0	2.353×10^{9}	1.786×10^{-3}
		^3S-^1P	16.636	1-1	3.267×10^{11}	1.355×10^{-2}
		^1S-^1P	19.262	0-1	4.718×10^{11}	2.624×10^{-2}
61	$1s^2$-$1s2p$	^1S-^3P	.31751	0-1	5.211×10^{15}	2.363×10^{-1}
		^1S-^1P	.31188	0-1	1.071×10^{16}	4.687×10^{-1}
	$1s2s$-$1s2p$	^3S-^3P	16.461	1-2	1.047×10^{12}	7.086×10^{-2}
			138.17	1-1	1.105×10^{9}	3.164×10^{-3}
			119.48	1-0	2.481×10^{9}	1.770×10^{-3}
		^3S-^1P	15.603	1-1	3.838×10^{11}	1.401×10^{-2}
		^1S-^1P	17.969	0-1	5.591×10^{11}	2.707×10^{-2}
62	$1s^2$-$1s2p$	^1S-^3P	.30674	0-1	5.596×10^{15}	2.368×10^{-1}
		^1S-^1P	.30112	0-1	1.139×10^{16}	4.647×10^{-1}
	$1s2s$-$1s2p$	^3S-^3P	15.412	1-2	1.231×10^{12}	7.303×10^{-2}
			135.31	1-1	1.129×10^{9}	3.099×10^{-3}
			115.76	1-0	2.623×10^{9}	1.756×10^{-3}
		^3S-^1P	14.642	1-1	4.503×10^{11}	1.447×10^{-2}
		^1S-^1P	16.776	0-1	6.616×10^{11}	2.791×10^{-2}

Table A.5. Continued

Z	Transition	Multiplet	λ(Å)	J_i - J_k	A_{ki} (s^{-1})	f_{ik}
63	$1s^2$-$1s2p$	^1S-^3P	.29649	0-1	6.001×10^{15}	2.373×10^{-1}
		^1S-^1P	.29087	0-1	1.211×10^{16}	4.607×10^{-1}
	$1s2s$-$1s2p$	^3S-^3P	14.438	1-2	1.445×10^{12}	7.525×10^{-2}
			132.63	1-1	1.151×10^{9}	3.035×10^{-3}
			112.24	1-0	2.768×10^{9}	1.743×10^{-3}
		^3S-^1P	13.748	1-1	5.274×10^{11}	1.495×10^{-2}
		^1S-^1P	15.675	0-1	7.812×10^{11}	2.878×10^{-2}
64	$1s^2$-$1s2p$	^1S-^3P	.28671	0-1	6.428×10^{15}	2.376×10^{-1}
		^1S-^1P	.28109	0-1	1.285×10^{16}	4.568×10^{-1}
	$1s2s$-$1s2p$	^3S-^3P	13.536	1-2	1.693×10^{12}	7.752×10^{-2}
			129.97	1-1	1.174×10^{9}	2.974×10^{-3}
			108.79	1-0	2.925×10^{9}	1.730×10^{-3}
		^3S-^1P	12.916	1-1	6.169×10^{11}	1.543×10^{-2}
		^1S-^1P	14.658	0-1	9.208×10^{11}	2.966×10^{-2}
65	$1s^2$-$1s2p$	^1S-^3P	.27739	0-1	6.875×10^{15}	2.379×10^{-1}
		^1S-^1P	.27177	0-1	1.363×10^{16}	4.529×10^{-1}
	$1s2s$-$1s2p$	^3S-^3P	12.698	1-2	1.982×10^{12}	7.984×10^{-2}
			127.40	1-1	1.198×10^{9}	2.914×10^{-3}
			105.47	1-0	3.090×10^{9}	1.718×10^{-3}
		^3S-^1P	12.140	1-1	7.204×10^{11}	1.592×10^{-2}
		^1S-^1P	13.718	0-1	1.083×10^{12}	3.057×10^{-2}
66	$1s^2$-$1s2p$	^1S-^3P	.26848	0-1	7.345×10^{15}	2.381×10^{-1}
		^1S-^1P	.26287	0-1	1.445×10^{16}	4.489×10^{-1}
	$1s2s$-$1s2p$	^3S-^3P	11.920	1-2	2.316×10^{12}	8.221×10^{-2}
			124.93	1-1	1.221×10^{9}	2.857×10^{-3}
			102.27	1-0	3.263×10^{9}	1.706×10^{-3}
		^3S-^1P	11.418	1-1	8.401×10^{11}	1.642×10^{-2}
		^1S-^1P	12.848	0-1	1.273×10^{12}	3.149×10^{-2}
67	$1s^2$-$1s2p$	^1S-^3P	.25998	0-1	7.838×10^{15}	2.383×10^{-1}
		^1S-^1P	.25436	0-1	1.529×10^{16}	4.450×10^{-1}
	$1s2s$-$1s2p$	^3S-^3P	11.197	1-2	2.702×10^{12}	8.463×10^{-2}
			122.54	1-1	1.244×10^{9}	2.800×10^{-3}
			99.174	1-0	3.446×10^{9}	1.694×10^{-3}
		^3S-^1P	10.744	1-1	9.782×10^{11}	1.693×10^{-2}
		^1S-^1P	12.042	0-1	1.492×10^{12}	3.244×10^{-2}
68	$1s^2$-$1s2p$	^1S-^3P	.25184	0-1	8.354×10^{15}	2.383×10^{-1}
		^1S-^1P	.24623	0-1	1.618×10^{16}	4.411×10^{-1}
	$1s2s$-$1s2p$	^3S-^3P	10.524	1-2	3.147×10^{12}	8.710×10^{-2}
			120.24	1-1	1.267×10^{9}	2.745×10^{-3}
			96.200	1-0	3.638×10^{9}	1.683×10^{-3}
		^3S-^1P	10.115	1-1	1.137×10^{12}	1.745×10^{-2}
		^1S-^1P	11.295	0-1	1.747×10^{12}	3.340×10^{-2}

Table A.5. Continued

Z	Transition	Multiplet	λ(Å)	J_i - J_k	A_{ki} (s^{-1})	f_{ik}
69	$1s^2$-$1s2p$	^1S-^3P	.24406	0-1	8.895×10^{15}	2.383×10^{-1}
		^1S-^1P	.23845	0-1	1.710×10^{16}	4.372×10^{-1}
	$1s2s$-$1s2p$	^3S-^3P	9.8976	1-2	3.661×10^{12}	8.962×10^{-2}
			118.02	1-1	1.289×10^{9}	2.691×10^{-3}
			93.333	1-0	3.839×10^{9}	1.671×10^{-3}
		^3S-^1P	9.5283	1-1	1.321×10^{12}	1.797×10^{-2}
		^1S-^1P	10.602	0-1	2.041×10^{12}	3.439×10^{-2}
70	$1s^2$-$1s2p$	^1S-^3P	.23661	0-1	9.460×10^{15}	2.382×10^{-1}
		^1S-^1P	.23099	0-1	1.805×10^{16}	4.333×10^{-1}
	$1s2s$-$1s2p$	^3S-^3P	9.3141	1-2	4.253×10^{12}	9.219×10^{-2}
			115.90	1-1	1.310×10^{9}	2.638×10^{-3}
			90.582	1-0	4.049×10^{9}	1.660×10^{-3}
		^3S-^1P	8.9801	1-1	1.531×10^{12}	1.851×10^{-2}
		^1S-^1P	9.9578	0-1	2.381×10^{12}	3.539×10^{-2}
71	$1s^2$-$1s2p$	^1S-^3P	.22947	0-1	1.005×10^{16}	2.381×10^{-1}
		^1S-^1P	.22385	0-1	1.905×10^{16}	4.294×10^{-1}
	$1s2s$-$1s2p$	^3S-^3P	8.7696	1-2	4.934×10^{12}	9.481×10^{-2}
			113.81	1-1	1.332×10^{9}	2.587×10^{-3}
			87.892	1-0	4.273×10^{9}	1.650×10^{-3}
		^3S-^1P	8.4673	1-1	1.773×10^{12}	1.906×10^{-2}
		^1S-^1P	9.3588	0-1	2.774×10^{12}	3.642×10^{-2}
72	$1s^2$-$1s2p$	^1S-^3P	.22263	0-1	1.067×10^{16}	2.379×10^{-1}
		^1S-^1P	.21701	0-1	2.009×10^{16}	4.254×10^{-1}
	$1s2s$-$1s2p$	^3S-^3P	8.2617	1-2	5.716×10^{12}	9.749×10^{-2}
			111.80	1-1	1.354×10^{9}	2.536×10^{-3}
			85.308	1-0	4.507×10^{9}	1.639×10^{-3}
		^3S-^1P	7.9878	1-1	2.051×10^{12}	1.961×10^{-2}
		^1S-^1P	8.8015	0-1	3.226×10^{12}	3.747×10^{-2}
73	$1s^2$-$1s2p$	^1S-^3P	.21606	0-1	1.132×10^{16}	2.376×10^{-1}
		^1S-^1P	.21044	0-1	2.116×10^{16}	4.215×10^{-1}
	$1s2s$-$1s2p$	^3S-^3P	7.7872	1-2	6.614×10^{12}	1.002×10^{-1}
			109.86	1-1	1.375×10^{9}	2.487×10^{-3}
			82.808	1-0	4.754×10^{9}	1.629×10^{-3}
		^3S-^1P	7.5389	1-1	2.369×10^{12}	2.018×10^{-2}
		^1S-^1P	8.2825	0-1	3.747×10^{12}	3.854×10^{-2}
74	$1s^2$-$1s2p$	^1S-^3P	.20976	0-1	1.199×10^{16}	2.373×10^{-1}
		^1S-^1P	.20414	0-1	2.227×10^{16}	4.175×10^{-1}
	$1s2s$-$1s2p$	^3S-^3P	7.3439	1-2	7.642×10^{12}	1.030×10^{-1}
			107.99	1-1	1.395×10^{9}	2.438×10^{-3}
			80.399	1-0	5.011×10^{9}	1.619×10^{-3}
		^3S-^1P	7.1185	1-1	2.732×10^{12}	2.076×10^{-2}
		^1S-^1P	7.7988	0-1	4.346×10^{12}	3.963×10^{-2}

Table A.5. Continued

Z	Transition	Multiplet	λ(Å)	J_i - J_k	A_{ki} (s^{-1})	f_{ik}
75	$1s^2$-$1s2p$	^1S-^3P	.20371	0-1	1.269×10^{16}	2.369×10^{-1}
		^1S-^1P	.19809	0-1	2.343×10^{16}	4.135×10^{-1}
	$1s2s$-$1s2p$	^3S-^3P	6.9291	1-2	8.821×10^{12}	1.058×10^{-1}
			106.15	1-1	1.416×10^{9}	2.391×10^{-3}
			78.047	1-0	5.287×10^{9}	1.609×10^{-3}
		^3S-^1P	6.7243	1-1	3.149×10^{12}	2.134×10^{-2}
		^1S-^1P	7.3472	0-1	5.034×10^{12}	4.074×10^{-2}
76	$1s^2$-$1s2p$	^1S-^3P	.19789	0-1	1.343×10^{16}	2.365×10^{-1}
		^1S-^1P	.19227	0-1	2.463×10^{16}	4.095×10^{-1}
	$1s2s$-$1s2p$	^3S-^3P	6.5410	1-2	1.017×10^{13}	1.087×10^{-1}
			104.38	1-1	1.435×10^{9}	2.345×10^{-3}
			75.783	1-0	5.574×10^{9}	1.600×10^{-3}
		^3S-^1P	6.3548	1-1	3.624×10^{12}	2.194×10^{-2}
		^1S-^1P	6.9258	0-1	5.823×10^{12}	4.187×10^{-2}
77	$1s^2$-$1s2p$	^1S-^3P	.19230	0-1	1.419×10^{16}	2.360×10^{-1}
		^1S-^1P	.18667	0-1	2.587×10^{16}	4.055×10^{-1}
	$1s2s$-$1s2p$	^3S-^3P	6.1775	1-2	1.171×10^{13}	1.116×10^{-1}
			102.68	1-1	1.454×10^{9}	2.299×10^{-3}
			73.596	1-0	5.875×10^{9}	1.590×10^{-3}
		^3S-^1P	6.0082	1-1	4.167×10^{12}	2.255×10^{-2}
		^1S-^1P	6.5320	1-1	6.727×10^{12}	4.303×10^{-2}
78	$1s^2$-$1s2p$	^1S-^3P	.18692	0-1	1.499×10^{16}	2.355×10^{-1}
		^1S-^1P	.18129	0-1	2.715×10^{16}	4.014×10^{-1}
	$1s2s$-$1s2p$	^3S-^3P	5.8370	1-2	1.347×10^{13}	1.146×10^{-1}
			101.06	1-1	1.471×10^{9}	2.253×10^{-3}
			71.494	1-0	6.187×10^{9}	1.580×10^{-3}
		^3S-^1P	5.6828	1-1	4.785×10^{12}	2.317×10^{-2}
		^1S-^1P	6.1639	0-1	7.761×10^{12}	4.420×10^{-2}
79	$1s^2$-$1s2p$	^1S-^1P	.18174	0-1	1.581×10^{16}	2.349×10^{-1}
		^1S-^1P	.17611	0-1	2.848×10^{16}	3.973×10^{-1}
	$1s2s$-$1s2p$	^3S-^3P	5.5177	1-2	1.547×10^{13}	1.177×10^{-1}
			99.493	1-1	1.488×10^{9}	2.208×10^{-3}
			69.458	1-0	6.515×10^{9}	1.571×10^{-3}
		^3S-^1P	5.3772	1-1	5.489×10^{12}	2.379×10^{-2}
		^1S-^1P	5.8194	0-1	8.943×10^{12}	4.540×10^{-2}
80	$1s^2$-$1s2p$	^1S-^3P	.17676	0-1	1.668×10^{16}	2.343×10^{-1}
		^1S-^1P	.17112	0-1	2.986×10^{16}	3.932×10^{-1}
	$1s2s$-$1s2p$	^3S-^3P	5.2182	1-2	1.775×10^{13}	1.208×10^{-1}
			28.029	1-1	1.501×10^{9}	2.162×10^{-3}
			67.511	1-0	6.852×10^{9}	1.561×10^{-3}
		^3S-^1P	5.0901	1-1	6.290×10^{12}	2.443×10^{-2}
		^1S-^1P	5.4970	0-1	1.029×10^{13}	4.662×10^{-2}

Table A.5. Continued

Z	Transition	Multiplet	λ(Å)	J_i - J_k	A_{ki} (s^{-1})	f_{ik}
81	$1s^2$-$1s2p$	^1S-^3P	.17195	0-1	1.757×10^{16}	2.337×10^{-1}
		^1S-^1P	.16632	0-1	3.128×10^{16}	3.891×10^{-1}
	$1s2s$-$1s2p$	^3S-^3P	4.9368	1-2	2.035×10^{13}	1.239×10^{-1}
			96.554	1-1	1.516×10^{9}	2.119×10^{-3}
			65.597	1-0	7.215×10^{9}	1.551×10^{-3}
		^3S-^1P	4.8199	1-1	7.202×10^{12}	2.508×10^{-2}
		^1S-^1P	5.1946	0-1	1.183×10^{13}	4.787×10^{-2}
82	$1s^2$-$1s2p$	^1S-^3P	.16732	0-1	1.850×10^{16}	2.330×10^{-1}
		^1S-^1P	.16168	0-1	3.274×10^{16}	3.849×10^{-1}
	$1s2s$-$1s2p$	^3S-^3P	4.6715	1-2	2.333×10^{13}	1.272×10^{-1}
			94.713	1-1	1.550×10^{9}	2.085×10^{-3}
			63.553	1-0	7.663×10^{9}	1.547×10^{-3}
		^3S-^1P	4.5647	1-1	8.242×10^{12}	2.575×10^{-2}
		^1S-^1P	4.9113	0-1	1.359×10^{13}	4.913×10^{-2}
83	$1s^2$-$1s2p$	^1S-^3P	.16286	0-1	1.947×10^{16}	2.323×10^{-1}
		^1S-^1P	.15721	0-1	3.425×10^{16}	3.807×10^{-1}
	$1s2s$-$1s2p$	^3S-^3P	4.4243	1-2	2.667×10^{13}	1.304×10^{-1}
			93.893	1-1	1.536×10^{9}	2.030×10^{-3}
			62.005	1-0	7.971×10^{9}	1.531×10^{-3}
		^3S-^1P	4.3267	1-1	9.411×10^{12}	2.641×10^{-2}
		^1S-^1P	4.6453	0-1	1.559×10^{13}	5.042×10^{-2}
84	$1s^2$-$1s2p$	^1S-^3P	.15855	0-1	2.047×10^{16}	2.315×10^{-1}
		^1S-^1P	.15290	0-1	3.581×10^{16}	3.765×10^{-1}
	$1s2s$-$1s2p$	^3S-^3P	4.1908	1-2	3.048×10^{13}	1.338×10^{-1}
			92.702	1-1	1.541×10^{19}	1.985×10^{-3}
			60.318	1-0	8.365×10^{9}	1.521×10^{-3}
		^3S-^1P	4.1016	1-1	1.074×10^{13}	2.709×10^{-2}
		^1S-^1P	4.3956	0-1	1.786×10^{13}	5.173×10^{-2}
85	$1s^2$-$1s2p$	^1S-^3P	.15439	0-1	2.151×10^{16}	2.307×10^{-1}
		^1S-^1P	.14874	0-1	3.741×10^{16}	3.723×10^{-1}
	$1s2s$-$1s2p$	^3S-^3P	3.9711	1-2	3.480×10^{13}	1.371×10^{-1}
			91.597	1-1	1.542×10^{9}	1.939×10^{-3}
			58.698	1-0	8.769×10^{9}	1.510×10^{-3}
		^3S-^1P	3.8894	1-1	1.225×10^{13}	2.778×10^{-2}
		^1S-^1P	4.1612	0-1	2.044×10^{13}	5.306×10^{-2}
86	$1s^2$-$1s2p$	^1S-^3P	.15038	0-1	2.259×10^{16}	2.298×10^{-1}
		^1S-^1P	.14472	0-1	3.907×10^{16}	3.680×10^{-1}
	$1s2s$-$1s2p$	^3S-^3P	3.7644	1-2	3.970×10^{13}	1.406×10^{-1}
			90.664	1-1	1.535×10^{9}	1.892×10^{-3}
			57.175	1-0	9.168×10^{9}	1.498×10^{-3}
		^3S-^1P	3.6896	1-1	1.396×10^{13}	2.849×10^{-2}
		^1S-^1P	3.9409	0-1	2.337×10^{13}	5.441×10^{-2}

Table A.5. Continued

Z	Transition	Multiplet	λ(Å)	J_i - J_k	A_{ki} (s^{-1})	f_{ik}
87	$1s^2$-$1s2p$	^1S-^3P	.14649	0-1	2.372×10^{16}	2.289×10^{-1}
		^1S-^1P	.14083	0-1	4.077×10^{16}	3.637×10^{-1}
	$1s2s$-$1s2p$	^3S-^3P	3.5693	1-2	4.525×10^{13}	1.441×10^{-1}
			89.657	1-1	1.533×10^{9}	1.847×10^{-3}
			55.649	1-0	9.607×10^{9}	1.487×10^{-3}
		^3S-^1P	3.5008	1-1	1.589×10^{13}	2.920×10^{-2}
		^1S-^1P	3.7334	0-1	2.670×10^{13}	5.579×10^{-2}
88	$1s^2$-$1s2p$	^1S-^1P	.14274	0-1	2.488×10^{16}	2.280×10^{-1}
		^1S-^1P	.13707	0-1	4.252×10^{16}	3.593×10^{-1}
	$1s2s$-$1s2p$	^3S-^3P	3.3856	1-2	5.154×10^{13}	1.476×10^{-1}
			88.824	1-1	1.522×10^{9}	1.801×10^{-3}
			52.214	1-0	1.004×10^{10}	1.474×10^{-3}
		^3S-^1P	3.3229	1-1	1.808×10^{13}	2.992×10^{-2}
		^1S-^1P	3.5382	0-1	3.047×10^{13}	5.719×10^{-2}
89	$1s^2$-$1s2p$	^1S-^3P	.13911	0-1	2.608×10^{16}	2.270×10^{-1}
		^1S-^1P	.13344	0-1	4.432×10^{16}	3.549×10^{-1}
	$1s2s$-$1s2p$	^3S-^3P	3.2124	1-2	5.864×10^{13}	1.512×10^{-1}
			88.089	1-1	1.507×10^{9}	1.753×10^{-3}
			52.837	1-0	1.048×10^{10}	1.462×10^{-3}
		^3S-^1P	3.1548	1-1	2.055×10^{13}	3.066×10^{-2}
		^1S-^1P	3.3544	0-1	3.475×10^{13}	5.862×10^{-2}
90	$1s^2$-$1s2p$	^1S-^3P	.13560	0-1	2.732×10^{16}	2.260×10^{-1}
		^1S-^1P	.12993	0-1	4.616×10^{16}	3.505×10^{-1}
	$1s2s$-$1s2p$	^3S-^3P	3.0489	1-2	6.668×10^{13}	1.549×10^{-1}
			87.437	1-1	1.488×10^{9}	1.706×10^{-3}
			51.510	1-0	1.093×10^{10}	1.449×10^{-3}
		^3S-^1P	2.9961	1-1	2.334×10^{13}	3.141×10^{-2}
		^1S-^1P	3.1812	0-1	3.959×10^{13}	6.006×10^{-2}
91	$1s^2$-$1s2p$	^1S-^3P	.13221	0-1	2.861×10^{16}	2.249×10^{-1}
		^1S-^1P	.12652	0-1	4.806×10^{16}	3.460×10^{-1}
	$1s2s$-$1s2p$	^3S-^3P	2.8946	1-2	7.575×10^{13}	1.586×10^{-1}
			86.947	1-1	1.462×10^{9}	1.656×10^{-3}
			50.253	1-0	1.137×10^{10}	1.435×10^{-3}
		^3S-^1P	2.8461	1-1	2.649×10^{13}	3.216×10^{-2}
		^1S-^1P	3.0179	0-1	4.507×10^{13}	6.153×10^{-2}
92	$1s^2$-$1s2p$	^1S-^3P	.12892	0-1	2.994×10^{16}	2.238×10^{-1}
		^1S-^1P	.12323	0-1	5.001×10^{16}	3.415×10^{-1}
	$1s2s$-$1s2p$	^3S-^3P	2.7488	1-2	8.600×10^{13}	1.624×10^{-1}
			86.565	1-1	1.430×10^{9}	1.607×10^{-3}
			49.047	1-0	1.181×10^{10}	1.420×10^{-3}
		^3S-^1P	2.7043	1-1	3.004×10^{13}	3.293×10^{-2}
		^1S-^1P	2.8638	0-1	5.126×10^{13}	6.303×10^{-2}

Table A.5. Continued

Z	Transition	Multiplet	λ(Å)	J_i - J_k	A_{ki} (s^{-1})	f_{ik}
93	$1s^2$-$1s2p$	^1S-^3P	.12573	0-1	3.132×10^{16}	2.227×10^{-1}
		^1S-^1P	.12004	0-1	5.200×10^{16}	3.370×10^{-1}
	$1s2s$-$1s2p$	^3S-^3P	2.6111	1-2	9.757×10^{13}	1.662×10^{-1}
			86.331	1-1	1.392×10^{9}	1.556×10^{-3}
			47.899	1-0	1.225×10^{10}	1.405×10^{-3}
		^3S-^1P	2.5702	1-1	3.404×10^{13}	3.371×10^{-2}
		^1S-^1P	2.7184	0-1	5.826×10^{13}	6.455×10^{-2}
94	$1s^2$-$1s2p$	^1S-^3P	.12265	0-1	3.274×10^{16}	2.215×10^{-1}
		^1S-^1P	.11694	0-1	5.405×10^{16}	3.324×10^{-1}
	$1s2s$-$1s2p$	^3S-^3P	2.4809	1-2	1.106×10^{14}	1.701×10^{-1}
			86.327	1-1	1.344×10^{9}	1.502×10^{-3}
			46.827	1-0	1.267×10^{10}	1.388×10^{-3}
		^3S-^1P	2.4433	1-1	3.855×10^{13}	3.450×10^{-2}
		^1S-^1P	2.5812	0-1	6.616×10^{13}	6.609×10^{-2}
95	$1s^2$-$1s2p$	^1S-^3P	.11965	0-1	3.422×10^{16}	2.203×10^{-1}
		^1S-^1P	.11394	0-1	5.614×10^{16}	3.278×10^{-1}
	$1s2s$-$1s2p$	^3S-^3P	2.3577	1-2	1.253×10^{14}	1.741×10^{-1}
			86.407	1-1	1.294×10^{9}	1.449×10^{-3}
			45.783	1-0	1.309×10^{10}	1.371×10^{-3}
		^3S-^1P	2.3231	1-1	4.364×10^{13}	3.531×10^{-2}
		^1S-^1P	2.4514	0-1	7.509×10^{13}	6.765×10^{-2}
96	$1s^2$-$1s2p$	^1S-^3P	.11675	0-1	3.573×10^{16}	2.191×10^{-1}
		^1S-^1P	.11103	0-1	5.828×10^{16}	3.232×10^{-1}
	$1s2s$-$1s2p$	^3S-^3P	2.2411	1-2	1.419×10^{14}	1.781×10^{-1}
			86.782	1-1	1.233×10^{9}	1.393×10^{-3}
			44.819	1-0	1.348×10^{10}	1.353×10^{-3}
		^3S-^1P	2.2093	1-1	4.936×10^{13}	3.612×10^{-2}
		^1S-^1P	2.3288	0-1	8.516×10^{13}	6.924×10^{-2}
97	$1s^2$-$1s2p$	^1S-^1P	.11393	0-1	3.730×10^{16}	2.178×10^{-1}
		^1S-^1P	.10821	0-1	6.047×10^{16}	3.185×10^{-1}
	$1s2s$-$1s2p$	^3S-^3P	2.1308	1-2	1.605×10^{14}	1.821×10^{-1}
			87.441	1-1	1.164×10^{9}	1.334×10^{-3}
			43.923	1-0	1.383×10^{10}	1.333×10^{-3}
		^3S-^1P	2.1016	1-1	5.580×10^{13}	3.694×10^{-2}
		^1S-^1P	2.2129	0-1	9.651×10^{13}	7.085×10^{-2}
98	$1s^2$-$1s2p$	^1S-^3P	.11121	0-1	3.892×10^{16}	2.165×10^{-1}
		^1S-^1P	.10547	0-1	6.271×10^{16}	3.137×10^{-1}
	$1s2s$-$1s2p$	^3S-^3P	2.0263	1-2	1.815×10^{14}	1.863×10^{-1}
			88.394	1-1	1.087×10^{9}	1.274×10^{-3}
			43.082	1-0	1.415×10^{10}	1.312×10^{-3}
		^3S-^1P	1.9994	1-1	6.304×10^{13}	3.778×10^{-2}
		^1S-^1P	2.1032	0-1	1.093×10^{14}	7.248×10^{-2}

Table A.5. Continued

Z	Transition	Multiplet	λ(Å)	J_i - J_k	A_{ki} (s^{-1})	f_{ik}
99	$1s^2$-$1s2p$	^1S-^3P	.10855	0-1	4.059×10^{16}	2.151×10^{-1}
		^1S-^1P	.10281	0-1	6.500×10^{16}	3.090×10^{-1}
	$1s2s$-$1s2p$	^3S-^3P	1.9274	1-2	2.052×10^{14}	1.904×10^{-1}
			89.866	1-1	9.986×10^{8}	1.209×10^{-3}
			42.345	1-0	1.438×10^{10}	1.289×10^{-3}
		^3S-^1P	1.9026	1-1	7.117×10^{13}	3.862×10^{-2}
		^1S-^1P	1.9994	0-1	1.237×10^{14}	7.414×10^{-2}
100	$1s^2$-$1s2p$	^1S-^3P	.10597	0-1	4.231×10^{16}	2.137×10^{-1}
		^1S-^1P	.10022	0-1	6.733×10^{16}	3.042×10^{-1}
	$1s2s$-$1s2p$	^3S-^3P	1.8336	1-2	2.317×10^{14}	1.947×10^{-1}
			91.813	1-1	9.035×10^{8}	1.142×10^{-3}
			41.673	1-0	1.457×10^{10}	1.264×10^{-3}
		^3S-^1P	1.8108	1-1	8.031×10^{13}	3.948×10^{-2}
		^1S-^1P	1.9011	0-1	1.399×10^{14}	7.582×10^{-2}

Table A.6. Wavelengths λ and transition probabilities A_{ki} for $1s^2$-$1s2s$ M1 and $1s^2$-$1s2p$ M2 transitions in He-like ions, from [A.4]

Z	Transition	Multiplet	λ(Å)	J_i - J_k	A_{ki} (s^{-1})	Type
2	$1s^2$-$1s2s$	^1S-^3S	625.48	0-1	1.266×10^{-4}	M1
	$1s^2$-$1s2p$	^1S-^3P	591.33	0-2	3.271×10^{-1}	M2
3	$1s^2$-$1s2s$	^1S-^3S	210.05	0-1	2.035×10^{-2}	M1
	$1s^2$-$1s2p$	^1S-^3P	202.30	0-2	3.498×10^{1}	M2
4	$1s^2$-$1s2s$	^1S-^3S	104.54	0-1	5.615×10^{-1}	M1
	$1s^2$-$1s2p$	^1S-^3P	101.68	0-2	6.171×10^{2}	M2
5	$1s^2$-$1s2s$	^1S-^3S	62.436	0-1	6.696	M1
	$1s^2$-$1s2p$	^1S-^3P	61.085	0-2	5.014×10^{-3}	M2
6	$1s^2$-$1s2s$	^1S-^3S	41.470	0-1	4.860×10^{1}	M1
	$1s^2$-$1s2p$	^1S-^3P	40.726	0-2	2.622×10^{4}	M2
7	$1s^2$-$1s2s$	^1S-^3S	29.533	0-1	2.537×10^{2}	M1
	$1s^2$-$1s2p$	^1S-^3P	29.081	0-2	1.030×10^{5}	M2
8	$1s^2$-$1s2s$	^1S-^3S	22.097	0-1	1.047×10^{3}	M1
	$1s^2$-$1s2p$	^1S-^3P	21.800	0-2	3.308×10^{5}	M2
9	$1s^2$-$1s2s$	^1S-^3S	17.152	0-1	3.621×10^{3}	M1
	$1s^2$-$1s2p$	^1S-^3P	16.947	0-2	9.155×10^{5}	M2
10	$1s^2$-$1s2s$	^1S-^3S	13.699	0-1	1.092×10^{4}	M1
	$1s^2$-$1s2p$	^1S-^3P	13.550	0-2	2.257×10^{6}	M2
11	$1s^2$-$1s2s$	^1S-^3S	11.191	0-1	2.952×10^{4}	M1
	$1s^2$-$1s2p$	^1S-^3P	11.080	0-2	5.079×10^{6}	M2
12	$1s^2$-$1s2s$	^1S-^3S	9.3141	0-1	7.293×10^{4}	M1
	$1s^2$-$1s2p$	^1S-^3P	9.2280	0-2	1.060×10^{7}	M2
13	$1s^2$-$1s2s$	^1S-^3S	7.8720	0-1	1.672×10^{5}	M1
	$1s^2$-$1s2p$	^1S-^3P	7.8037	0-2	2.080×10^{7}	M2
14	$1s^2$-$1s2s$	^1S-^3S	6.7402	0-1	3.598×10^{5}	M1
	$1s^2$-$1s2p$	^1S-^3P	6.6849	0-2	3.873×10^{7}	M2
15	$1s^2$-$1s2s$	^1S-^3S	5.8357	0-1	7.333×10^{5}	M1
	$1s^2$-$1s2p$	^1S-^3P	5.7900	0-2	6.895×10^{7}	M2
16	$1s^2$-$1s2s$	^1S-^3S	5.1014	0-1	1.426×10^{6}	M1
	$1s^2$-$1s2p$	^1S-^3P	5.0631	0-2	1.181×10^{8}	M2
17	$1s^2$-$1s2s$	^1S-^3S	4.4972	0-1	2.661×10^{6}	M1
	$1s^2$-$1s2p$	^1S-^3P	4.4645	0-2	1.955×10^{8}	M2
18	$1s^2$-$1s2s$	^1S-^3S	3.9941	0-1	4.787×10^{6}	M1
	$1s^2$-$1s2p$	^1S-^3P	3.9658	0-2	3.141×10^{8}	M2
19	$1s^2$-$1s2s$	^1S-^3S	3.5707	0-1	8.341×10^{6}	M1
	$1s^2$-$1s2p$	^1S-^3P	3.5459	0-2	4.914×10^{8}	M2
20	$1s^2$-$1s2s$	^1S-^3S	3.2110	0-1	1.412×10^{7}	M1
	$1s^2$-$1s2p$	^1S-^3P	3.1891	0-2	7.510×10^{8}	M2

Table A.6. Continued

Z	Transition	Multiplet	λ(Å)	J_i - J_k	A_{ki} (s^{-1})	Type
21	$1s^2$-$1s2s$	^1S-^3S	2.9028	0-1	2.328×10^7	M1
	$1s^2$-$1s2p$	^1S-^3P	2.8832	0-2	1.123×10^9	M2
22	$1s^2$-$1s2s$	^1S-^3S	2.6368	0-1	3.750×10^7	M1
	$1s^2$-$1s2p$	^1S-^3P	2.6191	0-2	1.648×10^9	M2
23	$1s^2$-$1s2s$	^1S-^3S	2.4056	0-1	5.913×10^7	M1
	$1s^2$-$1s2p$	^1S-^3P	2.3895	0-2	2.377×10^9	M2
24	$1s^2$-$1s2s$	^1S-^3S	2.2034	0-1	9.143×10^7	M1
	$1s^2$-$1s2p$	^1S-^3P	2.1886	0-2	3.373×10^9	M2
25	$1s^2$-$1s2s$	^1S-^3S	2.0255	0-1	1.389×10^8	M1
	$1s^2$-$1s2p$	^1S-^3P	2.0118	0-2	4.718×10^9	M2
26	$1s^2$-$1s2s$	^1S-^3S	1.8682	0-1	2.075×10^8	M1
	$1s^2$-$1s2p$	^1S-^3P	1.8554	0-2	6.510×10^9	M2
27	$1s^2$-$1s2s$	^1S-^3S	1.7284	0-1	3.053×10^8	M1
	$1s^2$-$1s2p$	^1S-^3P	1.7164	0-2	8.873×10^9	M2
28	$1s^2$-$1s2s$	^1S-^3S	1.6036	0-1	4.431×10^8	M1
	$1s^2$-$1s2p$	^1S-^3P	1.5923	0-2	1.195×10^{10}	M2
29	$1s^2$-$1s2s$	^1S-^3S	1.4917	0-1	6.346×10^8	M1
	$1s^2$-$1s2p$	^1S-^3P	1.4811	0-2	1.594×10^{10}	M2
30	$1s^2$-$1s2s$	^1S-^3S	1.3911	0-1	8.981×10^8	M1
	$1s^2$-$1s2p$	^1S-^3P	1.3809	0-2	2.104×10^{10}	M2
31	$1s^2$-$1s2s$	^1S-^3S	1.3002	0-1	1.257×10^9	M1
	$1s^2$-$1s2p$	^1S-^3P	1.2905	0-2	2.751×10^{10}	M2
32	$1s^2$-$1s2s$	^1S-^3S	1.2179	0-1	1.740×10^9	M1
	$1s^2$-$1s2p$	^1S-^3P	1.2086	0-2	3.567×10^{10}	M2
33	$1s^2$-$1s2s$	^1S-^3S	1.1430	0-1	2.385×10^9	M1
	$1s^2$-$1s2p$	^1S-^3P	1.1341	0-2	4.587×10^{10}	M2
34	$1s^2$-$1s2s$	^1S-^3S	1.0748	0-1	3.239×10^9	M1
	$1s^2$-$1s2p$	^1S-^3P	1.0661	0-2	5.855×10^{10}	M2
35	$1s^2$-$1s2s$	^1S-^3S	1.0124	0-1	4.360×10^9	M1
	$1s^2$-$1s2p$	^1S-^3P	1.0040	0-2	7.420×10^{10}	M2
36	$1s^2$-$1s2s$	^1S-^3S	.95525	0-1	5.822×10^9	M1
	$1s^2$-$1s2p$	^1S-^3P	.94710	0-2	9.341×10^{10}	M2
37	$1s^2$-$1s2s$	^1S-^3S	.90272	0-1	7.714×10^9	M1
	$1s^2$-$1s2p$	^1S-^3P	.89479	0-2	1.168×10^{11}	M2
38	$1s^2$-$1s2s$	^1S-^3S	.85434	0-1	1.015×10^{10}	M1
	$1s^2$-$1s2p$	^1S-^3P	.84661	0-2	1.453×10^{11}	M2
39	$1s^2$-$1s2s$	^1S-^3S	.80969	0-1	1.325×10^{10}	M1
	$1s^2$-$1s2p$	^1S-^3P	.80214	0-2	1.796×10^{11}	M2
40	$1s^2$-$1s2s$	^1S-^3S	.76840	0-1	1.719×10^{10}	M1
	$1s^2$-$1s2p$	^1S-^3P	.76099	0-2	2.208×10^{11}	M2
41	$1s^2$-$1s2s$	^1S-^3S	.73012	0-1	2.217×10^{10}	M1
	$1s^2$-$1s2p$	^1S-^3P	.72286	0-2	2.701×10^{11}	M2

Table A.6. Continued

Z	Transition	Multiplet	λ(Å)	J_i - J_k	A_{ki} (s^{-1})	Type
42	$1s^2$-$1s2s$	^1S-^3S	.69459	0-1	2.842×10^{10}	M1
	$1s^2$-$1s2p$	^1S-^3P	.68745	0-2	3.289×10^{11}	M2
43	$1s^2$-$1s2s$	^1S-^3S	.66153	0-1	3.622×10^{10}	M1
	$1s^2$-$1s2p$	^1S-^3P	.65451	0-2	3.985×10^{11}	M2
44	$1s^2$-$1s2s$	^1S-^3S	.63073	0-1	4.592×10^{10}	M1
	$1s^2$-$1s2p$	^1S-^3P	.62382	0-2	4.808×10^{11}	M2
45	$1s^2$-$1s2s$	^1S-^3S	.60198	0-1	5.792×10^{10}	M1
	$1s^2$-$1s2p$	^1S-^3P	.59517	0-2	5.775×10^{11}	M2
46	$1s^2$-$1s2s$	^1S-^3S	.57511	0-1	7.270×10^{10}	M1
	$1s^2$-$1s2p$	^1S-^3P	.56838	0-2	6.910×10^{11}	M2
47	$1s^2$-$1s2s$	^1S-^3S	.54996	0-1	9.083×10^{10}	M1
	$1s^2$-$1s2p$	^1S-^3P	.54331	0-2	8.237×10^{11}	M2
48	$1s^2$-$1s2s$	^1S-^3S	.52637	0-1	1.130×10^{11}	M1
	$1s^2$-$1s2p$	^1S-^3P	.51979	0-2	9.781×10^{11}	M2
49	$1s^2$-$1s2s$	^1S-^3S	.50423	0-1	1.399×10^{11}	M1
	$1s^2$-$1s2p$	^1S-^3P	.49772	0-2	1.157×10^{12}	M2
50	$1s^2$-$1s2s$	^1S-^3S	.48342	0-1	1.726×10^{11}	M1
	$1s^2$-$1s2p$	^1S-^3P	.47697	0-2	1.365×10^{12}	M2
51	$1s^2$-$1s2s$	^1S-^3S	.46384	0-1	2.120×10^{11}	M1
	$1s^2$-$1s2p$	^1S-^3P	.45744	0-2	1.605×10^{12}	M2
52	$1s^2$-$1s2s$	^1S-^3S	.44538	0-1	2.595×10^{11}	M1
	$1s^2$-$1s2p$	^1S-^3P	.43903	0-2	1.881×10^{12}	M2
53	$1s^2$-$1s2s$	^1S-^3S	.42796	0-1	3.164×10^{11}	M1
	$1s^2$-$1s2p$	^1S-^3P	.42166	0-2	2.197×10^{12}	M2
54	$1s^2$-$1s2s$	^1S-^3S	.41151	0-1	3.846×10^{11}	M1
	$1s^2$-$1s2p$	^1S-^3P	.40526	0-2	2.560×10^{12}	M2
55	$1s^2$-$1s2s$	^1S-^3S	.39596	0-1	4.658×10^{11}	M1
	$1s^2$-$1s2p$	^1S-^3P	.38974	0-2	2.974×10^{12}	M2
56	$1s^2$-$1s2s$	^1S-^3S	.38124	0-1	5.625×10^{11}	M1
	$1s^2$-$1s2p$	^1S-^3P	.37506	0-2	3.446×10^{12}	M2
57	$1s^2$-$1s2s$	^1S-^3S	.36729	0-1	6.771×10^{11}	M1
	$1s^2$-$1s2p$	^1S-^3P	.36115	0-2	3.983×10^{12}	M2
58	$1s^2$-$1s2s$	^1S-^3S	.35406	0-1	8.126×10^{11}	M1
	$1s^2$-$1s2p$	^1S-^3P	.34795	0-2	4.592×10^{12}	M2
59	$1s^2$-$1s2s$	^1S-^3S	.34151	0-1	9.725×10^{11}	M1
	$1s^2$-$1s2p$	^1S-^3P	.33542	0-2	5.281×10^{12}	M2
60	$1s^2$-$1s2s$	^1S-^3S	.32958	0-1	1.161×10^{12}	M1
	$1s^2$-$1s2p$	^1S-^3P	.32352	0-2	6.060×10^{12}	M2
61	$1s^2$-$1s2s$	^1S-^3S	.31824	0-1	1.382×10^{12}	M1
	$1s^2$-$1s2p$	^1S-^3P	.31220	0-2	6.939×10^{12}	M2

Table A.6. Continued

Z	Transition	Multiplet	λ(Å)	J_i - J_k	A_{ki} (s^{-1})	Type
62	$1s^2$-$1s2s$	^1S-^3S	.30744	0-1	1.640×10^{12}	M1
	$1s^2$-$1s2p$	^1S-^3P	.30143	0-2	7.927×10^{12}	M2
63	$1s^2$-$1s2s$	^1S-^3S	.29715	0-1	1.943×10^{12}	M1
	$1s^2$-$1s2p$	^1S-^3P	.29116	0-2	9.038×10^{12}	M2
64	$1s^2$-$1s2s$	^1S-^3S	.28735	0-1	2.295×10^{12}	M1
	$1s^2$-$1s2p$	^1S-^3P	.28138	0-2	1.028×10^{13}	M2
65	$1s^2$-$1s2s$	^1S-^3S	.27799	0-1	2.706×10^{12}	M1
	$1s^2$-$1s2p$	^1S-^3P	.27204	0-2	1.168×10^{13}	M2
66	$1s^2$-$1s2s$	^1S-^3S	.26906	0-1	3.183×10^{12}	M1
	$1s^2$-$1s2p$	^1S-^3P	.26312	0-2	1.324×10^{13}	M2
67	$1s^2$-$1s2s$	^1S-^3S	.26053	0-1	3.736×10^{12}	M1
	$1s^2$-$1s2p$	^1S-^3P	.25461	0-2	1.498×10^{13}	M2
68	$1s^2$-$1s2s$	^1S-^3S	.25237	0-1	4.376×10^{12}	M1
	$1s^2$-$1s2p$	^1S-^3P	.24646	0-2	1.691×10^{13}	M2
69	$1s^2$-$1s2s$	^1S-^3S	.24457	0-1	5.115×10^{12}	M1
	$1s^2$-$1s2p$	^1S-^3P	.23867	0-2	1.907×10^{13}	M2
70	$1s^2$-$1s2s$	^1S-^3S	.23709	0-1	5.968×10^{12}	M1
	$1s^2$-$1s2p$	^1S-^3P	.23121	0-2	2.146×10^{13}	M2
71	$1s^2$-$1s2s$	^1S-^3S	.22993	0-1	6.950×10^{12}	M1
	$1s^2$-$1s2p$	^1S-^3P	.22406	0-2	2.412×10^{13}	M2
72	$1s^2$-$1s2s$	^1S-^3S	.22307	0-1	8.079×10^{12}	M1
	$1s^2$-$1s2p$	^1S-^3P	.21721	0-2	2.706×10^{13}	M2
73	$1s^2$-$1s2s$	^1S-^3S	.21649	0-1	9.375×10^{12}	M1
	$1s^2$-$1s2p$	^1S-^3P	.21063	0-2	3.032×10^{13}	M2
74	$1s^2$-$1s2s$	^1S-^3S	.21017	0-1	1.086×10^{13}	M1
	$1s^2$-$1s2p$	^1S-^3P	.20432	0-2	3.392×10^{13}	M2
75	$1s^2$-$1s2s$	^1S-^3S	.20410	0-1	1.256×10^{13}	M1
	$1s^2$-$1s2p$	^1S-^3P	.19826	0-2	3.789×10^{13}	M2
76	$1s^2$-$1s2s$	^1S-^3S	.19827	0-1	1.450×10^{13}	M1
	$1s^2$-$1s2p$	^1S-^3P	.19243	0-2	4.226×10^{13}	M2
77	$1s^2$-$1s2s$	^1S-^3S	.19266	0-1	1.672×10^{13}	M1
	$1s^2$-$1s2p$	^1S-^3P	.18683	0-2	4.708×10^{13}	M2
78	$1s^2$-$1s2s$	^1S-^3S	.18727	0-1	1.925×10^{13}	M1
	$1s^2$-$1s2p$	^1S-^3P	.18145	0-2	5.237×10^{13}	M2
79	$1s^2$-$1s2s$	^1S-^3S	.18208	0-1	2.213×10^{13}	M1
	$1s^2$-$1s2p$	^1S-^3P	.17626	0-2	5.819×10^{13}	M2
80	$1s^2$-$1s2s$	^1S-^3S	.17708	0-1	2.540×10^{13}	M1
	$1s^2$-$1s2p$	^1S-^3P	.17127	0-2	6.457×10^{13}	M2

Table A.6. Continued

Z	Transition	Multiplet	λ(Å)	J_i - J_k	A_{ki} (s^{-1})	Type
81	$1s^2$-$1s2s$	^1S-^3S	.17226	0-1	2.913×10^{13}	M1
	$1s^2$-$1s2p$	^1S-^3P	.16645	0-2	7.157×10^{13}	M2
82	$1s^2$-$1s2s$	^1S-^3S	.16762	0-1	3.335×10^{13}	M1
	$1s^2$-$1s2p$	^1S-^3P	.16181	0-2	7.923×10^{13}	M2
83	$1s^2$-$1s2s$	^1S-^3S	.16314	0-1	3.813×10^{13}	M1
	$1s^2$-$1s2p$	^1S-^3P	.15734	0-2	8.761×10^{13}	M2
84	$1s^2$-$1s2s$	^1S-^3S	.15882	0-1	4.355×10^{13}	M1
	$1s^2$-$1s2p$	^1S-^3P	.15302	0-2	9.676×10^{13}	M2
85	$1s^2$-$1s2s$	^1S-^3S	.15465	0-1	4.968×10^{13}	M1
	$1s^2$-$1s2p$	^1S-^3P	.14885	0-2	1.068×10^{14}	M2
86	$1s^2$-$1s2s$	^1S-^3S	.15063	0-1	5.660×10^{13}	M1
	$1s^2$-$1s2p$	^1S-^3P	.14483	0-2	1.176×10^{14}	M2
87	$1s^2$-$1s2s$	^1S-^3S	.14673	0-1	6.442×10^{13}	M1
	$1s^2$-$1s2p$	^1S-^3P	.14094	0-2	1.295×10^{14}	M2
88	$1s^2$-$1s2s$	^1S-^3S	.14297	0-1	7.325×10^{13}	M1
	$1s^2$-$1s2p$	^1S-^3P	.13718	0-2	1.425×10^{14}	M2
89	$1s^2$-$1s2s$	^1S-^3S	.13933	0-1	8.319×10^{13}	M1
	$1s^2$-$1s2p$	^1S-^3P	.13354	0-2	1.565×10^{14}	M2
90	$1s^2$-$1s2s$	^1S-^3S	.13581	0-1	9.439×10^{13}	M1
	$1s^2$-$1s2p$	^1S-^3P	.13002	0-2	1.718×10^{14}	M2
91	$1s^2$-$1s2s$	^1S-^3S	.13241	0-1	1.070×10^{14}	M1
	$1s^2$-$1s2p$	^1S-^3P	.12662	0-2	1.884×10^{14}	M2
92	$1s^2$-$1s2s$	^1S-^3S	.12911	0-1	1.212×10^{14}	M1
	$1s^2$-$1s2p$	^1S-^3P	.12332	0-2	2.064×10^{14}	M2
93	$1s^2$-$1s2s$	^1S-^3S	.12592	0-1	1.371×10^{14}	M1
	$1s^2$-$1s2p$	^1S-^3P	.12012	0-2	2.260×10^{14}	M2
94	$1s^2$-$1s2s$	^1S-^3S	.12282	0-1	1.550×10^{14}	M1
	$1s^2$-$1s2p$	^1S-^3P	.11703	0-2	2.471×10^{14}	M2
95	$1s^2$-$1s2s$	^1S-^3S	.11982	0-1	1.751×10^{14}	M1
	$1s^2$-$1s2p$	^1S-^3P	.11402	0-2	2.700×10^{14}	M2
96	$1s^2$-$1s2s$	^1S-^3S	.11691	0-1	1.976×10^{14}	M1
	$1s^2$-$1s2p$	^1S-^3P	.11111	0-2	2.948×10^{14}	M2
97	$1s^2$-$1s2s$	^1S-^3S	.11408	0-1	2.228×10^{14}	M1
	$1s^2$-$1s2p$	^1S-^3P	.10829	0-2	3.216×10^{14}	M2
98	$1s^2$-$1s2s$	^1S-^3S	.11135	0-1	2.510×10^{14}	M1
	$1s^2$-$1s2p$	^1S-^3P	.10555	0-2	3.506×10^{14}	M2
99	$1s^2$-$1s2s$	^1S-^3S	.10868	0-1	2.827×10^{14}	M1
	$1s^2$-$1s2p$	^1S-^3P	.10288	0-2	3.819×10^{14}	M2
100	$1s^2$-$1s2s$	^1S-^3S	.10610	0-1	3.181×10^{14}	M1
	$1s^2$-$1s2p$	^1S-^3P	.10029	0-2	4.156×10^{14}	M2

Table A.7. Calculated energies (in 10^4 cm^{-1}) for autoionizing states of He-like ions: a - intermediate coupling, from [A.5], b - $\boldsymbol{LS}$ coupling, from [A.6]

Z		$2s^2_{1/2}[0]$ $2s^2\,{}^1S_0$	$2p^2_{1/2}[0]$ $2p^2\,{}^3P_0$	$2p_{1/2}2p_{3/2}[1]$ $2p^2\,{}^3P_1$	$2p^2_{3/2}[2]$ $2p^2\,{}^3P_2$	$2p_{1/2}2p_{3/2}[2]$ $2p^2\,{}^1D_2$
10	a	523.0469	514.1237	514.0086	513.8086	510.2570
	b	523.0461	514.1224	514.0074	513.8073	510.8073
20	a	2154.080	2134.131	2131.850	2129.186	2120.800
	b	2154.081	2134.120	2131.836	2129.171	2120.788
30	a	4926.783	4891.303	4874.393	4867.670	4839.358
	b	4926.754	4891.269	4874.331	4867.584	4839.330
40	a	8901.914	8849.200	8776.999	8674.415	8767.209
	b	8901.741	8849.116	8776.789	8674.354	8766.889
50	a	14171.42	14102.18	13893.39	13641.98	13881.33
	b	14170.84	14102.00	13892.79	13641.86	13880.37
60	a	20868.34	20783.18	20300.78	19763.02	20287.16
	b	20866.90	20782.89	20299.28	19762.80	20284.73
70	a	29183.61	29082.96	28108.52	27063.83	28094.143
	b	29180.95	29082.84	28105.13	27063.45	28088.61
80	a	39394.53	39278.57	37473.05	35576.76	37458.26
	b	39391.52	39280.61	37466.02	35576.13	37477.29
90	a	51916.50	51783.19	48637.64	45341.01	48612.161
	b	51918.96	51796.82	48611.87	45339.98	48589.93
100	a	67409.82	67231.14	61923.80	56403.68	61913.378
	b	67436.55	67304.59	61902.66	56401.92	61874.89
110	a	87028.3	86635.23	77974.30	68821.09	77967.949

Z		$2s_{1/2}2p_{3/2}[1]$ $2s2p\,{}^1P_1$	$2s_{1/2}2p_{1/2}[1]$ $2s2p\,{}^3P_1$	$2s_{1/2}2p_{1/2}[0]$ $2s2p\,{}^3P_0$	$2s_{1/2}2p_{3/2}[2]$ $2s2p\,{}^3P_2$	$2p^2_{3/2}[0]$ $2p^2\,{}^1S_0$
10	a	509.3738	520.8473	520.9551	520.6223	498.1145
	b	509.3710	520.8511	520.9553	520.6223	498.1165
20	a	2122.628	2149.416	2151.001	2145.361	2097.153
	b	2122.599	2149.439	2151.008	2145.356	2097.155
30	a	4863.189	4918.433	4924.346	4894.778	4809.529
	b	4863.098	4918.469	4924.321	4894.750	4809.529
40	a	8764.767	8889.010	8900.633	8804.134	8640.906
	b	8764.564	8888.974	8900.539	8804.028	8640.912
50	a	13879.08	14154.37	14171.47	13926.97	13603.37
	b	13878.64	14154.10	14171.15	13926.67	13603.38
60	a	20282.51	20848.06	20869.87	20340.15	19717.64
	b	20281.58	20847.32	20869.32	20339.49	19717.68
70	a	28083.37	29161.35	29187.28	28152.13	27010.25
	b	28081.83	29160.13	29186.66	28151.23	27010.34
80	a	37435.63	39371.99	39401.34	37517.17	35513.61
	b	37434.44	39371.71	39402.42	37517.45	35513.80
90	a	48563.11	51894.77	51926.56	48659.45	45266.86
	b	48568.48	51903.42	51938.05	48668.23	45267.20
100	a	61804.50	67378.75	67410.50	61805.27	56316.97
	b	61842.79	67430.14	67468.45	61964.38	56317.46
110	a	77700.85	86909.97	86940.27	77835.77	68720.10

Table A.8. Calculated autoionization probabilities A_a in (10^{13} s^{-1}) of He-like satellites to resonance lines in H-like ions, from [A.7]

Transition	Z=10	Z=14	Z=18	Z=26
$2p3d\,^3P_2$ - $1s2s\,^3S_1$	1.79×10^{-4}			
$2p3p\,^3P_2$ - $1s2s\,^3S_1$	6.74×10^{-1}	6.92×10^{-1}	$6.70\times10^{-}1$	5.95×10^{-1}
$2p3s\,^3P_1$ - $1s2s\,^3S_1$	6.21×10^{-1}	6.07×10^{-1}		
$2p3s\,^1P_1$ - $1s2s\,^1S_0$	9.43	9.31	8.78	8.83
$2s3d\,^3D_3$ - $1s2p\,^3P_2$	2.53×10^{-1}	2.50×10^{-1}	2.36×10^{-1}	1.20×10^{-1}
$2p3p\,^1D_2$ - $1s2p\,^3P_2$	3.46	4.09		1.10×10^{1}
$2p3p\,^3P_2$ - $1s2p\,^3P_1$	1.84	2.95		
$2p3p\,^3P_2$ - $1s2p\,^3P_2$	1.84	2.95	7.23×10^{-1}	1.57
$2p3p\,^3P_0$ - $1s2p\,^3P_1$	7.92×10^{-3}	4.51×10^{-2}		
$2s3d\,^1D_2$ - $1s2p\,^1P_1$	6.83	7.85	3.60	1.31
$2p3p\,^3D_3$ - $1s2p\,^3P_2$	1.41×10^{-3}	4.39×10^{-3}		1359×10^{-1}
$2p3d\,^3D_2$ - $1s2p\,^1P_1$	4.84	2.02	3.10	
$2p3p\,^1P_1$ - $1s2s\,^1S_0$	4.95×10^{-2}			7.65×10^{-1}
$2p3p\,^1P_1$ - $1s2p\,^1P_1$	2.20×10^{-4}			
$2p3d\,^3P_2$ - $1s3s\,^3S_1$	1.79×10^{-4}			
$2p3p\,^1S_0$ - $1s3p\,^1P_1$	5.86×10^{-1}	8.05×10^{-1}	1.15	2.29
$2p3d\,^1P_1$ - $1s3d\,^1D_2$	5.72×10^{-1}	4.36×10^{-1}	2.56×10^{-1}	3.63×10^{-2}
$2p3d\,^1F_3$ - $1s3d\,^3D_2$	2.60	2.58	2.52	2.23
$2p3d\,^1F_3$ - $1s3d\,^1D_2$	2.60	2.58	2.52	2.23
$2p3d\,^3P_0$ - $1s3d\,^3D_1$	2.44×10^{-4}	1.15×10^{-3}	3.83×10^{-3}	1.56×10^{-2}
$2p3d\,^3P_1$ - $1s3d\,^3D_1$	2.35×10^{-4}	1.45×10^{-3}	8.63×10^{-3}	2.49×10^{-2}
$2p3d\,^3P_1$ - $1s3d\,^3D_2$	2.35×10^{-4}	1.45×10^{-3}	8.63×10^{-3}	
$2p3d\,^3P_2$ - $1s3d\,^3D_2$	1.79×10^{-4}	8.13×10^{-4}	2.87×10^{-3}	
$2p3d\,^1D_2$ - $1s3p\,^1P_1$	6.83	7.85	3.60	1.31
$2p3s\,^1P_1$ - $1s3s\,^1S_0$	9.43	9.31	8.78	8.83
$2p3p\,^3S_1$ - $1s3p\,^3P_2$	9.13×10^{-3}	1.10×10^{-2}		
$2p3s\,^1P_1$ - $1s3d\,^1D_1$	9.43	9.31	8.78	8.83
$2p3d\,^3D_2$ - $1s3p\,^3P_1$	4.84	2.02		
$2p3d\,^3D_3$ - $1s3p\,^3P_2$	2.53×10^{-1}	2.50×10^{-1}	2.36×10^{-1}	1.20×10^{-1}
$2p3d\,^3D_2$ - $1s3p\,^3P_2$	4.84			1.92
$2p3p\,^1D_2$ - $1s3p\,^3P_1$	3.46			
$2p3d\,^3D_3$ - $1s3d\,^3D_3$	4.52×10^{-3}	2.45×10^{-2}	7.73×10^{-2}	2.64×10^{-1}
$2p3d\,^3D_1$ - $1s3p\,^3P_0$	2.52×10^{-1}	2.14×10^{-1}		
$2p3d\,^3D_1$ - $1s3p\,^3P_1$	2.52×10^{-1}		1.71×10^{-1}	
$2p3d\,^3D_2$ - $1s3d\,^3D_1$	1.09×10^{-3}	6.52×10^{-3}	2.57×10^{-2}	
$2p3d\,^3D_2$ - $1s3d\,^3D_2$	1.09×10^{-3}	6.52×10^{-3}	2.57×10^{-2}	
$2p3d\,^3D_1$ - $1s3d\,^3D_1$	3.22×10^{-2}	2.39×10^{-1}	7.98×10^{-1}	
$2p3d\,^3D_1$ - $1s3d\,^3D_2$	3.22×10^{-2}	2.39×10^{-1}	7.98×10^{-1}	3.12×10^{-2}

Table A.8. Continued

Transition	Z=10	Z=14	Z=18	Z=26
$2p3p\,^3P_2$ - $1s3p\,^3P_1$	1.84	2.95	7.23×10^{-1}	1.57
$2p3p\,^3P_2$ - $1s3p\,^3P_2$	1.84	2.95	7.23×10^{-1}	
$2p3p\,^3P_1$ - $1s3p\,^3P_0$	2.42×10^{-3}		8.06×10^{-2}	2.25×10^{-2}
$2p3p\,^3P_1$ - $1s3p\,^3P_1$	2.42×10^{-3}			
$2p3p\,^3P_1$ - $1s3p\,^3P_2$	2.42×10^{-3}	4.02×10^{-2}	8.06×10^{-2}	
$2p3p\,^3P_0$ - $1s3p\,^3P_1$	7.92×10^{-3}	4.51×10^{-2}	1.26×10^{-1}	2.13×10^{-1}
$2p3s\,^3P_2$ - $1s3s\,^3S_1$	1.95×10^{-2}	2.04×10^{-5}	2.01×10^{-2}	1.50×10^{-1}
$2p3p\,^3P_1$ - $1s3s\,^3S_1$	7.92×10^{-2}	1.31×10^{-1}	2.57×10^{-1}	9.30×10^{-1}
$2p3p\,^3P_2$ - $1s3s\,^3S_1$	6.74×10^{-1}	6.92×10^{-1}	6.70×10^{-1}	
$2p3s\,^3P_2$ - $1s3s\,^3S_1$	6.21×10^{-1}	6.07×10^{-1}	6.09×10^{-1}	1.67×10^{-1}
$2s3d\,^3D_2$ - $1s3p\,^1P_1$	4.84	2.02	3.10	1.92
$2p3d\,^1D_2$ - $1s3d\,^1D_2$	1.77×10^{-2}	5.58×10^{-2}	6.38×10^{-2}	
$2p3p\,^1D_2$ - $1s3p\,^1P_1$	3.46	4.09	9.31	
$2p3p\,^3P_2$ - $1s3p\,^1P_1$	1.84	2.95		
$2p3d\,^3F_4$ - $1s3d\,^3D_3$	5.34×10^{-1}	5.34×10^{-1}	5.34×10^{-1}	5.34×10^{-1}
$2p3d\,^3F_3$ - $1s3d\,^3D_2$	5.33×10^{-1}	5.33×10^{-1}	5.41×10^{-1}	6.44×10^{-1}
$2p3d\,^3F_3$ - $1s3d\,^1D_2$	5.33×10^{-1}	5.33×10^{-1}	5.41×10^{-1}	6.44×10^{-1}
$2p3d\,^3F_2$ - $1s3d\,^3D_1$	5.33×10^{-1}	5.33×10^{-1}	5.41×10^{-1}	6.44×10^{-1}
$2p3d\,^3F_2$ - $1s3d\,^3D_2$	5.15×10^{-1}	4.72×10^{-1}	4.44×10^{-1}	3.61×10^{-1}
$2p3d\,^3F_2$ - $1s3d\,^3D_2$	5.15×10^{-1}	4.72×10^{-1}	4.44×10^{-1}	3.61×10^{-1}
$2p3p\,^3D_3$ - $1s3p\,^3P_2$	1.41×10^{-3}	4.39×10^{-3}	1.09×10^{-2}	
$2p3p\,^3D_2$ - $1s3p\,^3P_1$	9.35×10^{-3}	5.96×10^{-2}	2.31×10^{-1}	1.16
$2p3p\,^3D_2$ - $1s3p\,^3P_2$	9.35×10^{-3}	5.96×10^{-2}	2.31×10^{-1}	1.16
$2p3p\,^3D_1$ - $1s3p\,^3P_0$	1.55×10^{-4}	3.08×10^{-5}	4.99×10^{-4}	
$2p3p\,^3D_1$ - $1s3p\,^3P_1$	1.55×10^{-4}	3.08×10^{-5}	4.99×10^{-4}	4.03×10^{-3}
$2p3p\,^3P_1$ - $1s3d\,^3D_2$	7.92×10^{-2}	1.31×10^{-1}		9.30×10^{-1}
$2p3p\,^3P_0$ - $1s3d\,^3D_1$	1.09×10^{-1}	1.84×10^{-1}	2.66×10^{-1}	3.70×10^{-1}
$2p3p\,^1P_1$ - $1s3p\,^1P_1$	1.20×10^{-4}	8.37×10^{-4}		1.95×10^{-2}
$2p3s\,^1S_0$ - $1s3p\,^1P_1$	1.77×10^{-1}	1.74×101	1.70×101	1.67×10^{-1}
$2s3p\,^1P_1$ - $1s3s\,^1S_0$	4.95×10^{-2}	5.46×10^{-2}	7.15×10^{-2}	7.65×10^{-1}
$2s3s\,^3S_0$ - $1s3p\,^3P_1$	8.31×10^{-2}	8.11×10^{-2}	7.77×10^{-2}	
$2s3s\,^3S_1$ - $1s3p\,^3P_2$	8.31×10^{-2}	8.11×10^{-2}	7.77×10^{-2}	6.78×10^{-2}

Table A.9. Comparison of autoionization probabilities (in 10^{13} s^{-1}) for $2\ell 4\ell'$ levels in He-like ions calculated by different approaches: (a) AUTOLSJ, (b) MZ, (c) MCDF, from [A.7]

Designation *LSJ*		Z=10	Z=18	Z=26	Z=36	Z=42
	a	2.25×10^{-2}	2.74×10^{-2}	2.95×10^{-2}	3.09×10^{-2}	3.14×10^{-2}
$2p4f\,^3G_5$	b	2.95×10^{-2}	2.95×10^{-2}	2.95×10^{-2}	2.95×10^{-2}	2.95×10^{-2}
	c	1.39×10^{-2}	2.09×10^{-2}	2.36×10^{-2}	2.46×10^{-2}	2.41×10^{-2}
	a	5.24×10^{-3}	2.34×10^{-2}	2.85×10^{-2}	3.09×10^{-2}	3.16×10^{-2}
$2p4f\,^3F_4$	b	8.93×10^{-3}	1.71×10^{-2}	1.94×10^{-2}	2.03×10^{-2}	2.05×10^{-2}
	c	4.89×10^{-3}	1.93×10^{-2}	2.41×10^{-2}	2.64×10^{-2}	2.64×10^{-2}
	a	1.94×10^{-2}	1.12×10^{-2}	1.05×10^{-2}	1.62×10^{-2}	2.50×10^{-2}
$2p4f\,^3G_4$	b	9.09×10^{-3}	2.53×10^{-3}	3.09×10^{-4}	2.58×10^{-3}	1.02×10^{-2}
	c	1.17×10^{-2}	8.54×10^{-3}	8.74×10^{-3}	1.52×10^{-2}	2.36×10^{-2}
	a	3.88×10^{-2}	4.27×10^{-2}	4.41×10^{-2}	3.98×10^{-2}	3.16×10^{-2}
$2p4f\,^1G_4$	b	4.47×10^{-2}	4.31×10^{-2}	4.28×10^{-2}	3.98×10^{-2}	3.20×10^{-2}
	c	2.65×10^{-2}	3.49×10^{-2}	3.79×10^{-2}	3.39×10^{-2}	2.55×10^{-2}
	a	1.62×10^{-1}	1.66×10^{-2}	1.97×10^{-5}	1.06×10^{-3}	1.45×10^{-3}
$2p4d\,^3F_4$	b	2.10×10^{-1}	1.65×10^{-2}	6.11×10^{-5}	1.24×10^{-3}	1.65×10^{-3}
	c	1.12×10^{-1}	1.50×10^{-2}	7.50×10^{-7}	6.94×10^{-4}	1.02×10^{-3}
	a	5.43×10^{-2}	2.26×10^{-1}	2.53×10^{-1}	2.59×10^{-1}	2.61×10^{-1}
$2s4f\,^3F_4$	b	6.85×10^{-2}	2.62×10^{-1}	2.78×10^{-1}	2.77×10^{-1}	2.77×10^{-1}
	c	3.47×10^{-2}	1.80×10^{-1}	2.15×10^{-1}	2.25×10^{-1}	2.30×10^{-1}
	a	1.91×10^{-3}	4.37×10^{-2}	8.29×10^{-2}	9.58×10^{-2}	9.91×10^{-2}
$2s4f\,^3F_4$	b	2.11×10^{-3}	5.04×10^{-2}	9.05×10^{-2}	1.02×10^{-1}	1.04×10^{-1}
	c	2.71×10^{-3}	4.30×10^{-2}	7.92×10^{-2}	9.05×10^{-2}	9.29×10^{-2}
	a	1.04×10^{-1}	1.78×10^{-2}	2.11×10^{-2}	2.26×10^{-2}	1.76×10^{-3}
$2s4d\,^3D_3$	b	1.34×10^{-1}	1.15×10^{-2}	1.35×10^{-2}	1.70×10^{-2}	1.90×10^{-2}
	c	8.94×10^{-2}	1.44×10^{-2}	1.79×10^{-2}	1.99×10^{-2}	2.04×10^{-2}
	a	3.36×10^{-3}	1.28×10^{-3}	2.44×10^{-3}	2.21×10^{-3}	3.46×10^{-2}
$2p4f\,^1F_3$	b	2.00×10^{-3}	4.52×10^{-4}	2.34×10^{-3}	2.02×10^{-3}	1.47×10^{-3}
	c	4.65×10^{-3}	3.52×10^{-4}	3.42×10^{-3}	4.46×10^{-3}	4.39×10^{-3}
	a	4.33×10^{-3}	8.09×10^{-2}	4.56×10^{-2}	3.63×10^{-2}	7.76×10^{-2}
$2p4f\,^3F_3$	b	2.49×10^{-4}	9.17×10^{-2}	5.00×10^{-2}	3.90×10^{-2}	3.63×10^{-2}
	c	4.90×10^{-3}	7.40×10^{-2}	4.15×10^{-2}	3.35×10^{-2}	3.18×10^{-2}
	a	1.71×10^{-2}	9.01×10^{-3}	8.07×10^{-3}	7.85×10^{-3}	3.94×10^{-4}
$2p4f\,^3G_3$	b	9.91×10^{-4}	1.63×10^{-4}	6.89×10^{-4}	3.50×10^{-3}	5.60×10^{-3}
	c	9.64×10^{-3}	6.82×10^{-3}	6.60×10^{-3}	6.61×10^{-3}	6.46×10^{-3}
	a	4.34×10^{-3}	2.81×10^{-4}	2.74×10^{-5}	2.60×10^{-4}	2.30×10^{-2}
$2p4f\,^3D_3$	b	4.33×10^{-3}	2.54×10^{-4}	1.74×10^{-5}	1.17×10^{-4}	1.43×10^{-4}
	c	4.35×10^{-3}	1.63×10^{-4}	5.90×10^{-5}	2.58×10^{-4}	3.34×10^{-4}
	a	4.33×10^{-1}	2.15×10^{-1}	2.91×10^{-1}	3.42×10^{-1}	3.61×10^{-1}
$2p4d\,^3F_3$	b	2.27×10^{-1}	2.61×10^{-1}	3.41×10^{-1}	3.52×10^{-1}	3.93×10^{-1}
	c	1.22×10^{-1}	1.98×10^{-1}	3.05×10^{-1}	3.92×10^{-1}	4.28×10^{-1}
	a	5.63×10^{-3}	7.60×10^{-2}	1.05×10^{-1}	1.22×10^{-1}	1.34×10^{-1}
$2p4d\,^3D_3$	b	1.33×10^{-2}	8.74×10^{-2}	1.08×10^{-1}	1.08×10^{-1}	1.08×10^{-1}
	c	7.79×10^{-3}	6.23×10^{-2}	7.08×10^{-2}	4.84×10^{-2}	3.50×10^{-2}
	a	4.57×10^{-2}	2.45×10^{-3}	1.39×10^{-2}	1.99×10^{-2}	1.71×10^{-2}
$2s4f\,^3F_3$	b	3.76×10^{-2}	1.65×10^{-3}	7.40×10^{-3}	1.35×10^{-2}	1.17×10^{-2}
	c	3.08×10^{-2}	4.80×10^{-6}	1.97×10^{-3}	4.21×10^{-3}	3.74×10^{-3}
	a	8.10×10^{-4}	8.50×10^{-2}	1.63×10^{-1}	2.81×10^{-1}	3.53×10^{-1}
$2s4f\,^1F_3$	b	4.58×10^{-2}	1.08×10^{-1}	1.91×10^{-1}	3.97×10^{-1}	4.16×10^{-1}
	c	1.37×10^{-5}	7.69×10^{-2}	1.53×10^{-1}	2.73×10^{-1}	3.58×10^{-1}

Table A.9. Continued

	a	1.03	1.03	8.92×10^{-1}	7.39×10^{-1}	1.71×10^{-2}
$2p4d\,^1F_3$	b	1.28	1.15	9.57×10^{-1}	7.70×10^{-1}	6.75×10^{-1}
	c	9.04×10^{-1}	9.27×10^{-1}	8.09×10^{-1}	6.54×10^{-1}	5.70×10^{-1}
	a	3.35×10^{-2}	4.94×10^{-1}	1.01	1.38	1.55
$2p4f\,^3D_2$	b	4.11×10^{-2}	3.43×10^{-1}	1.08	1.46	1.69
	c	3.77×10^{-2}	5.53×10^{-1}	1.12	1.56	1.75
	a	2.52	1.65×10^{-1}	4.95×10^{-1}	3.07×10^{-1}	2.38×10^{-1}
$2p4p\,^1D_2$	b	2.75	1.06	5.14×10^{-1}	3.19×10^{-1}	2.52×10^{-1}
	c	2.44	9.91×10^{-1}	4.19×10^{-1}	2.20×10^{-1}	1.59×10^{-1}
	a	1.04×10^{-1}	1.25×10^{-1}	1.49×10^{-1}	3.15×10^{-1}	3.17×10^{-1}
$2p4p\,^3P_2$	b	1.40×10^{-1}	1.04×10^{-1}	1.65×10^{-1}	9.67×10^{-3}	9.35×10^{-3}
	c	2.58×10^{-1}	1.41×10^{-1}	5.63×10^{-2}	4.52×10^{-3}	2.90×10^{-3}
	a	1.13×10^{-1}	2.09×10^{-3}	3.99×10^{-3}	6.72×10^{-3}	5.41×10^{-3}
$2s4d\,^3D_2$	b	1.40×10^{-1}	2.00×10^{-3}	4.80×10^{-3}	1.02×10^{-1}	2.93×10^{-1}
	c	1.11×10^{-1}	6.70×10^{-3}	4.95×10^{-2}	2.09×10^{-1}	2.02×10^{-1}
	a	1.67×10^{-3}	1.38×10^{-1}	1.22	1.82	1.95
$2p4f\,^3F_2$	b	1.07×10^{-3}	2.09×10^{-1}	1.34	1.90	2.02
	c	4.38×10^{-3}	1.19×10^{-1}	1.19	1.78	1.90
	a	1.52×10^{-1}	4.33	3.49	2.67	2.48
$2p4f\,^3D_2$	b	3.27	4.75	3.63	2.77	2.56
	c	4.95×10^{-11}	3.86	3.20	2.47	2.27
	a	1.80	5.63×10^{-2}	1.49×10^{-2}	4.56×10^{-3}	2.24×10^{-3}
$2p4d\,^1D_2$	b	2.20×10^{-1}	8.18×10^{-2}	1.59×10^{-2}	4.77×10^{-3}	2.39×10^{-3}
	c	1.48	5.09×10^{-2}	1.57×10^{-2}	4.89×10^{-3}	2.49×10^{-3}
	a	1.11	5.54×10^{-2}	1.75×10^{-2}	5.60×10^{-3}	2.68×10^{-3}
$2p4f\,^1D_2$	b	2.06×10^{-1}	5.51×10^{-2}	1.89×10^{-2}	6.14×10^{-3}	3.20×10^{-3}
	c	3.39×10^{-1}	6.07×10^{-2}	2.17×10^{-2}	7.34×10^{-3}	3.94×10^{-3}
	a	2.44×10^{-1}	2.42×10^{-1}	2.24×10^{-1}	2.17×10^{-1}	2.16×10^{-1}
$2s4p\,^3P_2$	b	2.84×10^{-1}	2.64×10^{-1}	2.38×10^{-1}	2.27×10^{-1}	2.25×10^{-1}
	c	2.31×10^{-1}	2.55×10^{-1}	2.43×10^{-1}	2.52×10^{-1}	2.62×10^{-1}
	a	5.28×10^{-4}	1.04	1.73×10^{-1}	1.81×10^{-1}	1.85×10^{-1}
$2p4s\,^3P_2$	b	1.95×10^{-3}	1.91×10^{-1}	1.91×10^{-1}	1.95×10^{-1}	1.98×10^{-1}
	c	1.44×10^{-2}	1.38×10^{-1}	1.55×10^{-1}	1.72×10^{-1}	1.82×10^{-1}
	a	1.59×10^{-1}	1.91×10^{-2}	6.77×10^{-3}	2.77×10^{-3}	2.15×10^{-3}
$2p4d\,^3F_2$	b	2.10×10^{-1}	2.08×10^{-2}	7.80×10^{-3}	3.89×10^{-3}	3.46×10^{-3}
	c	1.12×10^{-1}	1.61×10^{-2}	1.31×10^{-2}	1.03×10^{-2}	8.71×10^{-3}
	a	1.63×10^{-2}	2.19×10^{-3}	2.57×10^{-2}	2.17×10^{-2}	1.86×10^{-2}
$2p4d\,^1D_2$	b	1.81×10^{-2}	2.77×10^{-3}	2.66×10^{-2}	2.03×10^{-2}	1.71×10^{-2}
	c	1.23×10^{-2}	4.93×10^{-4}	1.68×10^{-2}	1.26×10^{-2}	1.00×10^{-2}
	a	4.64×10^{-4}	2.56×10^{-2}	3.10×10^{-2}	4.35×10^{-2}	4.69×10^{-2}
$2p4d\,^3D_2$	b	5.81×10^{-4}	2.82×10^{-2}	3.20×10^{-2}	4.43×10^{-2}	4.76×10^{-2}
	c	1.54×10^{-3}	1.63×10^{-2}	2.63×10^{-2}	4.51×10^{-2}	5.34×10^{-2}
	a	4.02×10^{-2}	4.62×10^{-3}	1.10×10^{-2}	1.33×10^{-2}	1.48×10^{-2}
$2s4f\,^3F_2$	b	5.02×10^{-2}	5.33×10^{-2}	5.80×10^{-2}	6.03×10^{-2}	5.97×10^{-2}
	c	2.48×10^{-2}	4.11×10^{-2}	4.74×10^{-2}	5.39×10^{-2}	5.72×10^{-2}
	a	2.48×10^{-5}	3.93×10^{-3}	1.00×10^{-2}	1.25×10^{-2}	1.40×10^{-2}
$2p4d\,^3P_2$	b	4.02×10^{-5}	4.62×10^{-3}	1.10×10^{-2}	1.33×10^{-2}	1.48×10^{-2}
	c	2.91×10^{-3}	2.19×10^{-3}	8.35×10^{-3}	1.05×10^{-2}	1.10×10^{-2}
	a	4.56×10^{-2}	4.55×10^{-2}	4.03×10^{-2}	3.51×10^{-2}	3.30×10^{-2}
$2s4s\,^3S_1$	b	5.79×10^{-2}	5.16×10^{-2}	4.38×10^{-2}	3.72×10^{-2}	3.47×10^{-2}
	c	6.24×10^{-2}	5.74×10^{-2}	5.06×10^{-2}	4.52×10^{-2}	4.27×10^{-2}
	a	4.96×10^{-5}	2.07×10^{-4}	2.11×10^{-3}	7.50×10^{-3}	1.05×10^{-2}
$2p4p\,^3D_1$	b	7.05×10^{-6}	3.64×10^{-4}	2.60×10^{-3}	8.10×10^{-3}	1.13×10^{-2}
	c	1.78×10^{-4}	3.16×10^{-4}	2.83×10^{-3}	9.74×10^{-3}	1.53×10^{-2}

Table A.9. Continued

	a	3.33×10^{-4}	6.43×10^{-3}	1.26×10^{-2}	1.69×10^{-2}	1.85×10^{-2}
$2p4p\,^1P_1$	b	4.01×10^{-4}	6.93×10^{-3}	1.15×10^{-2}	1.82×10^{-2}	1.38×10^{-2}
	c	6.22×10^{-4}	4.37×10^{-3}	4.89×10^{-3}	3.68×10^{-3}	3.09×10^{-3}
	a	1.44×10^{-3}	9.89×10^{-2}	1.10×10^{-1}	1.07×10^{-1}	1.05×10^{-1}
$2p4p\,^3P_1$	b	2.61×10^{-3}	1.14×10^{-1}	1.22×10^{-1}	1.19×10^{-1}	1.17×10^{-1}
	c	7.76×10^{-4}	9.54×10^{-2}	1.16×10^{-1}	1.23×10^{-1}	1.26×10^{-1}
	a	1.08×10^{-1}	2.00×10^{-2}	7.77×10^{-3}	9.98×10^{-3}	4.69×10^{-2}
$2s4d\,^3D_1$	b	1.37×10^{-1}	2.18×10^{-2}	8.53×10^{-3}	1.07×10^{-2}	1.22×10^{-2}
	c	9.99×10^{-2}	2.18×10^{-2}	9.93×10^{-3}	1.66×10^{-2}	2.19×10^{-2}
	a	4.21×10^{-3}	9.94×10^{-3}	4.56×10^{-2}	1.72×10^{-2}	1.68×10^{-2}
$2p4p\,^3S_1$	b	6.16×10^{-3}	1.20×10^{-2}	1.87×10^{-2}	1.86×10^{-2}	1.80×10^{-2}
	c	1.11×10^{-4}	4.34×10^{-3}	1.22×10^{-2}	1.35×10^{-2}	1.33×10^{-2}
	a	3.28×10^{-3}	2.76×10^{-4}	1.02×10^{-5}	1.47×10^{-4}	2.08×10^{-4}
$2p4f\,^3P_1$	b	3.26×10^{-3}	2.51×10^{-4}	1.83×10^{-5}	1.70×10^{-4}	2.31×10^{-4}
	c	2.87×10^{-3}	1.16×10^{-4}	7.07×10^{-5}	2.44×10^{-4}	2.86×10^{-4}
	a	4.17×10^{-2}	8.12×10^{-2}	3.10×10^{-2}	2.33×10^{-2}	1.14×10^{-2}
$2s4p\,^1P_1$	b	5.48×10^{-2}	8.85×10^{-2}	5.00×10^{-2}	2.28×10^{-2}	3.45×10^{-2}
	c	5.11×10^{-2}	6.96×10^{-2}	4.46×10^{-2}	9.05×10^{-2}	1.68×10^{-1}
	a	2.41×10^{-1}	3.64×10^{-1}	5.38×10^{-1}	7.72×10^{-1}	9.14×10^{-1}
$2s4p\,^3P_1$	b	2.78×10^{-1}	4.13×10^{-1}	5.90×10^{-1}	8.22×10^{-1}	9.57×10^{-1}
	c	2.25×10^{-1}	4.37×10^{-1}	6.71×10^{-1}	1.01	1.20
	a	1.75×10^{-2}	4.23×10^{-1}	1.49	1.87	1.86
$2p4s\,^3P_1$	b	3.04×10^{-2}	3.03×10^{-1}	1.58	1.92	1.90
	c	4.75×10^{-2}	4.73×10^{-1}	1.55	1.93	1.93
	a	8.25×10^{-2}	1.04×10^{-1}	8.62×10^{-3}	2.77×10^{-3}	5.62×10^{-2}
$2p4d\,^3D_1$	b	4.59×10^{-1}	1.08×10^{-1}	1.02×10^{-2}	3.26×10^{-2}	2.96×10^{-2}
	c	1.19×10^{-1}	1.07×10^{-1}	1.13×10^{-2}	5.50×10^{-2}	7.03×10^{-2}
	a	3.03	2.98	2.00	1.43	1.32
$2p4s\,^1P_1$	b	3.23	3.15	2.06	1.48	1.36
	c	3.08	2.86	1.89	1.40	1.31
	a	2.77×10^{-2}	3.41×10^{-2}	1.45×10^{-2}	1.39×10^{-2}	1.44×10^{-2}
$2p4d\,^3P_1$	b	1.93×10^{-2}	3.04×10^{-2}	1.57×10^{-2}	1.48×10^{-2}	1.51×10^{-2}
	c	2.07×10^{-2}	2.43×10^{-2}	1.14×10^{-2}	1.28×10^{-2}	1.42×10^{-2}
	a	3.33×10^{-1}	3.20×10^{-2}	6.34×10^{-3}	2.36×10^{-2}	2.71×10^{-2}
$2p4d\,^1P_1$	b	2.51×10^{-1}	2.46×10^{-2}	7.38×10^{-3}	2.42×10^{-2}	2.73×10^{-2}
	c	1.41×10^{-1}	8.09×10^{-3}	1.42×10^{-2}	3.29×10^{-2}	3.50×10^{-2}
	a	6.18	6.11	5.64	5.32	5.23
$2s4s\,^1S_0$	b	7.10	6.57	5.92	5.50	5.38
	c	4.90	5.45	5.43	5.60	5.85
	a	1.11×10^{-2}	1.08×10^{-1}	1.13×10^{-1}	7.07×10^{-2}	5.40×10^{-2}
$2p4p\,^3P_0$	b	1.42×10^{-2}	1.20×10^{-1}	1.28×10^{-1}	9.06×10^{-2}	8.56×10^{-2}
	c	1.22×10^{-2}	1.20×10^{-1}	1.84×10^{-1}	2.79×10^{-1}	4.07×10^{-1}
	a	1.98×10^{-1}	5.81×10^{-1}	1.20	1.67	1.81
$2p4p\,^1S_0$	b	2.10×10^{-1}	6.22×10^{-1}	1.27	1.73	1.86
	c	2.03×10^{-1}	5.42×10^{-1}	1.09	1.49	1.62
	a	2.41×10^{-1}	2.24×10^{-1}	2.15×10^{-1}	2.13×10^{-1}	2.13×10^{-1}
$2s4p\,^3P_0$	b	2.56×10^{-1}	2.34×10^{-1}	2.19×10^{-1}	2.10×10^{-1}	2.07×10^{-1}
	c	2.05×10^{-1}	2.37×10^{-1}	2.77×10^{-1}	3.65×10^{-1}	4.37×10^{-1}
	a	1.62×10^{-2}	3.56×10^{-2}	4.20×10^{-2}	4.21×10^{-2}	4.10×10^{-2}
$2p4s\,^3P_0$	b	3.05×10^{-2}	4.80×10^{-2}	5.36×10^{-2}	5.53×10^{-2}	5.64×10^{-2}
	c	5.50×10^{-2}	7.14×10^{-2}	8.85×10^{-2}	1.27×10^{-1}	1.67×10^{-1}
	a	1.59×10^{-6}	3.94×10^{-3}	1.31×10^{-2}	2.02×10^{-2}	2.24×10^{-2}
$2p4d\,^3P_0$	b	1.60×10^{-6}	4.59×10^{-3}	1.43×10^{-2}	2.29×10^{-2}	2.35×10^{-2}
	c	2.21×10^{-3}	1.74×10^{-3}	1.04×10^{-2}	1.84×10^{-2}	2.12×10^{-2}

A.3 Spectroscopic Data for X-Ray Lasers

Table A.10. Measured wavelengths (in Å) of $3p - 3s$ x-ray laser lines in neon-like ions (from [A.8]). Filled subshells, as for instance $2p^2_{1/2}$, have been dropped in the designation of the configurations

Z	$2p^3_{3/2}3p_{3/2}\,{}^1D_2$ - $2p^3_{3/2}3s_{1/2}\,{}^3P_1$	$2p_{1/2}3p_{3/2}\,{}^3P_2$ - $2p_{1/2}3s_{1/2}\,{}^1P_1$	$2p^3_{3/2}3p_{1/2}\,{}^3D_2$ - $2p^3_{3/2}3s_{1/2}\,{}^1P_1$	$2p^3_{3/2}3p_{3/2}\,{}^3D_1$ - $2p^3_{3/2}3s_{1/2}\,{}^1P_1$
16	859.93(5)	903.13(5)	924.06(5)	894.35(5)
17	754.22(6)	788.00(6)		783.43(6)
18	670.89(4)	697.83(4)	725.86(4)	696.49(4)
	670.8(2)	697.6(1)	725.8(3)	696.5(2)
	670.9(2)	697.5(2)	725.7(3)	696.5(2)
20	547.404(30)	565.317(10)	597.65(1)	568.698(10)
21	500.015(20)	515.045(20)	549.065(20)	520.191(20)
22	459.338(10)	472.088(10)	507.660(10)	478.72(1)
	459.52(10)	472.26(10)	507.82(10)	478.79(10)
		472.2(5)	507.6(5)	
23	423.92(2)	434.887(20)	471.884(20)	442.779(20)
24	392.81(2)	402.346(20)	440.772(20)	411.28(2)
		402.2(5)	440.7(5)	285.5(5)
25	365.169(20)	373.525(20)	413.382(20)	
26	340.40(3)	347.85(3)		358.24(3)
	340.47(5)	347.96(5)	389.25(5)	358.25(3)
		347.6(5)	388.9(5)	
27	318.0(5)	324.5(5)	367.3(5)	
28	297.90(5)	303.80(5)	348.05(5)	315.01(5)
		303.9(1)	348.0(1)	
	297.7(3)	303.6(3)	347.5(5)	314.8(5)
29	279.31(4)	284.67(4)		221.11(4)
	279.40(5)	284.70(5)	330.44(5)	296.07(5)
	279.3(3)	284.7(3)	331.5(10)	296.2(3)
30	262.32(6)	267.23(6)		212.17(6)
31	246.70(6)	251.11(6)		
32	232.24(4)	236.26(4)	286.46(4)	247.32(4)
33	218.84(6)	222.56(6)		
34	206.35(4)	209.73(4)		
	206.38(5)	209.78(5)	262.94(5)	220.28(5)
35	194.7(1)	197.8(1)	252.4(2)	207.9(1)
36	183.87(3)	186.79(3)		
	183.90(15)	186.70(10)	242.85(15)	196.30(20)
37	173.5(1)	176.1(1)	233.5(2)	185.2(1)
38	164.08(5)	166.49(5)		
	164.1(1)	166.5(1)	224.9(1)	175.1(1)
39	154.97(15)	157.14(15)		165.37(15)
40	146.6(1)	148.6(1)	209.6(2)	156.3(1)
41	138.6(1)	140.4(1)	202.5(2)	147.6(1)
42	131.0(1)	132.7(1)		139.4(1)
47	99.365(30)	100.377(30)		105.079(30)

Table A.10. Continued

Z	$2p_{1/2}3p_{1/2}\,^1S_0$ - $2p_{1/2}3s_{1/2}\,^1P_1$	$2p^3_{3/2}3p_{3/2}\,^3P_0$ - $2p^3_{3/2}3s_{1/2}\,^3P_1$	$2p_{1/2}3p_{1/2}\,^1S_0$ - $2p^3_{3/2}3s_{1/2}\,^3P_1$
16	608.39(5)	816.81(5)	
17	528.97(6)	711.74(6)	490.13(6)
18	468.5(2)	628.(3)	
20	383.269(30)	504.330(10)	344.785(10)
21	352.158(20)	456.417(20)	
22	326.30(10)	415.21(10)	285.08(10)
	326.5(3)		
	326.3(5)		
23	304.211(20)	379.694(20)	
24	285.375(20)	348.356(20)	
		240.2(5)	
26	254.87(3)	295.98(3)	204.65(3)
	254.75(5)		204.6(1)
	254.9(5)		204.2(5)
27	242.4(5)		
28		254.10(10)	176.01(5)
	231.1(2)		
29		163.6(1)	
	221.1(3)		
32	196.06(4)		
34	182.44(4)		
	182.43(5)		113.43(5)
35	176.3(1)		
36	170.55(5)		
37	165.0(1)		
40	150.4(1)	118.9(1)	
41	145.9(1)	112.5(1)	
42	141.6(1)	106.4(1)	
47	122.980(30)	81.563(30)	

References

Chapter 1

1.1 S. Svanberg: *Atomic and Molecular Spectroscopy* 2nd edn. Springer Ser. Atom. Plasm., Vol. 6 (Springer, Berlin, Heidelberg 1992)
1.2 H. Winick: *Synchrotron Radiation Sources – A Technical Primer* (World Scientific, River Edge 1994)
1.3 V.G. Pal'chikov, V.P. Shevelko: *Reference Data on Multicharged Ions* Springer Ser. Atom. Plasm., Vol. 16 (Springer, Berlin, Heidelberg 1995)
1.4 R.K. Janev (ed.): *Atomic and Molecular Processes in Fusion Edge Plasmas* (Plenum, New York 1995)
1.5 R.D. Deslattes, E.G. Kessler, W.C. Sauder, A. Henins: Ann. Phys. (NY) **129,** 378 (1980)

Chapter 2

2.1 K. Siegbahn: *Alpha-, Beta- and Gamma-Ray Spectroscopy* (North Holland, Amsterdam 1979)
2.2 B.K. Agarwal: *X-ray spectroscopy*, 2nd ed., Springer Ser. Opt. Sci., Vol. 15 (Springer, Berlin, Heidelberg 1991)
2.3 A.G. Michette, C.J. Buckley (eds.):*X-Ray Science and Technology* (IOP, Bristol 1993)
2.4 N.A. Dyson: *X Rays in Atomic and Nuclear Physics* (Longman, London 1973)
2.5 E.F. Kaelble (ed.): *Handbook of X Rays* (McGraw-Hill, New York 1967)
2.6 P. Schmor, K.-N. Leung, G. Dutto (eds.): *Proc. 6th Int'l Conf. on Ion Sources* Whistler, Canada 1995; Rev. Sci. Instrum. **67(3)**, (1996)
2.7 I.G. Brown (ed.): *The Physics and Technology of Ion Sources* (Wiley, New York 1989)
2.8 B. Wolf (ed.): *Handbook of Ion Sources* (CRC Press, New York 1995)
2.9 N. Angert: CERN Accelerator School, 5th General Accelerator Physics Course, ed. by S. Turner, Vol. 2 (Jyväskylä, Finland 1994) p. 619
2.10 I.I. Sobelman, L.A. Vainshtein, E.A. Yukov: *Excitation of Atoms and Broadening of Spectral Lines*, 2nd edn. Springer Ser. Atoms Plasmas, Vol. 15 (Springer, Berlin , Heidelberg 1995)
2.11 J.A. Bittencourt: *Fundamentals of Plasma Physics* (Pergammon, New York 1986)
2.12 F.F. Chen: *Introduction to Plasma Physics* (Plenum, New York 1974)
2.13 S.C. Brown: *Introduction to Electrical Discharges in Gases* (Wiley, New York 1966)

2.14 I.H. Hutchinson: *Principles of Plasma Diagnostics* (Cambridge Univ. Press, Cambridge 1987)
2.15 H. Winter: In *Experimental Methods in Heavy Ion Physics*, ed. by K. Bethge, Lecture Notes Phys., 83 (Springer, Berlin, Heidelberg 1978)
2.16 M. Lamoureux: Adv. At. Mol. Phys. **31,** 233 (1993)
2.17 A. Girard: Rev. Sci. Instrum. **63**, 2676 (1992)
2.18 P. Debye, W. Hückel: Physikal. Z. **24**, 185 (1923)
2.19 T.A. Carlson, C.W. Nestor, Jr., N. Wasserman, J.D. McDowell: At. Data Nucl. Data Tables **2**, 63 (1970)
2.20 A. Müller: In *Physics of Ion Impact Phenomena*, ed. by D. Mathur, Springer Ser. Chem. Phys., Vol. 54 (Springer, Berlin, Heidelberg 1991)
2.21 V.G. Pal'chikov, V.P. Shevelko: *Reference Data on Multicharged Ions*, Springer Ser. Atom. Plasm., Vol. 16 (Springer, Berlin, Heidelberg 1995)
2.22 W. Lotz: Z. Phys. **206**, 205 (1967); ibid **216**, 241 (1968); ibid **220**, 466 (1969); ibid **232**, 101 (1970)
2.23 V.P. Shevelko, H. Tawara: J. Phys. B **24,** L589 (1995)
2.24 C. Bélenger, P. Defrance, E. Salzborn, V.P. Shevelko, H. Tawara, D.B. Uskov: J. Phys. B **30,** 1 (1997)
2.25 D.A. Knapp: In *Physics with Multiply Charged Ions*, ed. by D. Liesen (Plenum, New York 1995)
2.26 L. Bex: Third European Particle Accelerator Conference, ed. by H. Henke, H. Homeyer, C. Petit-Jean-Genaz, (Editions Frontièrs, Singapore 1992) p. 252
2.27 V.B. Kutner: Rev. Sci. Instrum. **65**, 1039 (1994)
2.28 D.J. Chivers: Rev. Sci. Instrum. **63,** 2501 (1992)
2.29 F.M. Penning: Physica **4**, 71 (1937)
2.30 A.T. Finkelstein: Rev. Sci. Instrum. **11**, 94 (1940)
2.31 R.S. Livingston: J. Appl. Phys. **15**, 2 (1944)
2.32 H. Heil: Z. Phys. **120**, 212 (1943)
2.33 R.S. Livingston, R.J. Jones: Rev. Sci. Instrum. **25**, 552 (1954)
2.34 R.E. Jones, A. Zucker: Rev. Sci. Instrum. **25**, 562 (1954)
2.35 A. Guthrie, R.K. Wackerling: *The Characteristics of Electrical Discharges in Magnetic Fields* (McGraw-Hill, New York 1949)
2.36 J.R.J. Bennett: IEEE Trans. NS-**19**, 48 (1972)
2.37 B.H. Wolf: IEEE Trans. NS-**19**, 74 (1972)
2.38 T.S. Green: Rep. Prog. Phys. **37**, 1257 (1974)
2.39 L. Vályi: *Atom and Ion Sources* (Wiley, London 1977)
2.40 B.F. Gavin: In *The Physics and Technology of Ion Sources* ed. by I.G. Brown (Wiley, New York 1989)
2.41 B.H. Wolf: In *Handbook of Ion Sources* ed. by B.H. Wolf (CRC Press, New York 1995) p. 69
2.42 B.H. Wolf: Rev. Sci. Instrum. **65**, 1248 (1994)
2.43 J.R. Pierce: *Theory and Design of Electron Beams* (Van Nostrand, Toronto 1954)
2.44 J.D. Schneider, D.D. Armstrong: IEEE Trans. NS-**30**, 2844 (1983)
2.45 R. Keller: In *The Physics and Technology of Ion Sources*, ed. by I.G. Brown (Wiley, New York 1989)
2.46 P. Spädtke: In *Handbook of Ion Sources*, ed. by B.H. Wolf (CRC Press, New York 1995)
2.47 C.D. Child: Phys. Rev. **32**, 492 (1911)
2.48 I. Langmuir, K.T. Compton: Rev. Mod. Phys. **3**, 251 (1931)
2.49 R. Keller, P. Spädtke, H. Emig: Vacuum **36**, 833 (1986)
2.50 C.E. Anderson, K.W. Ehlers: Rev. Sci. Instrum. **27**, 809 (1956)

2.51 B. Gavin: Nucl. Instrum. Methods. **64**, 73 (1968)
2.52 J. Consolino, R. Geller, C. Lerot: Proc. 1st Int'l. Conf. on Ion Sources, (Saclay, 1969) p. 537
2.53 H. Postma: Phys. Lett. A **31**, 196 (1970)
2.54 P. Apard, S. Bliman, R. Geller, B. Jacquot, C. Jacquot: Proc. 2nd Int'l. Conf. on Ion Sources, (Wien 1971) p. 632
2.55 M. Sekiguchi, T. Nakagawa (ed.): *Proc. 12th Int'l Workshop on ECR Ion Sources* Vol. INS-J-182 (Riken, Japan 1995)
2.56 B.H. Wolf: In *Handbook of Ion Sources*, ed. by B.H. Wolf (CRC Press, New York 1995) p. 121
2.57 R. Geller: *Electron Cyclotron Resonance Ion Sources* (IOP, Bristol 1996)
2.58 G. Melin, F. Bourg, P. Briand, M. Delaunay, G. Gaudart, A. Girard, D. Hitz J.P. Klein, P. Ludwig, T.K. Nguyen, M. Pontonnier, Y. Su: Rev. Sci. Instrum. **65**, 1051 (1994)
2.59 A.G. Drentje: Rev. Sci. Instrum. **65**, 1045 (1994)
2.60 R. Geller: Proc. Int'l Conf. Phys. Highly Charged Ions, Gießen, 1990; Suppl. Z. Phys. D **21**, S117 (1991)
2.61 R. Geller: Annu. Rev. Nucl. Part. Sci. **40**, 15 (1990)
2.62 C.M. Lyneis, T.A. Antaya: Rev. Sci. Instrum. **61**, 221 (1990)
2.63 R. Geller: IEEE Trans. NS-**26**, 2120 (1979)
2.64 J. Arianer, R. Geller: Annu. Rev. Nucl. Part. Sci. **31**, 19 (1981)
2.65 D. Hitz, F. Bourg, P. Ludwig, G. Melin, M. Pontonnier, T.K. Nguyen: In *12th Int'l Workshop on ECR Ion Sources*, ed. by M. Sekiguchi, T. Nakagawa, Vol. INS-J-182 (INS, Riken, Japan 1995) p. 126
2.66 D. Hitz, F. Bourg, M. Delaunay, P. Ludwig, G. Melin, M. Pontonnier, T.K. NGuyen: Rev. Sci. Instrum. **67,** 883 (1996)
2.67 P. Briand, R. Geller, B. Jacquot, C. Jacquot: Nucl. Instrum. Methods. **131,** 407 (1975)
2.68 B. Jacquot, P. Briand, F. Bourg, R. Geller: Nucl. Instrum. Methods. A **269,** 1 (1988)
2.69 T.A. Antaya, S. Gammino: Rev. Sci. Instrum. **65**, 1723 (1994)
2.70 P. Sortais et al.: Int'l Conf. Ion Sources, East Lansing, USA, 1987, Rept. NSCL-MSUCP-47, p. 334
2.71 M. Liehr, R. Tassel, M. Schlapp, E. Salzborn: Rev. Sci. Instrum. **63**, 2541 (1992)
2.72 G.D. Alton, D.N. Smithe: Rev. Sci. Instrum. **65**, 775 (1994)
2.73 G.D. Alton, D.N. Smithe: In *12th Int'l Workshop on ECR Ion Sources*, ed. by M. Sekiguchi, T. Nakagawa, Vol. INS-J-182 (INS, Riken, Japan 1995) p. 100
2.74 P. Mak, G. King, T.A. Grotjohn, J. Asmussen: J. Vac. Sci. Tech. A **10**, 1281 (1992)
2.75 H. Amemiya, M. Maeda: In *12th Int'l Workshop on ECR Ion Sources*, ed. by M. Sekiguchi, T. Nakagawa, Vol. INS-J-182 (INS, Riken, Japan 1995) p. 53
2.76 M.A. Liebermann, A.J. Lichtenberg: Plasma Phys. **14**, 1073 (1972); ibid **15**, 125 (1973)
2.77 O. Eldridge: Phys. Fluids **15**, 676 (1972)
2.78 R. Geller: In *Proc. 12th Int'l Workshop on ECR Ion Sources*, ed. by M. Sekiguchi, T. Nakagawa, Vol. INS-J-182 (Riken, Japan 1995) p. 4
2.79 G. Ciavola, S. Gammino, P. Briand, G. Melin, P. Seyfert: Rev. Sci. Instrum. **65**, 1057 (1994)

2.80 G. Ciavola, S. Gammino, F. Bourg, P. Briand, R. Langnier, G. Melin, P. Seyfert, G. Gaggero, M. Losasso, R. Penco: In *12th Int'l Workshop on ECR Ion Sources*, ed. by M. Sekiguchi, T. Nakagawa, Vol. INS-J-182 (INS, Riken, Japan 1995) p. 131

2.81 G. Ciavola, S. Gammino, M. Cafici, M. Castro, F. Chines, S. Marletta, F. Alessandra, F. Bourg, P. Briand, G. Melin, R. Langnier, P. Seyfert, G. Gaggero, M. Losasso, R. Penco: Rev. Sci. Instrum. **67,** 889 (1996)

2.82 T. Nakagawa, T. Kageyama: Jpn. J. Appl. Phys. **30**, L1588 (1991)

2.83 T. Nakagawa, T. Kageyama, M. Kase, A. Goto, Y. Yano: Jpn. J. Appl. Phys. **31**, L1129 (1992)

2.84 G. Melin, C. Barué, F. Bourg, P. Briand, J. Debernardi, M. Delaunay, R. Geller, A. Girard, K.S. Golovanivsky, D. Hitz, B. Jacquot, P. Ludwig, J.M. Marthonnet, T.K. Nguyen, L. Pin, M. Pontonnier, J.C. Rocco: In *10th Int'l Workshop on ECR Ion Sources*, ORNLCONF-9011136 (Oakridge, USA 1990) p. 1

2.85 Z.Q. Xie, C.M. Lyneis: In *12th Int'l Workshop on ECR Ion Sources*, ed. by M. Sekiguchi, T. Nakagawa, Vol. INS-J-182 (INS, Riken, Japan 1995) p. 24

2.86 Z.Q. Xie, C.M. Lyneis: Rev. Sci. Instrum. **67,** 886 (1996)

2.87 R. Geller: Proc. 8th Workshop on Electron Cyclotron Resonance Ion Sources, NSCL Rept. MSUCP-47 (East Lansing 1987) p. 1

2.88 T. Antaya: J. Phys. C **1**, 707 (1989)

2.89 G. Shirkov: Rev. Sci. Instrum. **63**, 2894 (1992)

2.90 H.W. Zhao, A.A. Efremov, V.B. Kutner: Nucl. Instrum. Methods. B **98**, 545 (1995)

2.91 P. Briand, R. Geller, G. Melin: Nucl. Instrum. Methods. A **294**, 673 (1992)

2.92 P. Sortais: Rev. Sci. Instrum. **63**, 2801 (1992)

2.93 M.P. Bougarel, C.E. Hill, H. Haseroth, K. Langbein, E. Tanke: In *12th Int'l Workshop on ECR Ion Sources*, ed. by M. Sekiguchi, T. Nakagawa, Vol. INS-J-182 (INS, Riken, Japan 1995) p. 193

2.94 C. Mühle, U. Ratzinger, W. Bleuel, G. Jöst, K. Leible, S. Schennach, B.H. Wolf: Rev. Sci. Instrum. **65**, 1078 (1994)

2.95 C. Barué, P. Briand, A. Girard, G. Melin, G. Brifford: Rev. Sci. Instrum. **63**, 2844 (1992)

2.96 C. Barué, M. Lamoureux, P. Briand, A. Girard, G. Melin: J. Appl. Phys. **63**, 2662 (1992)

2.97 R. Friedlein, S. Heprich, U. Lehnert, H. Tyrroff, H. Wirth, C. Zippe, G. Zschornack: Nucl. Instrum. Methods. B **98**, 585 (1995)

2.98 E.D. Donets: USSR Inventor's Certificate No. 248860, 16 March 1967, Byull. OIPOTZ No. 23, (1969) p. 65

2.99 E.D. Donets, V.I. Ilushchenko, V.A. Alpert: In *Premiére Conf. sur les Sources d'Ions* INSTM, (Saclay, France 1969) p. 625

2.100 R.E. Marrs, M.A. Levine, D.A. Knapp, J.R. Henderson: Phys. Rev. Lett. **60**, 1715 (1988)

2.101 E.D. Donets: In *The Physics, Technology of Ion Sources*, ed. by I.G. Brown (Wiley, New York 1989)

2.102 E.D. Donets: Rev. Sci. Instrum. **67,** 873 (1996)

2.103 R. Becker: In *Handbook of Ion Sources*, ed. by B. Wolf (CRC Press, New York 1995)

2.104 M.P. Stöckli: Proc. Int'l Conf. Phys. Highly Charged Ions, Gießen, 1990; Suppl. Z. Phys. D **21**, S111 (1991)

2.105 R.E. Marrs, P. Beiersdorfer, D. Schneider: Phys. Today **47**, 27 (October 1994)

2.106 M.P. Stöckli, R.M. Ali, C.L. Cocke, M.L.A. Raphaelian, P. Richard: Rev. Sci. Instrum. **63**, 2822 (1992)

2.107 J. Arianer, A. Cabrespine, C. Goldstein, T. Junquera, A. Courtois, G. Deschamps, M. Olivier: Nucl. Instrum. Methods. **198**, 175 (1982)
2.108 R. Becker, M. Kleinod: Rev. Sci. Instrum. **65**, 1063 (1994)
2.109 G. Shirkov, E. Donets: Rev. Sci. Instrum. **63**, 2819 (1992)
2.110 R. Becker, M. Kleinod, H. Thomae, E.D. Donets: In *Proc. HCI-92*, ed. by P. Richard, M. Stöckli, C.L. Cocke, C.D. Lin, AIP Conf. Proc. **274**, 686 (AIP, New York, 1993)
2.111 Y.S. Kim, R.H. Pratt: Phys. Rev. A **27**, 2913 (1983)
2.112 R.W. Schmieder: In *Physics of Highly-Ionized Atoms*, ed. by R. Marrus, NATO ASI Series, Vol. 201 (Plenum, New York 1989) p. 321
2.113 L. Spitzer: *Physics of Ionized Gases* (Wiley, New York 1962)
2.114 M.A. Levine, R.E. Marrs, J.R. Henderson, D.A. Knapp, M.B. Schneider: Phys. Scr. **T22**, 157 (1988)
2.115 B.M. Penetrante, D. Schneider, R.E. Marrs, J.N. Bardsley: Rev. Sci. Instrum. **63**, 2806 (1992)
2.116 P. Beiersdorfer, V. Decaux, K. Widmann: Nucl. Instrum. Methods. B **98**, 566 (1995)
2.117 R.E. Marrs, S.R. Elliott, D.A. Knapp: Phys. Rev. Lett. **72**, 4082 (1994)
2.118 J.H. Scofield: Phys. Rev. A **40**, 3054 (1989)
2.119 G.S. Hurst, M.G. Payne, S.D. Kramer, J.P. Young: Rev. Mod. Phys. **51**, 767 (1979)
2.120 F. Scheerer, V.N. Fedoseyev, H.-J. Kluge, V.I. Mishin, V.S. Letokhov, H.L. Ravn, Y. Shirakabe, S. Sundell, O. Tengblad: Rev. Sci. Instrum. **63**, 2831 (1992)
2.121 H.-J. Kluge: Phys. Scr. **T22**, 85 (1988)
2.122 L.J. Radziemski, D.A. Cremers (eds.): *Laser-Induced Plasmas and Applications* (Dekker, New York 1989)
2.123 T.P. Hughes: *Plasmas and Laser Light* (Hilger, London 1975)
2.124 R. Fedosejevs, R. Ottmann, R. Siegel, G. Kühnle, S. Szatmári, F.P. Schäfer: Appl. Phys. B **50**, 79 (1990)
2.125 J.C. Kieffer, P. Audebert, M. Chaker, J.P. Matte, H. Pépin, T.W. Johnston, P. Maine, D. Meyerhofer, J. Delettrez, D. Strickland, P. Bado, G. Mourou: Phys. Rev. Lett. **62**, 760 (1989)
2.126 U. Teubner, J. Bergmann, B. van Wonterghem, F.P. Schäfer, R. Sauerbrey: Phys. Rev. Lett. **70**, 794 (1993)
2.127 J.S. Pearlman, G.H. Dahlbacka: Appl. Phys. Lett. **31**, 414 (1977)
2.128 G.A. Doschek, U. Feldman, P.G. Burkhalter, T. Finn, W.A. Feibelman: J. Phys. B **10**, L745 (1977)
2.129 W. Mróz, J. Wolowski, E. Woryna, B. Králiková, J. Krása, L. Láska, K. Mašek, J. Skála, K. Rohlena, O.B. Shamaev, B.Y. Sharkov, H. Haseroth, K.L. Langbein, T.R. Sherwood: Rev. Sci. Instrum. **65**, 1272 (1994)
2.130 V.B. Kutner, Y.A. Bykovsky, V.P. Gusev, Y.P. Kozyrev, V.D. Peklenkov: Rev. Sci. Instrum. **63**, 2835 (1992)
2.131 Yu.A. Bykovsky, V.P. Gusev, Yu.P. Kozyrev, I.V. Kolesov, V.B. Kutner, A.S. Pasyuk, V.D. Peklenkov, S.G. Stetsenko, K.G. Suvorov, D.A. Uzienko: Joint Institute for Nuclear Research, Rept. No. P9-86-2 (Dubna, 1986)
2.132 N.J. Peacock, R.S. Pease: Brit. J. Appl. Phys. (J. Phys. D) **2**, 1705 (1969)
2.133 Yu.A. Bykovsky, Yu.P. Kozyrev, S.V. Ryzhikh, S.M. Sil'nov, V.F. Elesin, V.I. Dimovich: *Injector of Highly Charged Ions* Sov. Patent No. 324938 (8 June 1969)
2.134 T.R. Sherwood: Rev. Sci. Instrum. **63**, 2789 (1992)
2.135 T. Henkelmann, J. Sellmair, G. Korschinek: Nucl. Instrum. Methods. B **56**, 1152 (1991)

2.136 T. Henkelmann, G. Korschinek, G. Belayev, V. Dubenkov, A. Golubev, S. Latyshev, B. Sharkov, A. Shumshurov, B. Wolf: Rev. Sci. Instrum. **63**, 2828 (1992)
2.137 B.Y. Sharkov, A.V. Shumshurov, V.P. Dubenkov, O.B. Shamaev, A.A. Golubev: Rev. Sci. Instrum. **63**, 2841 (1992)
2.138 Y. Amdidouche, H. Haseroth, A. Kuttenberger, K. Langbein, J. Sellmair, T.R. Sherwood, B. Williams, B. Sharkov, O. Sharmaev: Rev. Sci. Instrum. **63,** 2838 (1992)
2.139 G. Korschinek, J. Sellmair: Rev. Sci. Instrum. **57,** 745 (1986)
2.140 R.H. Hughes, R.J. Anderson, C.K. Manka, M.R. Carruth, L.G. Gray, J.P. Rosenfeld: J. Appl. Phys. **51,** 4088 (1980)
2.141 V. Dubenkov, B. Sharkov, P. Spädtke, H. Emig, D.H.H. Hoffmann, J. Jacoby, K. Leible, B. Wolf, R. Friedlein, J. Jensen: GSI Sci. Rept. 1992, GSI 93-1, (GSI Darmstadt 1993) p. 431
2.142 J.D. Lindl, R.L. McCrory, E.M. Campbell: Phys. Today **9**, 32 (September 1992)
2.143 J.D. Lindl: Il Nuovo Cimento **106 A,** 1467 (1993)
2.144 G.B. Zimmerman, W.L. Kruer: Comments Plasma Phys. and Controlled Fusion **2,** 51 (1975)
2.145 E. Förster, K. Gäbel, I. Uschmann: Rev. Sci. Instrum. **63**, 5012 (1992)
2.146 E. Förster, E.E. Fill, K. Gäbel, H. He, T. Missala, O. Renner, I. Uschmann: J. Quant. Spectrosc. Radiat. Transfer **51**, 101 (1994)
2.147 J.D. Kmetec, C.L. Gordon, J.J. Macklin, B.E. Lemoff, S.G. Brown, S.E. Harris: Phys. Rev. Lett. **68**, 1527 (1992)
2.148 G. Jamelot: In *Physics with Multiply Charged Ions*, ed. by D. Liesen (Plenum, New York 1995)
2.149 J.S. Wark, A. Djaoui, S.J. Rose, H. He, O. Renner, T. Missalla, E. Förster: Phys. Rev. Lett. **72**, 1826 (1994)
2.150 H. Wiedemann: *Particle Accelerator Physics* Vol. 1 and 2 (Springer, Berlin, Heidelberg 1993,1995);
2.151 M. Reiser: *Theory and Design of Charged Particle Beams* (Wiley, New York 1994)
2.152 P.J. Bryant, K. Johnsen: *The Principles of Circular Accelerators and Storage Rings* (Cambridge Univ. Press, Cambridge 1993)
2.153 S. Turner (ed.): *CERN Accelerator School, 5th General Accelerator Physics Course* Vol. 1 and 2 (CERN, Geneva 1994)
2.154 J.D. Cockcroft, E.T.S. Walton: Proc. R. Soc. London A **136,** 619 (1932)
2.155 R.J. Van de Graaff: Phys. Rev.**38,** 1919 (1931)
2.156 G. Ising: Arkiv för Mathematik, Astronomi och Fysik **18,** 1 (1924)
2.157 R. Wideroe: Arch. Elektrotechnik **21,** 387 (1928)
2.158 L.W. Alvarez: Phys. Rev. **70,** 799 (1946)
2.159 E.O. Lawrence, N.E. Edlefsen: Science **72,** 376 (1930)
2.160 E.M. McMillan: Phys. Rev. Letter to the Editor **68,** 1434 (1945)
2.161 V. Veksler: J. Phys. USSR **9,** 153 (1945)
2.162 P.J. Bryant: CERN Accelerator School, 5th General Accelerator Physics Course, ed. by S. Turner, Vol. 1 (CERN, Jyväskylä, Finland 1994) p. 1
2.163 T. Roser: Fourth European Particle Accelerator Conference, ed. by V. Suller and C. Petit-Jean-Genaz, Vol. 1 (World Scientific, Singapore 1994) p. 151
2.164 H. Haseroth: Fourth European Particle Accelerator Conference, ed. by V. Suller and C. Petit-Jean-Genaz, Vol. 1 (World Scientific, Singapore 1994) p. 138
2.165 H.H.J. Bongartz: *Particle Accelerators Around the World* http://elsar1.physik.uni-bonn.de/accelerator_list.html

2.166 K. Blasche, B. Franczak: Third European Particle Accelerator Conference, ed. by H. Henke, H. Homeyer and C. Petit-Jean-Genaz, Vol. 1 (Editions Frontièrs, Singapore 1992) p. 9
2.167 K. Blasche, B. Franzke: Fourth European Particle Accelerator Conference, ed. by V. Suller and C. Petit-Jean-Genaz, Vol. 1 (World Scientific, Singapore 1994) p. 133
2.168 N. Bohr, J. Lindhard: K. Dan. Vidensk. Selsk. Mat.-Fys. Medd. **28,** No. 7 (1954)
2.169 H.D. Betz: Rev. Mod. Phys. **44,** 465 (1972)
2.170 H.D. Betz, L. Grodzins: Phys. Rev. Lett. **25,** 211 (1970)
2.171 E. Baron, C. Ricaud: European Particle Accelerator Conference, Rome 1988, ed. by S. Tazzari, Vol. 2 (World Scientific, Singapore 1989) p. 839
2.172 R. Anholt, W.E. Meyerhof: Phys. Rev. A **33,** 1556 (1986)
2.173 A. Magel, H. Geissel, B. Voss, P. Armbruster, T. Aumann, M. Bernas, B. Blank, T. Brohm, H.-G. Clerc, S. Czajkowski, H. Folger, A. Grewe, E. Hanelt, A. Heinz, H. Irnich, M. de Jong, A. Junghans, F. Nickel, M. Pfützner, A. Piechaczek, C. Röhl, C. Scheidenberger, K.-H. Schmidt, W. Schwab, S. Steinhäuser, K. Sümmerer, W. Trinder, H. Wollnik, G. Münzenberg: Nucl. Instrum. Methods. B **94,** 548 (1994)
2.174 C. Scheidenberger, H. Geissel, Th. Stöhlker, H. Folger, H. Irnich, C. Kozhuharov, A. Magel, P.H. Mokler, R. Moshammer, G. Münzenberg, F. Nickel, M. Pfützner, P. Rymuza, W. Schwab, J. Ullrich, B. Voss: Nucl. Instrum. Methods. B **90,** 36 (1994)
2.175 Th. Stöhlker, H. Geissel, H. Folger, C. Kozhuharov, P.H. Mokler, G. Münzenberg, D. Schardt, T. Schwab, M. Steiner, H. Stelzer, K. Sümmerer: Nucl. Instrum. Methods. B **61,** 408 (1991)
2.176 C. Scheidenberger, H. Geissel, H.H. Mikkelsen, F. Nickel, T. Brohm, H. Folger, H. Irnich, A. Magel, M.F. Mohar, G. Münzenberg, M. Pfützner, E. Roeckl, I. Schall, D. Schardt, K.-H. Schmidt, W. Schwab, M. Steiner, Th. Stöhlker, K. Sümmerer, D.J. Vieira, B. Voss, M. Weber: Phys. Rev. Lett. **73,** 50 (1994)
2.177 S. Bashkin (ed.): *Beam-Foil Spectroscopy,* Topics in Curr. Phys., Vol. 1 (Springer, Berlin, Heidelberg 1976)
2.178 J.H. Andrä: In *Progress in Atomic Spectroscopy,* ed. by Kleinpoppen (Plenum, New York 1979) Vol. B
2.179 S. Bashkin, J.O. Stoner, Jr.: *Atomic Energy Levels and Grotrian Diagrams* Vols. 1–4 (North Holland, Amsterdam 1975-1982)
2.180 H.F. Beyer, P. Indelicato, K. Finlayson, D. Liesen, R.D. Deslattes: Phys. Rev. A **43,** 223 (1991)
2.181 H.F. Beyer: In *Physics with Multiply Charged Ions,* ed. by D. Liesen (Plenum, New York 1995)
2.182 H.F. Beyer, R. Mann: In *Progress in Atomic Spectroscopy,* ed. by H.J. Beyer and H. Kleinpoppen, Part C (Plenum, New York 1984)
2.183 R. Mann, H.F. Beyer: Comments At. Mol. Phys. **12,** 149 (1982)
2.184 G.I. Budker: Symp. Int'l sur les Anneaux de Collisions, Saclay, France 1966 (Presses Universitaires de France, 1966) I-1-1
2.185 D. Möhl, G. Petrucci, L. Thorndahl, S. van der Meer: Phys. Rep. **58** (2) (1980)
2.186 G. Arnison, UA1 collaboration at CERN: Phys. Lett. B **122,** 103 (1983)
2.187 G.I. Budker, N.S. Dikansky, V.I. Kudelainen, I.N. Meshkov, V.I. Parkhomchuk, A.N. Skrinsky, B.N. Sukhina: Part. Accel. **7,** 197 (1976)
2.188 M. Bell, J. Chaney, H. Herr, F. Krienen, P. Møller-Petersen, G. Petrucci: Nucl. Instrum. Methods. **190,** 237 (1981)

2.189 B. Franzke: Third European Particle Accelerator Conference, ed. by H. Henke, H. Homeyer and C. Petit-Jean-Genaz, Vol. 1 (Editions Frontièrs, Singapore 1992) p. 367
2.190 S.R. Møller: Fourth European Particle Accelerator Conference, ed. by V. Suller and C. Petit-Jean-Genaz (World Scientific, Singapore 1994) Vol. 1, p. 173
2.191 A. Müller: Proc. 16th Int'l Conference on X-Ray and Innershell Processes, (Debrecen, Hungary 1993)
2.192 R. Schuch: X-Ray and Innershell Processes, ed. by T.A. Carlson, M.O. Krause and S.T. Manson, AIP Conf. Proc. **215**, 174 (AIP New York, 1990)
2.193 T. Quinteros, R. Schuch, M. Pajek, P. Sigray, H. Cederquist, H. Danared, L. Bagge, A. Filevich, J. Jeansson, A. Källberg, A. Paál: Nucl. Instrum. Methods. A **333,** 288 (1993)
2.194 M. Larsson, H. Danared, J.R. Mowat, P. Sigray, G. Sundström, L. Broström, A. Filevich, A. Källberg, S. Mannervik, K.G. Rensfelt, S. Datz: Phys. Rev. Lett. **70,** 430 (1993)
2.195 H. Danared: Nucl. Instrum. Methods. A **335,** 397 (1993)
2.196 B. Hochadel, F. Albrecht, M. Grieser, D. Habs, D. Schwalm, E. Szmola, A. Wolf: Nucl. Instrum. Methods. A **343,** 401 (1994)
2.197 W. Petrich, M. Grieser, R. Grimm, A. Gruber, D. Habs, H.-J. Miesner, D. Schwalm, B. Wanner, H. Wernøe, A. Wolf: Phys. Rev. A **48,** 2127 (1993)
2.198 P. Forck, C. Broude, M. Grieser, D. Habs, J. Kenntner, J. Liebmann, R. Repnow, Z. Amitay, D. Zajfman: Phys. Rev. Lett. **72,** 2002 (1994)
2.199 P. Forck, M. Grieser, D. Habs, A. Lampert, R. Repnow, D. Schwalm, A. Wolf, D. Zajfman: Phys. Rev. Lett. **70,** 426 (1993)
2.200 L.H. Andersen, D. Mathur, H.T. Schmidt, L. Vejby-Christensen: Phys. Rev. Lett. **74,** 892 (1994)
2.201 J.S. Hangst, M. Kristensen, J.S. Nielsen, O. Poulsen, J.P. Schiffer, P. Shi: Phys. Rev. Lett. **67,** 1238 (1991)
2.202 L.H. Andersen, J.H. Posthumus, O. Vahtras, H. Ågren, N. Elander, A. Nunez, A. Scrinzi, M. Natiello, M. Larsson: Phys. Rev. Lett. **71,** 1812 (1993)
2.203 J.S. Hangst, J.S. Nielsen, O. Poulsen, P. Shi, J.P. Schiffer: Phys. Rev. Lett. **74,** 4432 (1995)
2.204 P. Lefèvre, D. Möhl: Workshop on Beam Cooling and Related Topics, ed. by J. Bosser, CERN Rept. 94-03 (Montreux, Switzerland 1993) p. 411
2.205 B. Franzke, K. Beckert, F. Bosch, H. Eickhoff, A. Gruber, O. Klepper, F. Nolden, U. Schaaf, P. Spädtke, M. Steck: Third European Particle Accelerator Conference, ed. by H. Henke, H. Homeyer and C. Petit-Jean-Genaz, Vol. 1 (Editions Frontiérs, Singapore 1992) p. 444
2.206 H. Eickhoff, K. Beckert, F. Bosch, B. Franzke, F. Nolden, R. Raabe, H. Reich, M. Steck, P. Spädtke: Workshop on Beam Cooling and Related Topics, ed. by J. Bosser, CERN Rept. 94-03 (Montreux, Switzerland 1993) p. 307
2.207 M. Grieser, M. Blum, D. Habs, R.V. Hahn, B. Hochadel, E. Jaeschke, C.M. Kleffner, M. Stampfer, M. Steck, A. Noda: In *Cooler Rings and their Applications*, ed. by T. Katayama and A. Noda (World Scientific, Singapore 1991) p. 190
2.208 N.S. Dikansky, I.N. Meshkov, A.N. Skrinsky: Nature **276** (1978)
2.209 A.H. Sørensen, E. Bonerup: Nucl. Instrum. Methods. **215,** 27 (1983)
2.210 A. Wolf: Phys. Scr. **T22,** 55 (1987)
2.211 H. Poth: Phys. Rep. **196,** 135 (1990)
2.212 V.V. Parkhomchuk, A.N. Skrinsky: Rep. Prog. Phys. **54,** 919 (1991)
2.213 J.D. Jackson: *Classical Electrodynamics* (Wiley, New York 1967)
2.214 Y. Debrenev, A.N. Skrinsky: Part. Accel. **8,** 1 (1977)

2.215 I. Hofmann: Workshop on Beam Cooling and Related Topics, ed. by J. Bosser, CERN Ret. 94-03 (Montreux, Switzerland 1993) p. 330
2.216 A. Wolf, C. Ellert, M. Grieser, D. Habs, B. Hochadel, R. Repnov, D. Schwalm: Workshop on Beam Cooling and Related Topics, ed. by J. Bosser, CERN Rept. 94-03 (Montreux, Switzerland 1993) p. 416
2.217 M. Steck, K. Beckert, F. Bosch, H. Eickhoff, B. Franzke, O. Klepper, R. Moshammer, F. Nolden, P. Spädtke, T. Winkler: Fourth European Particle Accelerator Conf., ed. by V. Suller and C. Petit-Jean-Genaz, Vol. 2 (World Scientific, Singapore 1994) p. 1197
2.218 T. Winkler, K. Beckert, F. Bosch, H. Eickhoff, B. Franzke, O. Klepper, F. Nolden, H. Reich, B. Schlitt, P. Spädtke, M. Steck: Hyp. Interact. **99,** 277 (1996)
2.219 A. Pivinski: 9th Int'l Conference on High-Energy Accelerators (Stanford, 1974) p. 405
2.220 H. Danared, G. Andler, L. Bagge, C.J. Herrlander, J. Hilke, J. Jeansson, A. Källberg, A. Nielsson, A. Paál, K.-G. Rensfelt, U. Rosengård, J. Starker, M. af Ugglas: Phys. Rev. Lett. **72,** 3775 (1994)
2.221 M. Grieser, F. Albrecht, D. Habs, R. von Hahn, B. Hochadel, C.-M. Kleffner, J. Liebmann, J. Kenntner, H.-J. Miesner, S. Pastuszka, R. Repnov, U. Schramm, D. Schwalm, A. Wolf, M. Steck: Fourth European Particle Accelerator Conference, ed. by V. Suller and C. Petit-Jean-Genaz, Vol. 1 (World Scientific, Singapore 1994)
2.222 V.L. Alperovich, Yu.B. Bolkhovityanov, A.G. Paulish, A.S. Terekhov: Nucl. Instrum. Methods. A **340,** 429 (1994)
2.223 J. Bosser, R. Ley, G. Molinari, G. Tranquille, F. Varenne, I. Meshkov, V. Polyakov, A. Smirnov, E. Syresin: Fourth European Particle Accelerator Conference, ed. by V. Suller and C. Petit-Jean-Genaz, Vol. 2 (World Scientific, Singapore 1994)
2.224 D. Möhl: CERN Accelerator School: Advanced Accelerator Physics, CERN Rept. 87-03 Vol 2 (CERN, Geneva, 1987)
2.225 F.T. Cole, F.E. Mills: Annu. Rev. Nucl. Part. Sci. **31,** 295 (1981)
2.226 D. Möhl: CERN Accelerator School: Antiprotons for Colliding Beam Facilities, CERN Rep. 84-15 (CERN, Geneva 1984)
2.227 S. Chattopadhyay: IEEE Trans. NS-**30,** 2646 (1983)
2.228 N.S. Dikansky, D.V. Pestrikov: Sov. Phys. Tech. Phys. **29,** 271 (1984)
2.229 D.J. Wineland, R.E. Drullinger, F.L. Walls: Phys. Rev. Lett. **40,** 1639 (1978)
2.230 W. Neuhauser, M. Hohenstatt, P. Toschek, H. Dehmelt: Phys. Rev. Lett. **41,** 233 (1978)
2.231 S.V. Andreev, V.I. Balykin, V.S. Letokhov, V.G. Minogin: JETP Lett. **34,** 442 (1981)
2.232 W.D. Phillips, H. Metcalf: Phys. Rev. Lett. **48,** 596 (1982)
2.233 S. Schröder, R. Klein, N. Boos, M. Gerhard, R. Grieser, G. Huber, A. Karafillidis, M. Krieg, N. Schmidt, T. Kühl, R. Neumann, V. Balykin, M. Grieser, D. Habs, E. Jaeschke, D. Krämer, M. Kristensen, M. Music, W. Petrich, D. Schwalm, P. Sigray, M. Steck, B. Wanner, A. Wolf: Phys. Rev. Lett. **64,** 2901 (1990)
2.234 T. Kühl, R. Neumann, D. Marx, H. Poth, K. Boos, R. Grieser, G. Huber, R. Klein, S. Schröder, V. Balykin, D. Habs, W. Petrich, B. Wanner, A. Wolf, D. Schwalm: Nucl. Instrum. Methods. B **56/57,** 1124 (1991)
2.235 T. Kühl: Phys. Scr. **T40,** 49 (1992)
2.236 H. Okamoto, A.M. Sessler, D. Möhl: Phys. Rev. Lett. **72,** 3977 (1994)
2.237 A. Aspect, J. Dalibard, A. Heidmann, C. Salomon, C. Cohen-Tannoudji: Phys. Rev. Lett. **57,** 1688 (1986)

2.238 R. Grimm, Yu.B. Ovchinnikov, A.I. Sidorov, V.S. Letokhov: Phys. Rev. Lett. **65,** 1415 (1990)

2.239 J. Söding, R. Grimm, J. Kowalski, Yu.B. Ovchinnikov, A.I. Sidorov: Europhys. Lett. **20,** 101 (1992)

2.240 L.S. Brown, G. Gabrielse: Rev. Mod. Phys. **58,** 233 (1986)

2.241 W. Paul: Rev. Mod. Phys. **62,** 531 (1990)

2.242 H. Dehmelt: Rev. Mod. Phys. **62,** 525 (1990)

2.243 R.S. Van Dyck Jr., P.B. Schwinberg, H.G. Dehmelt: Phys. Rev. Lett. **59,** 26 (1987)

2.244 P.B. Schwinberg, R.S. Van Dyck Jr., H.G. Dehmelt: Phys. Lett. A **81,** 119 (1981)

2.245 G. Gabrielse, X. Fei, L.A. Orozco, R.L. Tjoelker, J. Haas, H. Kalinowsky, T.A. Trainor, W. Kells: Phys. Rev. Lett. **65,** 1317 (1990)

2.246 G. Gabrielse, D. Phillips, W. Quint, H. Kalinowsky, G. Rouleau: Phys. Rev. Lett. **74,** 3544 (1995)

2.247 T. Otto, G. Bollen, G. Savard, L. Schweikhard, H. Stolzenberg, G. Audi, R.B. Moore, G. Rouleau, J. Szerpo, Z. Patyk: Nucl. Phys. A **567,** 281 (1994)

2.248 F. DiFilippo, V. Natarajan, K.R. Boyce, D.E. Pritchard: Phys. Rev. Lett. **73,** 1481 (1994)

2.249 F. DiFilippo, V. Natarajan, M. Bradley, F. Palmer, S. Rusinkiewicz, D.E. Pritchard: IEEE Trans. IM-**44,** 550 (1995)

2.250 H.-J. Kluge: Nucl. Instrum. Methods. B **98,** 500 (1995)

2.251 W. Quint: Phys. Scr. **T59,** 203 (1995)

2.252 M.H. Prior, R. Marrus, C.R. Vane: Phys. Rev. A **28,** 141 (1983)

2.253 D.A. Church, S.D. Kravis, B.M. Johnson, Y. Azuma, J. Levin, I.A. Sellin, M. Meron, K.W. Jones, M. Druetta, N. Mansour, H.G. Berry, R.T. Short: Nucl. Instrum. Methods. B **56/57,** 417 (1991)

2.254 H.F. Beyer, G. Bollen, F. Bosch, P. Egelhof, B. Franzke, R.W. Hasse, H.-J. Kluge, C. Kozhuharov, T. Kühl, D. Liesen, R. Mann, P.H. Mokler, A. Müller, G. Münzenberg, H. Poth, L. Schweikhard, R. Schuch, G. Werth: *HITRAP A Facility for Experimentation with Trapped Highly Charged Ions at GSI*, GSI-90-20 (GSI Darmstadt 1990)

2.255 J.J. Bollinger, J.S. Wells, D.J. Wineland, W.M. Itano: Phys. Rev. A **31,** 2711 (1985)

2.256 D. Wineland, H. Dehmelt: J. Appl. Phys. **46,** 919 (1975)

2.257 W.M. Itano, D.J. Wineland: Phys. Rev. A **25,** 35 (1982)

2.258 I. Bergström, H. Borgenstrand, C. Carlberg, G. Rouleau, R. Schuch, B. Smith: Phys. Scr. **47,** 475 (1993)

2.259 C. Carlberg, I. Bergström, G. Bollen, H. Borgenstrand, R. Jertz, H.-J. Kluge, G. Rouleau, R. Schuch, T. Schwarz, L. Schweikhard, P. Senne, F. Söderberg: IEEE Trans. IM-**44,** 553 (1995)

2.260 C. Carlberg: In *Physics with Multiply Charged Ions* ed. by D. Liesen (Plenum, New York 1995)

2.261 G. Bollen, S. Becker, H.-J. Kluge, M. König, R.B. Moore, T. Otto, H. Raimbault-Hartmann, G. Savard, L. Schweikhard, H. Stolzenberg, the ISOLDE Collaboration: Nucl. Instrum. Methods. A **368,** 675 (1996)

2.262 G. Bollen, R.B. Moore, G. Savard, H. Stolzenberg: J. Appl. Phys. **68,** 4355 (1990)

2.263 G. Gräff, H. Kalinowsky, J. Traut: Z. Phys. A **297,** 35 (1980)

2.264 D. Schneider, D.A. Church, G. Weinberg, J. Steiger, B. Beck, J. McDonald, E. Magee, D. Knapp: Rev. Sci. Instrum. **65,** 3472 (1994)

2.265 D. Church, G. Weinberg, J. Steiger, B. Beck, D. Schneider: EBIT Annual Rept. 1994, UCRL-ID-121572, ed. by D. Schneider (Lawrence Livermore National Laboratory, 1995) p. 82
2.266 J. Steiger, B. Beck, J. McDonald, E. McGee, D. Schneider: EBIT Annual Rept. 1994, UCRL-ID-121572, ed. by D. Schneider (Lawrence Livermore National Laboratory, 1995) p. 84

Chapter 3

3.1 A.H. Gabriel: Mon. Not. R. Astron. Soc. **160,** 99 (1972)
3.2 D.R. Plante, W.R. Johnson, J. Sapirstein: Phys. Rev. A **49**, 3519 (1994)
3.3 I.I. Sobelman: *Atomic Spectra and Radiative Transitions*, 2nd edn. Springer Ser. Atoms Plasmas, Vol.12 (Springer, Berlin, Heidelberg 1992)
3.4 H.A. Bethe, E.E. Salpeter: *Quantum Mechanics of One- and Two-Electron Atoms* (Plenum, New York 1977)
3.5 E. Lindroth: Nucl. Instrum. Methods. **98**, 1 (1995)
3.6 W.R. Johnson, G. Soff: At. Data Nucl. Data Tables **33**, 405 (1985)
3.7 V.A. Boiko, V.G. Pal'chikov, I.Yu. Skobelev, A.Ya. Faenov: *The Spectroscopic Constants of Atoms and Ions* (CRC, Boca Raton 1994)
V.G. Pal'chikov: Unpublished (1996)
3.8 G.W. Drake: Can. J. Phys. **66**, 586 (1988)
3.9 L.A. Vainshtein, U.I. Safronova: Phys. Scr. **31**, 519 (1985)
3.10 P. Indelicato, O. Gorceix, J.P. Desclaux: J. Phys. B **20**, 651 (1987)
3.11 J. Hata, I.P. Grant: J. Phys. B **16**, 523 (1983); *ibid* **17**, 931 (1984)
3.12 P. Indelicato: Nucl. Instrum. Methods. B **31**, 14 (1988)
3.13 P. Indelicato, J.P. Desclaux: Phys. Rev. A **42**, 5139 (1990)
3.14 M.H. Chen, K.T. Chen, W.R. Johnson: Phys. Rev. A **47**, 3692 (1993)
3.15 W.R. Johnson, J. Sapirstein: Phys. Rev. A **46**, 2197 (1992)
3.16 I. Lindgren, H. Person, S. Salomonson, P. Sunnergren: Phys. Scr. **T46**, 125 (1993)
3.17 W.R. Johnson, C.D. Lin: Phys. Rev. A **14**, 565 (1976)
3.18 W.C. Martin: Phys. Scr. **24**, 725 (1981)
3.19 P. Beiersdorfer, A. Osterheld, S.R. Elliott, M.H. Chen, D. Knapp, K. Reed: Phys. Rev. A **52,** 2693 (1995)
3.20 Th. Stöhlker, A.E. Livingston: Acta Pysica Polonica B **27,** 441 (1996)
3.21 Th. Stöhlker, P. Mokler, H. Geissel, R. Moshammer, P. Rymuza, E.M. Berstein, C.L. Cocke, C. Kozhuharov, G. Munzenberg, F. Nickel, C. Scheidenberger, Z. Stachura, J. Ullrich, A. Warczak: Phys. Lett. A **168**, 285 (1992)
3.22 S. Martin, A. Denis, M.C. Buchet-Poulizac, J.P. Buchet, J. Desesquelles: Phys. Rev. A **42**, 6570 (1990)
3.23 H.G. Berry, R.W. Dunford, A.E. Livingston: Phys. Rev. A **47**, 698 (1993)
3.24 P.H. Mokler, Th. Stöhlker, C. Kozhuharov, R. Mashammer, P. Rymuza, F. Bosch, T. Kandler: Phys. Scr. **T51**, 28 (1994)
3.25 E.G. Myers, J.K. Thompson, E.P. Gavathas, N.R. Claussen, J.D. Silver, D.J.H. Howie: Phys. Rev. Lett. **75**, 3637 (1994)
3.26 J. Suleiman, H.G. Berry, R.W. Dunford, D.S. Gemmel, E.P. Kanter, S. Cheng, L.J. Curtis: Phys. Rev. A **51**, 1905 (1995)
3.27 K.W. Kukla, A.E. Livingston, J. Suleiman, H.G. Berry, R.W. Dunford, D.S. Gemmel, E.P. Kanter, S. Chang, L.J. Curtis: Phys. Rev. A **51**, 1905 (1995)

3.28 A.E. Livingston, K.W. Kukla, C.M. Vogel Vogt, H.G. Berry, R.W. Dunford, D.S. Gemmel, E.P. Kanter, J. Suleiman, R. Ali, S. Cheng, L.J. Curtis: Nucl. Instrum. Methods. **98**, 28 (1995)
3.29 R.E. Marrs, S.R. Elliott, Th. Stöhlker: Phys. Rev. A **52**, 3577 (1995)
3.30 K. Widmann, P. Beiersdorfer, V. Desclaux: Nucl. Instrum. Methods. B **98**, 45 (1995)
3.31 E. Hinnov, the TFTR Team, B. Denne, the JET Team: Phys. Rev. A **40**, 4357 (1989)
3.32 J. Schweppe, A. Belkacen, L. Blumenfeld, N. Claytor, B. Feinberg, H. Gould, V.E. Kostroun, L. Levy, S. Misawa, J.R. Mowat, M.H. Prior: Phys. Rev. Lett **66**, 1434 (1991)
3.33 P. Beiersdorfer, D. Knapp, R.E. Marrs, S.R. Elliot, M.H. Chen: Phys. Rev. Lett. **71**, 3939 (1993)
3.34 R. Büttner: Spectroscopic Determination of the QED Contributions to the $2s-2p$ Transition Energies of Li-like Heavy Ions. Thesis, Astrophysical Institute, Gießen, Germany 1996; R. Büttner, U. Staude, M. Nicolai, J. Brankhoff, K.-H. Schartner, F. Folkmann, P.H. Mokler: Nucl. Instrum. Methods. B **98**, 41 (1995)
3.35 V. Decaux, P. Beiersdorfer, A. Osterheld: Nucl. Instrum. Methods. B **98**, 129 (1995)
3.36 S.A. Blundell: Phys. Rev. **46**, 5139 (1990); *ibid* **47**, 1790 (1993)
3.37 W.R. Johnson, J. Sapirstein, K.T. Cheng: Phys. Rev. A **51**, 297 (1995)
3.38 H. Nusbaumer, P.J. Storey: J. Phys. B **12**, 1647 (1979)
3.39 A.K. Bhatia, H.E. Mason: Astron. Astrophys. **103**, 324 (1981)
3.40 W.F. Perger, B.P. Das: J. Phys. B **20**, 6651 (1987)
3.41 E. Avgoustoglou, W.R. Johnson, Z.W. Liu, J. Sapirstein: Phys. Rev. A **51**, 1196 (1995)
3.42 E.P. Ivanova, A.V. Gulov: At. Data Nucl. Data Tables **49**, 1 (1991)
3.43 C. Jupen, L.J. Curtis: Phys. Scr. **53**, 312 (1996)
3.44 J.F. Seely, C.M. Brown, U. Feldman, J.O. Ekberg, C.J. Keane, B.J. MacGowan, D.R. Kania, W.E. Behling: At. Data Nucl. Data Tables **47**, 1 (1991)
3.45 M.H. Chen, K.T. Chen, W.R. Johnson, J. Sapirstein: Phys. Rev. A **50**, 247 (1994)
3.46 D.R. Plante, W.R. Johnson, J. Sapirstein: Unpublished (1994)
3.47 R.D.Deslattes, H.F. Beyer, F. Folkmann: J. Phys. B **17**, L689 (1984)
3.48 J.P. Briand, J.P. Mosse, P. Indelicato, P. Chevallier, D. Girard-Verhet, A. Chetioui: Phys. Rev. A **28**, 1413 (1983)
3.49 J.P. Briand, M. Tavernier, R. Marrus, J.P. Desclaux: Phys. Rev. A **29**, 3143 (1984)
3.50 S. MacLaren, P. Beiersdorfer, D.A. Vogel, D. Knapp, R.E. Marrus, K. Wong, R. Zasadzinski: Phys. Rev. A **45**, 329 (1992)
3.51 P. Indelicato, J.P. Briand, M. Tavernier, D. Liesen: Z. Phys. D **2**, 249 (1986)
3.52 J.P. Briand, P. Indelicato, A. Simionovici, V. San Vicente, D. Liesen, D. Dietrich: Europhys. Lett. **9**, 225 (1989)
3.53 J.P. Briand, P. Chevallier, P. Indelicato, K.P. Ziock, D. Dietrich: Phys. Rev. Lett. **65**, 2761 (1990)
3.54 P. Beiersdorfer, M. Bitter, S. von Goeler, K.W. Hill: Phys. Rev. A **40**, 150 (1989)
3.55 J. Fleming, A. Hibbert: Phys. Scr. **51**, 339 (1995)
3.56 M.H. Chen, K.T.Chen, W.R. Johnson, J. Sapirstein: Phys. Rev. **52**, 266 (1995)
3.57 A. Ynnerman, J. James, I. Lindgren, H. Persson, S. Salomonson: Phys. Rev. A 50, 4671 (1994)

3.58 B. Edlen: Phys. Scr. 28, 51 (1983)
3.59 W.C. Martin, R. Zalubas, A. Musgrove: J. Phys. Chem. Ref. Data **14,** 751 (1985)
3.60 J. Sugar, C. Corliss: J. Phys. Chem. Ref. Data **14,** Suppl. No. 2 (1985)
3.61 J. Reader, J. Sugar, N. Acquista, R. Bahr: J. Opt. Soc. Am. B **11,** 1930 (1994)
3.62 K.G. Widing, J.D. Purcell: Astrophys. J. **204,** L151 (1976)
3.63 P. Beiersdorfer: Bull. Am. Phys. Soc. **39**, 1208 (1994)
3.64 V.M. Shabaev: Opt. Spectrosc. **56,** 244 (1984)
3.65 P. Raghavan: Atom. Data Nucl. Data Tables, **42,** 189 (1989)
3.66 V.M. Shabaev: J. Phys. B **27,** 5825 (1994)
3.67 V.M. Shabaev, M.B. Shabaeva, I.I. Tupitsyn: Phys. Rev. A **52,** 3686 (1995)
3.68 M.B. Shabaeva, V.M. Shabaev: Phys. Rev A **52,** 2811 (1995)
3.69 I. Klaft, S. Borneis, T. Engel, B. Fricke, R. Grieser, G. Huber, T. Kühl, D. Marx, R. Neumann, S. Schröder, P. Seelig, L. Völker: Phys. Rev. Lett. **73,** 2425 (1994)
3.70 H. Persson, S.M. Schneider, W. Greiner, G. Soff, I. Lindgren: Phys. Rev. Lett. **76,** 1433 (1996)
3.71 M. Tomaselli, S.M. Schneider, E. Kankeleit, T. Kühl: Phys. Rev. C **51,** 2989 (1995)
3.72 H. Persson, S. Salomonson, P. Sunnergren, I. Lindgren, M. Gustavson: Hyp. Interact. **108**, 3 (1997)
3.73 Th. Stöhlker, P.H. Mokler, K. Beckert, F. Bosch, H. Eickhoff, B. Franzke, M. Jung, T. Kandler, O. Kleppner, C. Kozhuharov, R. Moshammer, F. Nolden, H. Reich, P. Rymuza, P. Spädtke, M. Steck: Phys. Rev. Lett. **71,** 2184 (1993)
3.74 H.F. Beyer: IEEE Trans. IM-**44,** 510 (1995)
3.75 H.F. Beyer, G. Menzel, D. Liesen, A. Gallus, F. Bosch, R.D. Deslattes, P. Indelicato, Th. Stöhlker, O. Klepper, R. Moshammer, F. Nolden, H. Eickhoff, B. Franzke, M. Steck: Z. Phys. D **35,** 169 (1995)
3.76 W.E. Lamb, R.C. Retherford: Phys. Rev. **72,** 241 (1947)
W.E. Lamb: Rep. Prog. Phys. **14,** 23 (1951)
3.77 A.I. Akhiezer, V.B. Berestetskii: *Quantum Electrodynamics* (Wiley, New York 1965)
3.78 J.R. Sapirstein, D.R. Yennie: In *Quantum Electrodynamics*, ed. by T. Kinoshita, (World Scientific, Singapore 1990)
3.79 P. Mohr: In *Atomic, Molecular and Optical Physics Reference Book*, ed. by G.W.F. Drake, (AIP, New York 1996)
3.80 P.J. Mohr: Nucl. Instrum. Methods. B **87,** 232 (1994)
3.81 H.F. Beyer: In *Physics with Multiply Charged Ions*, ed. by D. Liesen, NATO ASI Series B, Vol. 348 (Plenum, New York 1995)
3.82 P. Mohr: At. Data Nucl. Data Tables **29,** 453 (1983)
3.83 T. Franosch, G. Soff: Z. Phys. D **18,** 219 (1991)
3.84 J.D. Zumbro, E.B. Shera, Y. Tanaka, J. C.E. Bemis, R.A. Naumann, M.V. Hoehn, W. Reuter, R.M. Steffen: Phys. Rev. Lett. **53,** 1888 (1984)
3.85 V.M. Shabaev: J. Phys. B **26,** 1103 (1993)
3.86 V.M. Shabaev: Theor. Math. Phys. **63,** 588 (1985)
3.87 A.N. Artemyev, V.M. Shabaev, V.A. Yerokhin: Phys. Rev. A **52,** 1884 (1995)
3.88 P.J. Mohr: Ann. Phys. (N.Y). **88,** 26 (1974); ibid. p. 52
3.89 P.J. Mohr: Phys. Rev. A **46,** 4421 (1992)
3.90 P.J. Mohr, G. Soff: Phy. Rev. Lett. **70,** 158 (1993)
3.91 G. Soff, P.J. Mohr: Phys. Rev. A **38,** 5066 (1988)

3.92 N.L. Manakov, A.A. Nekipelov, A.G. Fainshtein: Sov. Phys. JETP **68,** 673 (1989)
3.93 H. Persson, I. Lindgren, S. Salomonson, P. Sunnergren: Phys. Rev. A **48,** 2772 (1993)
3.94 H.F. Beyer, K.D. Finlayson, D. Liesen, P. Indelicato, C.T. Chantler, R.D. Deslattes, J. Schweppe, F. Bos, M. Jung, O. Klepper, W. König, R. Moshammer, K. Beckert, H. Eickhoff, B. Franzke, A. Gruber, F. Nolden, P. Spädtke, M. Steck: J. Phys. B **26**, 1557 (1993)
3.95 P.H. Mokler, T. Stöhlker, C. Kozhuharov, R. Moshammer, P. Rymuza, Z. Stachura, A. Warczak: J. Phys. B **28,** 617 (1995)
3.96 H.F. Beyer, D. Liesen, F. Bosch, K.D. Finlayson, M. Jung, O. Klepper, R. Moshammer, K. Beckert, H. Eickhoff, B. Franzke, F. Nolden, P. Spädtke, M. Steck, G. Menzel, R.D. Deslattes: Phys. Lett. A **184**, 435 (1994)
3.97 S.A. Blundell, P.J. Mohr, W.R. Johnson, J. Sapirstein: Phys. Rev. A **48,** 2615 (1993)
3.98 I. Lindgren, H. Persson, S. Salomonson, L. Labzowsky: Phys. Rev. A **51,** 1167 (1995)
3.99 P. Indelicato: Phys. Rev. A **51,** 1132 (1995)
3.100 H. Persson, S. Salomonson, P. Sunnergren, I. Lindgren: Phys. Rev. Lett. **76,** 204 (1996)
3.101 S.R. Lundeen, F.M. Pipkin: Phys. Rev. Lett. **46**, 232 (1981)
3.102 V.G. Pal'chikov, Yu.L. Sokolov, V.P. Yakovlev: JETP Lett. **28,** 418 (1983)
3.103 E.W. Hagley, F.M. Pipkin: Phys. Rev. Lett. **72**, 1172 (1994)
3.104 S. Bourzeix, B. de Beauvoir, F. Nez, M.D. Plimmer, F. de Tomasi, L. Julien, F. Biraben: Phys. Rev. Lett. **76**, 384 (1996)
3.105 A. van Wijgaarten, J. Kwela, G.W.F. Drake: Phys. Rev. A **43,** 3325 (1991)
3.106 M.S. Dewey, R.W. Dunford: Phys. Rev. Lett. **60,** 2014 (1988)
3.107 M. Leventhal: Phys. Rev A **11,** 427 (1975)
3.108 H.W. Kugel, M. Leventhal, D.E. Murnick: Phys. Rev A **6,** 1306 (1972)
3.109 B. Curnutte, C.L. Cocke, R.D. Dubois: Nucl. Instrum. Methods. **202,** 119 (1982)
3.110 G.P. Lawrence, C.Y. Fan, S. Bashkin: Phys. Rev. Lett. **28,** 1612 (1972)
3.111 M. Leventhal, D.E. Murnick: Phys. Rev. Lett. 28, 1609 (1972)
3.112 H.W. Kugel, M. Leventhal, D.E. Murnick, C.K.N. Patel, O.R. Wood II: Phys. Rev. Lett. **35,** 647 (1975)
3.113 P. Pellegrin, Y. El Masri, L. Palffy: Phys. Rev. A **31,** 5 (1985)
3.114 D. Müller, J. Gassen, L. Kremer, H.-J. Pross, F. Scheuer, H.-D. Strater, P. von Brentano, A. Pape, J.C. Sens: Europhys. Lett. **5**, 503 (1988)
3.115 P. von Brentano, D. Platte, D. Budelsky, L. Kremer, H.J. Pross, F. Scheuer: Phys. Scr. **T46,** 162 (1993)
3.116 J. Gassen, D. Müller, D. Budelsky, L. Kremer, H.-J. Pross, F. Scheuer, P. von Brentano, A. Pape, J.C. Sens: Phys. Lett. A **147,** 385 (1990)
3.117 V. Zacek, H. Bohn, H. Brum, T. Faestermann, F. von Feilitzsch, G. Giorginis, P. Kienle, S. Schuhbeck: Z. Phys. A **318,** 7 (1986)
3.118 A.P. Georgiadis, D. Muller, H.-D. Sträter, J. Gassen, P. von Brentano, J.C. Sens, A. Pape: Phys. Lett. A **115,** 108 (1986)
3.119 O.R. Wood II, C.K.N. Patel, D.E. Murnick, E.T. Nelson, M. Leventhal, H.W. Kugel, Y. Niv: Phys. Rev. Lett. **35,** 647 (1975)
3.120 O.R. Wood II, C.K.N. Patel, D.E. Murnick, E.T. Nelson, M. Leventhal: Phys. Rev. Lett. **48**, 398 (1982)
3.121 H. Gould, R. Marrus: Phys. Rev. Lett. **41**, 1457 (1978)
3.122 H. Gould, R. Marrus: Phys. Rev. A **28,** 2001 (1983)

3.123 G. Plunien, Th. Beier, G. Soff, H. Persson, I. Lindgren, L.N. Labzowsky: In *GSI Sci. Rept. 1995*, GSI-96-01, GSI Darmstadt (1996) p.107
3.124 F. Nez, M.D. Plimmer, S. Bourzeix, L. Julien, F. Biraben, R. Felder, Y. Millerioux, P. De Natale: Europhys. Lett. **24,** 635 (1993)
3.125 M. Weitz, A. Huber, F. Schmidt-Kaler, D. Liebfried, T.W. Hänsch: Phys. Rev. Lett. **72**, 328 (1994)
3.126 D.J. Berkeland, E.A. Hinds, M.G. Boshier: Phys. Rev. Lett. **75**, 2470 (1995)
3.127 L. Schleinkofer, F. Bell, H.-D. Betz, G. Trollmann, J. Rothermel: Phys. Scr. **25**, 917 (1982)
3.128 E. Källne, J. Källne, P. Richard, M. Stöckli: J. Phys. B **17**, L115 (1984)
3.129 P. Richard, M. Stöckli, R.D. Deslattes, P. Cowan, R.E. LaVilla, B. Johnson, K. Jones, M. Meron, R. Mann, K.-H. Schartner: Phys. Rev. A **29**, 2939 (1984)
3.130 R.D. Deslattes, R. Schuch, E. Justiniano: Phys. Rev. A **32**, 1911 (1985)
3.131 H.F. Beyer, R.D. Deslattes, F. Folkmann, R.E. LaVilla: J. Phys. B **18**, 207 (1985)
3.132 H.-D. Dohmann, D. Liesen, E. Pfeng: GSI Sci. Rept. 1982, GSI Darmstadt (1983) p. 155
3.133 J.P. Briand, M. Tavernier, P. Indelicato, R. Marrus, M. Gould: Phys. Rev. Lett. **50**, 832 (1983)
3.134 A.F. McClelland: Development of a novel Technique for the Measurements of the $1s_{1/2}$ Lamb Shift in High Z Hydrogenic Ions by Lyman-α / Balmer-β Wavelength Intercomparison, PhD Thesis, St. Catherine's College, Oxford, UK 1989
3.135 H.F. Beyer, P. Indelicato, K.D. Finlayson, D. Liesen, R.D. Deslattes: Phys. Rev. A **43**, 223 (1991)
3.136 M. Tavernier, J.P. Briand, P. Indelicato, D. Liesen, P. Richard: J. Phys. B **18**, L327 (1985)
3.137 P. Mokler, Th. Stöhlker, C. Kozhuharov, R. Moshammer, P. Rymuza, Z. Stachura, A. Warczak: J. Phys. B **28**, 617 (1995)
3.138 G. Soff : Unpublished (1993)
3.139 J.H. Lupton, D.D. Dietrich, C.J. Hailey, R.E. Stewart, K.P. Ziock: Phys. Rev. A **50**, 2150 (1994)

Chapter 4

4.1 I.I. Sobelman: *Atomic Spectra and Radiative Transitions*, 2nd edn. Springer Ser. Atoms Plasmas, Vol. 12 (Springer, Berlin, Heidelberg 1992)
4.2 A.I. Akhiezer, V.B. Berestetskii: *Quantum Electrodynamics*, Interscience Monographs and Texts in Physics and Astronomy, (Wiley, New York 1965)
4.3 J.J. Sakurai: *Advanced Quantum Mechanics*, 10th edn. (Addison-Wesley, Ontario 1984)
4.4 R. Marrus, P.J. Mohr: Adv. At. Mol. Phys. **14**, 181 (1978)
4.5 W.R. Johnson: Phys. Rev. Lett. **29**, 1123 (1972)
4.6 F.A. Parpia, W.R. Johnson: Phys. Rev. A **26**, 1142 (1982)
4.7 S.P. Goldman, G.W.F. Drake: Phys. Rev. A **24**, 183 (1981)
4.8 S.P. Goldman: Phys. Rev. A **40**, 1185 (1989)
4.9 V.G. Pal'chikov, V.P. Shevelko: *Reference Data on Multicharged Ions*, Springer Ser. Atoms Plasmas, Vol. 16 (Springer, Berlin, Heidelberg 1995)
4.10 R.W. Dunford, M. Hass, E. Bakke, H.G. Berry, C.J. Liu, M.L.A. Raphaelian, C.J. Curtis: Phys. Rev. Lett. **62**, 2809 (1989)

4.11 V.A. Boiko, V.G. Pal'chikov, I.Yu. Skobelev, A.Ya. Faenov: *The Spectroscopic Constants of Atoms and Ions* (CRC, Boca Raton 1994)
4.12 G.W.F. Drake: Nucl. Instrum. Methods. B **9**, 465 (1985)
4.13 G.W.F. Drake: Phys. Rev. A **34**, 2871 (1986)
4.14 N.M. Cann, A.J. Thakkar: Phys. Rev A **46**, 5397 (1992)
4.15 W.R. Johnson, C.D. Lin: Phys. Rev.A **14**, 565 (1976)
4.16 C.D. Lin, W.R. Johnson, A. Dalgarno: Phys. Rev.A **15**, 154 (1977)
4.17 J. Hata, I.P. Grant: J. Phys. B **14**, 2111 (1981)
4.18 J. Krause: Phys. Rev.A **34**, 3692 (1986)
4.19 E. Lindroth, S. Salomonson: Phys. Rev.A **41**, 4659 (1990)
4.20 G. Feinberg, J. Sucher: Phys. Rev.Lett. **26**, 681 (1971)
4.21 W.R. Johnson, C.D. Lin: Phys. Rev.A **9**, 1486 (1974)
4.22 W.R. Johnson, D.R. Plante, J. Sapirstein: Adv. At. Mol. Opt. Phys. **35**, 255 (1995)
4.23 V.G. Pal'chikov: J. Appl. Spectrosc. **52**, 488 (1990)
4.24 R. Ali, I. Ahmad, H.G. Berry, R.W. Dunford, D.S. Gemmel, E.P. Kanter, P.H. Mokler, A.E. Livingston, S. Cheng, L.J. Curtis: Nucl. Instrum. Methods. **98**, 69 (1995)
4.25 R.C. Grossley: Phys. Scr. **T8**, 117 (1984)
4.26 I. Martinson, L.J. Curtis: Contemp. Phys. **30**, 173 (1989)
4.27 I. Martinson: Nucl. Instrum. Methods. **43**, 323 (1989)
4.28 L.J. Curtis, S.T. Maniak, R.W. Ghzist, R.E. Irving, D.G. Ellis, M. Henderson, M.H. Kacher, E. Träbert, J. Granzow, P. Bengtsson, L. Engström: Phys. Rev. A **51**, 4575 (1995)
4.29 X.-W. Zhu, K.T. Chang: Phys. Scr. **52**, 654 (1995)
4.30 J. Fleming, N. Vaeck, A. Hibbert, K.L. Bell, M.R. Godefroid: Phys. Scr. **53**, 446 (1996)
4.31 L.J. Curtis, D.G. Ellis: J. Phys. B **29**, 645 (1996)
4.32 E. Träbert, E.H. Pinnington, J.A. Kernahan, J. Dorfert, J. Granzow, P.H. Heckmann, R. Hutton: J. Phys. B **29,** 2647 (1996)
4.33 P. Quinet, J.E. Hansen: J. Phys. B **29,** 1879 (1996)
4.34 D.H. Baik, Y.G. Ohr, K.S. Kim, J.M. Lee, P. Indelicato, Y.K. Kim: At. Data Nucl. Data Tables **47,** 177 (1991)
4.35 A. Hibbert, M. Le Dourneuf, M. Mohan: At. Data Nucl. Data Tables **53,** 23 (1993)
4.36 V.P. Shevelko: *Atoms and Their Spectroscopic Properties*, Springer Ser. Atoms Plasmas, Vol. 18 (Springer, Berlin, Heidelberg 1996)
4.37 E. Träbert: Phys. Scr. **48**, 699 (1993)
4.38 J. Doerfert, E. Träbert, J. Granzow, A. Wolf, J. Kenntner, D. Habs, M. Grieser, T. Schüssler, U. Schramm, P. Forck: Nucl. Instrum. Methods. B **98,** 53 (1995)
4.39 J. Doerfert, E. Träbert, A. Wolf: Hyp. Interact. **99,** 155 (1996)
4.40 J.D. Gillaspy, E. Aglitskiy, E.W. Bell, C.M. Brown, C.T. Chantler, R.D. Deslattes, U. Feldmann, L.T. Hudson, J.M. Laming, E.S. Meyer, C.A. Morgan, A.I. Pipkin, J.R. Roberts, L.P. Ratliff, F.G. Serpa, J. Sugar, E. Tacács: Phys. Scr. **T59,** 392 (1995)
4.41 A. Simionovici, B.B. Birkett, J.P. Briand, P. Charles, D.D. Dietrich, K. Finlayson, P. Indelicato, D. Liesen, R. Marrus: Phys. Rev. A **48**, 1695 (1993)
4.42 R. Marrus, V. San Vincente, P. Charles, J.P. Briand, F. Bosch, D. Liesen, I. Varga: Phys. Rev. Lett. **56**, 1683 (1986)
4.43 L. Engström, C. Jupén, B. Denne, S. Huldt, W.T. Meng, P. Kaijser, J.O. Ekberg, U. Litzén, I. Martinson: Phys. Scr. **22**, 570 (1981)
4.44 T.W. Tunnel, C.P. Bhalla, C. Can: Phys. Lett. A **75,** 195 (1980)

4.45 I.P. Grant, B.J. McKenzie, P.H. Norrington, D.F. Mayers, N.C. Pyper: Comput. Phys. Commun. **21**, 207 (1980)
4.46 P. Indelicato, F. Parente, R. Marrus: Phys. Rev. A **56**, 1683 (1986)
4.47 K. Ando, Y. Zon, T. Kambara, Y. Nakai, Y. Kanai, M. Oura, Y. Awaya, T. Tonuma: Phys. Scr. **53**, 33 (1996)
4.48 R. Marrus, P. Charles, P. Indelicato, L. de Billy, C. Tazi, J.P. Briand, A. Simionovici, D. Dietrich, F. Bosch, D. Liesen: Phys. Rev. A **39**, 3725 (1989)
4.49 H. Gould, R. Marrus: Phys. Rev. A **28,** 2001 (1983)
4.50 C.T. Munger, H. Gould: Phys. Rev. Lett. **57,** 2927 (1986)
4.51 M. Westerlind, N. Reistad, C. Jupén, L. Engström: Nucl. Instrum. Methods. B **31,** 300 (1988)
4.52 C.A. Morgan, F.G. Serpa, E. Takács, E.S. Meyer, J.D. Gillaspy, J. Sugar, J.R. Roberts, C.W. Brown, U. Feldman: Phys. Rev. Lett. **74,** 1716 (1995)
4.53 L. LaJohn, T.M. Luke: Phys. Scr. **47**, 542 (1993)
4.54 E. Träbert, P.H. Heckmann, J. Doerfert, J. Granzow: Phys. Scr. **47**, 780 (1993)
4.55 J. Granzow, P.H. Heckmann, E. Träbert: Phys. Scr. **49**, 148 (1994)
4.56 J. Fleming, A. Hibbert: Phys. Scr. **51**, 339 (1995)
4.57 M. Finkbeiner, B. Fricke, W.-D. Sepp: Phys. Scr. **52**,258 (1995)
4.58 L.P. Presnyakov: Sov. Phys. - Usp. **19**, 387 (1976)
4.59 R.K. Janev, L.P. Presnyakov, V.P. Shevelko: *Physics of Highly Charged Ions* Springer Ser. Atoms Plasmas, Vol. 13 (Springer, Berlin, Heidelberg 1985)
4.60 E.V. Aglitsky, U.I. Safronova: *Spectroscopy of Autoionizing States of Multicharged Ions* (Energoatomizdat, Moscow 1985) (in Russian)
4.61 A. Müller: In *Physics of Ion Impact Phenomena*, ed. D.Mathur, Springer Ser. Chem. Phys., Vol. 54 (Springer, Berlin, Heidelberg 1991)
4.62 Y. Hahn: In *Atomic and Molecular Processes in Fusion Edge Plasmas*, ed. R.K. Janev (Plenum, New York 1995)
4.63 E.H.S. Burhop: *The Auger Effect and other Radiationless Transitions* (Cambridge Univ. Press, Cambridge 1952)
4.64 A. Temkin (ed.): *Autoionization* (Plenum, New York 1985)
4.65 P.B. Ivanov, U. Safronova: Phys. Scr. **49**, 408 (1994)
4.66 K.R. Karim: Phys. Scr. **49**, 565 (1994)
4.67 U. Safronova, M.S. Safronova, N.J. Snyderman, V.G. Pal'chikov: Phys. Scr. **50**, 29 (1994)
4.68 U. Safronova, R. Bruch: Phys. Scr. **50**, 45 (1994)
4.69 U. Safronova, M.S. Safronova, R. Bruch: J. Phys. B **28**, 2803 (1995)
4.70 U. Safronova, M.S. Safronova, R. Bruch, L.A. Vainshtein: Phys. Scr. **51**, 471 (1995)
4.71 U. Safronova, A.S. Shlyaptseva, I.E. Golovkin: Phys. Scr. **52**, 277 (1995)
4.72 J. Nilsen, U. Safronova, M.S. Safronova: Phys. Scr. **51**, 589 (1995)
4.73 P. Feldman, R. Novick: Phys. Rev. **160**, 143 (1967)
4.74 Ch. Froese-Fischer: *The Hartree-Fock Method for Atoms* (Wiley, New York 1977)
4.75 F. Bely-Dubau, J. Dubau, P. Faucher, L.J. Steenman-Clark: J. Phys. B **14**, 3313 (1981)
4.76 L.A. Vainshtein, U.I. Safronova: Atom. Data Nucl. Data Tables **21**, 49 (1978)
4.77 L.A. Vainshtein, U.I. Safronova: Phys. Scr. **31**, 519 (1985)
4.78 M. Cornille, J. Dubau, F. Bely-Dubau, P. Faucher, I.A. Ivanov, U.I. Safronova, J. Nilsen: J. Phys. B **23**, 4451 (1990)
4.79 J. Nilsen, M. Cornille, J. Dubau, F. Bely-Dubau, P. Faucher, I.A. Ivanov, U.I. Safronova, : J. Quant. Spectrosc. Radiat. Transfer **48**, 61 (1992)

Chapter 5

5.1 R.K. Janev, L.P. Presnyakov, V.P. Shevelko: *Physics of Highly Charged Ions* Springer Ser. Electrophys., Vol. 13 (Springer, Berlin, Heidelberg 1985)
5.2 A. Müller: Hyp. Interact. **99,** 31 (1996)
5.3 H.A. Bethe, E.E. Salpeter: *Quantum Mechanics of One- and Two-Electron Atoms* (Plenum, New York 1977)
5.4 U. Fano, J.W. Cooper: Rev. Mod. Phys. **40**, 441 (1968)
5.5 D.H. Sampson: *Atomic Photoionization* (Springer, Berlin, Heidelberg 1982)
5.6 M.Ya. Amusia: *Atomic Photoeffect* (Plenum, New York 1990)
5.7 V. Schmidt: Rep. Progr. Phys. **55**, 1483 (1992)
5.8 T.-N. Chang (ed.): *Many-body Theory of Atomic Structure and Photoionization* (World Scientific, Singapore 1993)
5.9 D.M.P. Holland, K. Codling, J.B. West, G.V. Marr: J. Phys. B **12**, 2465 (1979)
5.10 N. Saito, I.H. Suzuki: Int'l J. Mass. Spectr. and Ion Processes **115,** 157 (1992)
5.11 D.E. Osterbrock: *Astrophysics of Gaseous Nebulae and Active Galactic Nuclei* (University Science, Mill Valley, CA, 1989) p. 389
5.12 V.P. Shevelko, L.A. Vainshtein: *Atomic Physics for Hot Plasmas* (IOP, Bristol 1993)
5.13 J.M. Harriman: Phys. Rev. **101**, 594 (1956)
5.14 S.-C. Li: *Physics of High Temperature Radiation and the Quantum Theory of Radiation* (Publishing House of the National Defence Industry, Beijing 1992) (in Chinese)
5.15 W.J. Karzas, R. Latter: Astrophys. J. Suppl., **6**, 167 (1961)
5.16 K. Omidvar, A.M. McAllister: Phys. Rev. A **51**, 1063 (1995)
5.17 I.L.Beigman, L.A. Vainshtein, B.N. Chichkov: Sov. Phys. - JETP **53**, 541 (1981)
5.18 M. Pajek, R. Schuch: Phys. Rev. A **45**, 7894 (1992)
5.19 N.B. Delone, S.P. Goreslavsly, V.P. Krainov: J. Phys. B **27**, 4403 (194)
5.20 H.A. Kramers: Philos. Mag. **46**, 836 (1923)
5.21 M. Stobbe: Ann. Phys. (Leipzig) **7**, 661 (1930)
5.22 V.I. Kogan, A.B. Kukushkin, V.S. Lisitsa: Phys. Rep. **213,** 1 (1992)
5.23 I. Stohr (ed.): *X-Ray Absorption: Principles, Applications, Techniques of EXAFS, SEXAFS and XANES* (Wiley, New York 1988)
5.24 E. Jannitti, P. Nicolosi, G. Tondello: Phys. Scr. **36**, 93 (1987)
5.25 E. Jannitti, P. Nicolosi, G. Tondello: Phys. Lett. **131**, 186 (1988)
5.26 E. Jannitti, P. Nicolosi, G. Tondello: Photoionization Measurements of Light Ions in the Soft X-Ray Region, in *X-Ray and Inner-Shell Processes*, ed. by T.A. Carlson, M.O. Krause, S.T. Manson (AIP, New York 1990)
5.27 P. van Kampen, L. Kiernan, J.T. Costello, E.T. Kennedy, J.A.M. van der Mullen, G. O'Sullivan: J. Phys. B **290**, 4771 (1995)
5.28 R.F. Reilman, S.T. Manson: Astrophys. J. Suppl. Ser. **40**, 815 (1979)
5.29 K.L. Bell, A. Kingston: J. Phys. B **4**, 1308 (1971)
5.30 R.E. Clark, R.D. Cowan, F.W. Bobrowicz: At. Data Nucl. Data Tables **34,** 415 (1986)
5.31 K.S. Baliyan, R.H.G. Reid, A.E. Kingston: Phys. Scr. **46**, 115 (1992)
5.32 Y. Yu., M.J. Seaton: J. Phys. B **20**, 6409 (1987)
5.33 K. Butler, C. Mendoza, C.J. Zeippen: J. Phys. B **26,** 4409 (1993)
5.34 J.M. Bizau, D. Cubaynes, M. Richter, F. Wuilleumier, J. Obert, C.J. Putaux: Rev. Sci. Instrum. **63**, 1389 (1992)
5.35 M.J. Seaton: J. Phys. B **20**, 6363 (1987)

5.36 C. Mendoza: Atomic and Molecular Data for Space Astronomy: Needs, Analysis and Availability, *Lecture Notes Phys.*, 407, 85 ed. by P.L. Smith, W.L. Wiese (Springer Berlin, Heidelberg 1992)
5.37 Opacity Project Group: *Opacity Project*, Vols. 1 and 2 (IOP, Bristol 1996)
5.38 C.J. Zeippen: Phys. Scr. **T58**, 43 (1995)
5.39 A. Ron, I.B. Goldberg, J. Stein, S.T. Manson, R.H. Pratt, R.Y. Yin: Phys. Rev. A **50**, 1312 (1994)
5.40 A.F. Starace: In *Handbuch der Physik*, Vol. 31, ed. W. Mehlhorn (Springer, Berlin, Heidelberg 1979)
5.41 J. Berkowitz: *Photoabsorption, Photoionization and Photoelectron Spectroscopy* (Academic, New York 1979)
5.42 R.H. Pratt, A. Ron, H.K. Tseng: Rev. Mod. Phys. **45**, 273 (1973)
5.43 H.F. Beyer, D. Liesen, F. Bosch, K.D. Finlayson, M. Jung, O. Klepper, R. Moshammer, K. Beckert, H. Eickhoff, B. Franzke, F. Nolden, P. Spädtke, M. Steck, G. Menzel, R.D. Deslattes: Phys. Lett. A **184,** 435 (1994)
5.44 D. Liesen, H.F. Beyer, K.D. Finlayson, F. Bosch, M. Jung, O. Klepper, R. Moshammer. K. Beckert, H. Eickhoff, B. Franzke, F. Nolden, P. Spädtke, M. Steck, G. Menzel, R.D. Deslattes: Z. Phys. D **30,** 307 (1994)
5.45 P.F. Dittner, S. Datz: Early Measurements of Dielectronic Recombination: Multiply Charged Ions, in *Recombinaton of Atomic Ions*, ed. by W.G. Graham, W. Fritsch, Y. Hahn, J.A. Tanis, NATO ASI Series B: Physics, Vol. 296 (Plenum, New York 1992) pp. 133–141
5.46 H. Danared, G. Ander, L. Bagge, C.J. Herrlander, J. Hilke, J. Jeansson, A. Kallberg, A. Nilsson, A. Paal, K.G. Rensfelt, U. Rosengart, J. Starker, M. af Ugglas: Phys. Rev. Lett. **72**, 3 (1994)
5.47 L.H. Andersen, J. Bolko, P.I. Kvistgard: Phys. Rev. Lett. **64**, 729 (1990); L.H. Andersen, J. Bolko: J. Phys. B **23**, 3167 (1990); L.H. Andersen, J. Bolko: Phys. Rev. A **42**, 1184 (1990)
5.48 L.H. Andersen, G.-Y. Pan, H.T. Schmidt: J. Phys. B **25**, 277 (1992); H.T. Schmidt, G.-Y. Pan, L.H. Andersen: J. Phys. B **25**, 3165 (1992);
5.49 D.J. McLaughlin, Y. Hahn: Phys. Rev. A **43**, 1313 (1991)
5.50 Y. Hahn: Electron-Ion Recombination Processes in Plasmas, in: *Atomic and Molecular Processes in Fusion Edge Plasmas*, ed. by R.K. Janev (Plenum, New York 1995)
5.51 W. Spies, A. Müller, O. Uwira, C. Kozhuharov, M. Zimmermann, N. Grün, W. Scheid, M.S. Pindzola, N.R. Badnell: In Annual Rept. 1995, Institut für Strahlenphysik, ed. by R. Kunz and H.H. Pitz (Universität Stuttgart, 1995) p. 36
5.52 A. Müller: Nucl. Instrum. Methods. B **99**, 58 (1995)
5.53 H. Gao, D.R. DeWitt, R. Schuch, W. Zong, S. Asp, M. Pajek: Phys. Rev. Lett. **75**, 4381 (1995)
5.54 K. Omidvar, P.T. Guimaraes: Astrophys. J. Suppl. **73**, 555 (1990)
5.55 K. Omidvar: Phys. Rev. A **48**, 1997 (1993)
5.56 M.S. Adams, M.V. Fedorov, V.P. Krainov, D.D. Meyerhofer: Phys. Rev. A **52**, 125 (1995)
5.57 A. Burgess: Mon. Not. R. Astron. Soc. **118**, 477 (1958)
5.58 M.J. Seaton: Mon. Not. R. Astron. Soc. **119**, 81 (1959)
5.59 W.J. Boardman: Astrophys. J. Suppl. **9**, 185 (1964)
5.60 S.L. Haan, V.L. Jakobs: Phys. Rev. A **40**, 80 (1989)
5.61 O. Uwira, W. Spies, A. Frank, J. Linkemann, T. Cramer, C. Brandau, A. Müller, C. Kozhuharov, J. Klabunde, N. Angert, P.H. Mokler, R. Becker,M. Kleinod: In Annual Rept. 1995, Institut für Strahlenphysik, ed. by R. Kunz and H.H. Pitz (Universität Stuttgart, 1995) p. 38

5.62 D. Liesen, H.F. Beyer: GSI Rept. No. GSI-ESR/86-04 (Gesellschaft für Schwerionenforschung, Darmstadt 1986)
5.63 H.F. Beyer: J. Phys. (Paris) **50,** Cl-471 (1989)
5.64 H.F. Beyer: IEEE Trans. IM-**44,** 510 (1995)
5.65 H.F. Beyer, G. Menzel, D. Liesen, A. Gallus, F. Bosch, R.D. Deslattes, P. Indelicato, T. Stöhlker, O. Klepper, R. Moshammer, F. Nolden, H. Eickhoff, B. Franzke, M. Steck: Z. Phys. D **35,** 169 (1995)
5.66 D. Liesen: HIghly Charged Ions in Storage Rings, Proc. 17th Int'l. Conf. X-Ray Inner-Shell Phenomena, Sept. 9–13, Hamburg 1996, ed. by R.L. Johnson (AIP, New York 1996)
5.67 P.H. Mokler, Th. Stöhlker: *The Physics of Highly-Charged Ions Revealed by Storage/Cooler Rings*, Adv. At. Mol. Opt. Phys. **37**, 297 (1996)
5.68 M.S. Pindzola, F.J. Robicheaux, N.R. Badnell, M.H. Chen, M. Zimmermann: Phys. Rev. A **52**, 420 (1995)
5.69 A. Müller, S. Schennach, M. Wagner, J. Hasenbauer, O. Uwira, W. Spies: Phys. Scr. **T37**, 62 (1991)
5.70 A. Wolf, J. Berger, M. Bock, D. Hals, B. Hochadel, G. Kilgus, G. Neuzeither, U. Schzamm, D. Schwalm, E. Szmola, A. Müller, M. Wagner, R. Schuch: Z. Phys. D **21**, S69 (1991)
5.71 S. Schenach, A. Müller, O. Uwira, J. Haselbauer, W. Spies, A. Frank, M. Wagner, R. Becker, M. Kleinod, E. Jennewein, N. Angert, P.H. Mokler, N.R. Badnell, M.S. Pindzola: Z. Phys. D **30,** 291 (1994)
5.72 A. Frank, A. Müller, J. Haselbauer, S. Schenach, W. Spies, O. Uwira, M. Wagner, R. Becker, M. Kleinod, E. Jennewein, N. Angert, P.H. Mokler: *Proc. 6th Int'l. Conf. on the Physics of Highly Charged Ions*, ed. by P.Richard, M.Stöckli, C.L.Cocke, C.D.Lin, AIP Conf. Proc. **274**, 532 (AIP, New York, 1993)
5.73 Y. Hanh, P. Kristic: J. Phys. B **27**, L509 (1994)
5.74 N.R. Badnell, M.S. Pindzola: Phys. Rev. A **45**, 2820 (1992)
5.75 V.M. Shabaev: Phys. Rev. A **50**, 4521 (1994)
5.76 A. Sommerfeld: *Atombau und Spectrallinien* (Vieweg und Sohn, Braumschweig 1939)
5.77 W. Heitler: *Quantum Theory of Radiation*, 3rd edn. (Oxford Univ. Press, Oxford 1954)
5.78 L.D. Landau, L.M. Lifshitz: *The Classical Theory of Fields*, 3rd edn. (Pergamon, New York 1971)
5.79 J.D. Jackson: *Classical Electrodynamics*, 2nd edn. (Wiley, New York 1975)
5.80 J.M. Jauch, R. Rohrlich: *The Theory of Photons and Electrons*, 2nd ed. (Springer, Berlin, Heidelberg 1980)
5.81 R.H. Pratt, I.J. Fend: Electron-Atom Bremsstarhlung, in *Atomic Inner-Shell Physics*, ed. by B. Crasemann (Plenum, New York 1985) pp. 533–580
5.82 V.N. Tsytovitch, I.M. Oiringel (eds.): *Rolarization Bremsstrahlung of Particles and Atoms* (Nauka, Moscow 1987) (in Russian)
5.83 M.Ya. Amusia: Phys. Rep. **1620**, 249 (1988)
5.84 H.A. Bethe, W. Heitler: Proc. Roy. Soc. (London) A **146**, 83 (1934)
5.85 H.A. Bethe, L. Maximon: Phys. Rev. **93**, 768 (1954)
5.86 H.W. Koch, J.W. Motz: Rev. Mod. Phys. **31**, 920 (1959)
5.87 H. McMaster: Rev. Mod. Phys. **33**, 8 (1961)
5.88 L. Kissel, A. Quarles, R.H. Pratt: At. Data Nucl. Data Tables, **28**, 381 (1983)
5.89 M. Zeltzer, M.J. Berger: Nucl. Instrum. Methods. B **12**, 95 (1985)
M. Zeltzer, M.J. Berger: At. Data Nucl. Data Tables **35**, 345 (1986)
5.90 A. Dubois, A. Maquet: Phys. Rev. A **40**, 4288 (1989)
5.91 A.A. Al-Beteri, D.E. Raeside: Nucl. Instum. Methods. B **44**, 149 (1989)

5.92 L.C. Maximon, A. de Miniac, T. Aniel, E. Ganz: Phys. Rep. **147**, 189 (1987)
5.93 T.C. Chu, K. Ishii, A. Yamadera, M. Sebata, S. Morita: Phys. Rev. A **24** 1720 (1980)
5.94 H.W. Schnopper, J.P. Delvaille, K. Kalata, A.R. Sohval, M. Abdulwahab, K.W. Jones, H.E. Wegner: Phys. Lett. A **47**, 61 (1974)
5.95 F. Folkmann, G. Gaarde, T. Huus, K. Kemp: Nucl. Instrum. Methods. **116**, 487 (1974)
5.96 K. Ishii, S. Morita, H. Tawara: Phys. Rev. A **13**, 131 (1976)
5.97 K. Ishii, S. Morita: Phys. Rev. A **30**, 2278 (1984)
5.98 K. Ishii, S. Morita: Int. J. PIXE A **1**, 1 (1990)
5.99 K. Ishii: Nucl. Instrum. Methods. B **99**, 163 (1995)
5.100 P.C. Fischer: Phys. Rev. **92,** 420 (1953)
5.101 C.D. Curtis: Phys. Rev. **89,** 123 (1953)
5.102 E. Haug: Solar Phys. **71**, 77 (1981)
5.103 M. Lamoureux: Adv. At. Mol. Phys. **31**, 233 (1993)
5.104 A.M. Urnov: *Spectroscopic Diagnostics of Nouthermal Electrons in Hot Astrophusical and Laboratory Plasmas*, Rept. LLNL B239718 (Livermore 1994)
5.105 J. Dubau, M.K. Inal, A.M. Urnov: Phys. Scr. **T65**, 179 (1996)
5.106 J.C. Kieffer, J.P. Matte, H. Pepin, M. Chaker, Y. Beaudoin, T.W. Johnston, C.Y. Chien, S. Coe, G. Mourou, J. Dubau: Phys. Rev. Lett. **68**, 480 (1992)
5.107 J.C. Kieffer, J.P. Matte, M. Chaker, Y. Beaudoin, C.Y. Chien, S. Coe, G. Mourou, J. Dubau, M.K. Inal: Phys. Rev. E **48**, 4648 (1993)
5.108 M. Lamoureux, L. Jacquet, R.H. Pratt: Phys. Rev. A **39**, 6323 (1989)
5.109 R.K. Kirkwood, I.H. Hutchinson, S.C. Luckhardt, J.P. Squire: Nucl. Fusion **30**, 431 (1990)
5.110 R. Beier, C. Bachmann, R. Burhenn: J. Phys. D **14**, 643 (1981)
5.111 I.P. Tindo, V.D. Ivanov, S.L. Mandelshtam, A.I. Shurigin: Solar Phys., **24**, 429 (1970)
5.112 M. Siarkowski, I. Sylwester, G. Bromboszsz, V.V. Korneev, S.L. Mandelshtam, S.W. Oparin, A.M. Urnov, I.A. Zhitnik: Solar Phys., **81**, 63 (1982)
5.113 V.V. Korneev, S.L. Mandelshtam, S.N. Oparin, A.M. Urnov, I.A. Zhitnik: Adv. Space Res., **2**, No.11, 139 (1983)
5.114 S.L. Mandelshtam, A.M. Urnov, I.A. Zhitnik: Adv. Space Res., **4**, No.7, 87 (1984)
5.115 D.R. Kania, L.A. Jones: Phys. Rev. Lett., **53**, 166 (1984); ibid **55**, 1993 (1985)
5.116 S. von Goeler, J. Stevens, W. Stodiek, S. Bernabet, M. Bitter, K.W. Hill, D. Hillis, W. Hooke, F. Jobes, G. Lenner, E. Meservey, R. Motley, N. Sauthoff, S. Sesnic, F. Tenney: Rept. PPPL-2010, Princeton (1983)
5.117 L. Nocera, Yu.I. Skrynnikov, B.V. Somov: Solar Phys. **97**, 81 (1985)
5.118 C.M. Brown, U. Feldman, G.A. Doschek, J.F. Seely, R.E. LaVilla, V.L. Jacobs, J.R. Henderson, M.A. Levine: Phys. Rev. A **40**, 4089 (1989)
5.119 J.P. Apruzcsc, P.G. Burkhalter, J.E. Rogerson, J. Davis, J.F. Seely, C.M. Brown, D.A. Newman, R.W. Clark, J.P. Knauer, D.K. Bardley: Phys. Rev. A **39**, 5697 (1989)
5.120 J.R. Henderson, P. Beiersdorfer, C.L. Bennet, S. Chantrenne, D.A. Knapp, R.E. Marrs, M.B. Schneider, K.L. Wong, G.A. Doschek, J.F. Seely, C.M. Brown, R.E. LaVilla, J. Dubau, M.A. Levine: Phys. Rev. Lett. **65**, 705 (1990)
5.121 N.J. Fisch: Rev. Mod. Phys. **59**, 175 (1987)
5.122 V.V. Krutov, V.V. Korneev, S.L. Mandelshtam, I.P. Tindo, A.M. Urnov: Preprint P.N. Lebedev Physics Institute, Vol.133 (Moscow 1981)
5.123 K. Akita, K. Tanaka, T. Watanabe: Solar Phys. **86**, 101 (1983)

5.124 T. Fujimoto, F. Koike, K. Sakimoto, R. Okasawa, K. Kawasaki, K. Takiama, T. Oda, T. Kato: Int'l. Conf. Atomic Processes Relevant to Polarization Plasma Spectroscopy, NIFS-Data 16 (Nagoya, Japan 1992) Abstr.
5.125 I.C. Percival, M.J. Seaton: Philos. Trans. R. Soc. A **251**, 113(1958)
5.126 U. Fano, J.H. Macek: Rev. Mod. Phys. **45**, 533 (1973)
5.127 M.K. Inal, J. Dubau: J. Phys. B **20**, 4221 (1987); J. Phys. B **22**, 3329 (1989)
5.128 A.S. Shyaptseva, A.M. Urnov, A.V. Vinogradov: In *Proc. P.N. Lebedev Physics Institute*, Vol. 195, p. 85 (Nova Science, New York 1989)
5.129 Yu.P. Garbuzov: Polarization of X-Ray Lines in Multicharged Ions, Thesis, Moscow Physico-Technical Institute, Moscow (1990)
5.130 J.H. Scofield, M.H. Chen: Phys. Rev. A **52**, 2057 (1995)
5.131 K. Blum: *Density Matrix Theory and Applications* (Plenum, New York 1981)
5.132 W. Eissner, M.J. Seaton: J. Phys. B **5**, 2187 (1972)
5.133 H.E. Saraph: Comput. Phys. Commun. **15**, 247 (1978)
5.134 J. Mitroy: Phys. Rev. A **37**, 649 (1988); J. Mitroy, D.W. Nocross: Phys. Rev. A **37**, 3755 (1988)
5.135 H.L. Zhang, D.H. Sampson, R.E.H. Clark: Phys. Rev. A **41**, 198 (1990)
5.136 K.J. Reed, M.H. Reed: Phys. Rev. A **48**, 3644 (1993)
5.137 J.H. Scofield: Phys. Rev. A **40**, 3054 (1989)
5.138 D. Laundy: Nucl. Instrum. Methods. A **290**, 248 (1990)
5.139 C. Laulké, E. Jackuet, G. Cremer, J. Pascale, P. Boduch, G. Rieger, D. Lecler, M. Chantepie, J.L. Cojan: Phys. Rev. A **52,** 3803 (1995)
5.140 Z. Stachura, F. Bosch, F.J. Hambsch, B. Liu, D. Maor, P.H. Mokler, W.A. Schönfeldt, H. Wahl, B. Cleff, M. Brussermann, J. Wigger: J. Phys. B **17**, 835 (1984)
5.141 R.N. Badnell: Phys. Rev. A **42**, 3795 (1990)
5.142 D.L. Matthews, P.L. Hagelstein, M.D. Rosen, M.J. Eckart, N.M. Ceglio, A.U. Hazi, H. Medecki, B.J. MacGowan, J.E. Trebes, B.L. Whitten, E.M. Campbell, C.W. Hatcher, A.M. Hawryluk, R.L. Kauffman, L.D. Pleasance, G. Rambach, J.H. Scofield, G. Stone, T.A. Weaver: Phys. Rev. Lett. **54**, 110 (1985)
5.143 S. Suckewer, C.H. Skinner, H. Milchberg, C. Keane, D. Vorhees: Phys. Rev. Lett. **55**, 1753 (1985)
5.144 R.C. Elton: *X-ray Lasers* (Academic, New York 1990)
5.145 C.H. Skinner: Phys. Fluids B **3**, 2420 (1991)
5.146 E.E. Fill (ed.): *Proc. 3rd Int'l. Conf. on X-ray Lasers*, Schliersee, Germany, IOP Conf. Proc. **125** (IOP, Bristol 1992)
5.147 L.B. Da Silva, B.J. MacGowan, J.A. Koch, S.J. Mrowka, D.L. Matthews, D.C. Eder, R.A. London: *Short-Pulse High-Intensity Lasers and Applications II*, SPIE Proc. **1860**, 152 (1993)
5.148 D.L. Matthews: Nucl. Instrum. Methods. B **98**, 91 (1995)
5.149 E.Ya. Kononov, A.E. Kramida, L.I. Podobedova, E.N. Ragozin, V.A. Chirkov: Phys. Scr. **28**, 496 (1983)
5.150 E.N. Ragozin, S.S. Churilov, E.Ya. Kononov, A.N. Rabtsev, Yu.F. Zaykin: Phys. Scr. **37**, 742 (1988)
5.151 R.R. Gayazov, A.E. Kramida, L.I. Podobedova, E.N. Ragozin, V.A. Chirkov: *Proc. P.N. Lebedev Physics Institute* (Nova Science, New York 1988) Vol. 179
5.152 J. Nilsen, J.H. Scofield: Phys. Scr. **49**, 588 (1994)
5.153 G. Jamelot: X-ray Lasers, in *Physics with Multiply Charged Ions*, ed. by D. Liesen, NATO ASI Series B: Physics, Vol. 348 (Plenum, New York 1995) p. 291
5.154 N.M. Ceglio, P. Dhez (eds.): *Multilayer Structures and Laboratory X-ray Laser Research*, SPIE Proc. **688**, 76 (1986)

5.155 N.M. Ceglio: X-ray Sci. Technol. **1**, 7 (1989)
5.156 P. Chakraborty: Int. J. Mod. Phys. B **5**, 2133 (1991)
5.157 E. Spiller: *Soft X-Ray Optics* (SPIE Optical Engineering Press, Bellingham, Washington 1994)
5.158 J.J. Rocca, V. Shlyaptsev, F.G. Tomasel, O.D. Cortazar, D. Hartshorn, J.L.A. Chilla: Phys. Rev. Lett. **16**, 2192 (1994)
5.159 G.P. Collins: Phys. Today **22**, 19 (October 1994)

Chapter 6

6.1 A.H. Gabriel: Mon. Not. R. Astron. Soc. **160,** 99 (1972)
6.2 J. Dubau, S. Volonté: Rep. Prog. Phys. **43,** 199 (1980)
6.3 A. Müller: In *Physics of Ion Impact Phenomena*, ed. by D. Mathur, Springer Series in Chemical Physics, Vol. 54 (Springer, Berlin, Heidelberg 1991)
6.4 H.S.W. Massey, D.R. Bates: Rep. Prog. Phys. **9,** 62 (1942)
6.5 A. Burgess: Astrophys. J. **139,** 776 (1964)
6.6 C. Jordan: Mon. Not. R. Astron. Soc. **142,** 499 (1969)
6.7 H. Summers: Mon. Not. R. Astron. Soc. **169,** 663 (1974)
6.8 V.L. Jacobs, J. Davis, P.C. Kepple, M. Blaha: Astrophys. J. **211,** 605 (1977); *ibid* **215,** 106 (1977)
6.9 I.L. Beigman, L.A. Vainshtein, R. Sunyaev: Sov. Phys. - Usp. **11,** 411 (1968)
6.10 J.B.A. Mitchell, C.T. Ng, J.L. Forand, D.P. Levac, R.E. Mitchell, A. Sen, D.B. Miko, J.W. McGowan: Phys. Rev. Lett. **50,** 335 (1983)
6.11 D.S. Belić, G.H. Dunn, T.J. Morgan, D.W. Müller, C. Timmer: Phys. Rev. Lett. **50,** 339 (1983)
6.12 P.F. Dittner, S. Datz, S.D. Miller, C.D. Moak, P.H. Stelson, C. Bottcher, W.B. Dress, G.D. Aston, N. Nestković, C.M. Fou: Phys. Rev. Lett. **51,** 31 (1983)
6.13 K. LaGattuta, I. Nasser, Y. Hahn: Phys. Rev. A **33,** 2782 (1986)
6.14 C. Bottcher, D.C. Griffin, M.S. Pindzola: Phys. Rev. A **34,** 860 (1986)
6.15 A. Müller, D.S. Belić, B.D.D. Paola, N. Djurić, G.H. Dunn: Phys. Rev. A **36,** 599 (1987)
6.16 K. Widmann, P. Beiersdorfer, V. Decaux, M. Bitter: Phys. Rev. A **53,** 2200 (1996)
6.17 A.H. Gabriel: In *Highlights of Astronomy*, ed. by C. de Jager (Reidel, Dordrecht 1971)
6.18 L.P. Presnyakov: Sov. Phys. - Usp. **19**, 387 (1976)
6.19 U. Feldman: Phys. Scr. **24**, 681 (1981)
6.20 R.D. Cowan: *The Theory of Atomic Structure and Spectra* (Univ. California Press, Berkeley 1981)
6.21 E.V. Aglitsky, U.I. Safronova: *Spectroscopy of Autoionizing States of Multicharged Ions* (Energoatomizdat, Moscow 1985) (in Russian)
6.22 A.M. Urnov: *Spectroscopic Diagnostics of Nonthermal Electrons in Hot Astrophysical and Laboratory Plasmas*, Lawrence Livermore Lab., Rept. LLNL No. B239718 (Livermore, 1994)
6.23 L.A. Vainshtein, U.I. Safronova: At. Data Nucl. Data Tables **21**, 49 (1978)
6.24 L.A. Vainshtein, U.I. Safronova: At. Data Nucl. Data Tables **25**, 311 (1980)
6.25 J. Nilsen: At. Data Nucl. Data Tables **37**, 191 (1987)
6.26 J. Nilsen: At. Data Nucl. Data Tables **38**, 339 (1988)
6.27 J. Nilsen, U.I. Safronova, M.S.Safronova: J. Quantit. Spectrosc. Radiat. Transfer, **43**, 445 (1990)

6.28 A. Dasgupta, K.G. Whitney: At. Data Nucl. Data Tables **58**, 77 (1994)
6.29 R.H. Bell, M.J. Seaton: J. Phys. B **18**, 1589 (1985)
6.30 Y. Hahn: Adv. At. Mol. Phys. **21**, 123 (1985)
6.31 D.C. Griffin: Phys. Scr. **T28**, 17 (1989)
6.32 D.R. Bates, H.S.W. Massey: Phil. Trans. Roy. Soc. A **239**, 269 (1943)
6.33 P. Zimmerer, N. Grün, W. Scheid: Z. Phys. D **21**, 215 (1991)
6.34 P. Zimmerer, N. Grün, W. Scheid: J. Phys. B **24**, 2633 (1991)
6.35 P. Zimmerer, N. Grün, W. Scheid: Phys. Lett. A **148**, 457 (1990)
6.36 G. Kilgus, D. Habs, D. Schwalm, A. Wolf, N.R. Badnell, A. Müller: Phys. Rev. A **46**, 5730 (1992)
6.37 D.C. Griffin, M.S. Pindzola: Phys. Rev. A **35**, 2821 (1987)
6.38 N.R. Badnell, M.S. Pindzola: Phys. Rev. A **47**, 2937 (1993)
6.39 N.R. Badnell, M.S. Pindzola: Phys. Rev. A **45**, 2820 (1992)
6.40 A. Lampert, A. Wolf, D. Habs, J. Kenntner, G. Kilgus, D. Schwalm, M.S. Pindzola, N.R. Badnell: Phys. Rev. A **53**, 1413 (1996)
6.41 A. Dasgupta, K.G. Whitney: J. Phys. B **28**, 515 (1995); At. Data Nucl. Data Tables **58**, 77 (1994)
6.42 S. Datz: Nucl. Instrum. Methods. B **24/25**, 3 (1987)
6.43 L.H. Andersen, J. Bolko, P. Kvistgaard: Phys. Rev. A **41**, 1293 (1990)
6.44 L.H. Andersen, G.-Y. Pan, H.T. Schmidt, N.R. Badnell, M.S. Pindzola: Phys. Rev. A **45**, 7868 (1992)
6.45 L.H. Andersen, P. Hvelplund, H. Knudsen, P. Kvistgaard: Phys. Rev. Lett. **62**, 2656 (1989)
6.46 L.H. Andersen, G.-Y. Pan, H.T. Schmidt, M.S. Pindzola, N.R. Badnell: Phys. Rev. A **45**, 6332 (1992)
6.47 O. Uwira, A. Müller, W. Spies, A. Frank, J. Linkemann, L. Empacher, P.H. Mokler, R. Becker, M. Kleinod, S. Ricz: Nucl. Instrum. Methods. B **98**, 162 (1995)
6.48 S. Schenach, A. Müller, O. Uwira, J. Haselbauer, W. Spies, A. Frank, M. Wagner, R. Becker, M. Kleinod, E. Jennewein, N. Angert, P.H. Mokler, N.R. Badnell, M.S. Pindzola: Z. Phys. D **30**, 291 (1994)
6.49 R. Ali, C.P. Bhalla, C.L. Cocke, M. Stöckli: Phys. Rev. Lett. **64**, 633 (1990)
6.50 R. Ali, C.P. Bhalla, C.L. Cocke, M. Schulz, M. Stöckli: Phys. Rev. A **44**, 223 (1991)
6.51 A. Wolf, J. Berger, M. Bock, D. Habs, B. Hochadel, G. Kilgus, G. Neureither, U. Schramm, D. Schwalm, E. Szmola, A. Müller, M. Wagner, R. Schuch: Z. Phys. D **21**, S69 (1991)
6.52 J. Linkemann, A. Müller, W. Spies, O. Uwira, A. Frank, T. Cramer, J. Kenntner, A. Wolf, C. Broude, D. Habs, D. Schwalm, G.H. Dunn, N.R. Badnell, M.S. Pindzola: Annual Rept. 1995, ed. by R. Kunz and H.H. Pitz, Institut für Strahlenphysik, Universität Stuttgart (1996) p. 32
6.53 G. Kilgus, J. Berger, P. Blatt, M. Grieser, D. Habs, B. Hochadel, E. Jaeschke, D. Krämer, R. Neumann, G. Neureither, W. Ott, D. Schwalm, M. Steck, R. Stokstad, E. Szmola, A. Wolf, R. Schuch, A. Müller, M. Wagner: Phys. Rev. Lett. **64**, 737 (1990)
6.54 R. Schuch, W. Zong, D.R. DeWitt, H. Gao, S. Asp, J. Hvarfner, E. Lindroth, H. Danared, A. Källberg: Hyp. Interact. **99**, 317 (1996)
6.55 D.R. DeWitt, R. Schuch, S. Asp, C. Biedermann, H. Gao, W. Zong: Nucl. Instrum. Methods. B **99**, 86 (1995)
6.56 J. Kenntner, J. Linkemann, N.R. Badnell, C. Broude, D. Habs, G. Hofmann, A. Müller, M.S. Pindzola, E. Salzborn, D. Schwalm, A. Wolf: Nucl. Instrum. Methods. B **98**, 142 (1995)

6.57 J. Linkemann, J. Kenntner, A. Müller, A. Wolf, D. Habs, D. Schwalm, W. Spies, O. Uwira, A. Frank, A. Liedtke, G. Hofmann, E. Salzborn, N.R. Badnell, M.S. Pindzola: Nucl. Instrum. Methods. B **98,** 154 (1995)
6.58 P. Beiersdorfer, S. Chantrenne, M.H. Chen, R.E. Marrs, D.A. Vogel, K.L. Wong, R. Zasadzinski: Z. Phys. D **21,** S209 (1991)
6.59 P. Beiersdorfer, T.W. Phillips, K.L. Wong, R.E. Marrs, D.A. Vogel: Phys. Rev. A **46,** 3812 (1992)
6.60 A.J. Smith, P. Beiersdorfer, V. Decaux, K. Widmann, A. Osterheld, M. Chen: Phys. Rev. A **51,** 2808 (1995)
6.61 D.A. Knapp, R.E. Marrs, M.A. Levine, C.L. Bennett, M.H. Chen, J.R. Henderson, M.B. Schneider, J.H. Scofield: Phys. Rev. Lett. **62,** 2104 (1989)
6.62 D.A. Knapp, R.E. Marrs, M.B. Schneider, M.H. Chen, M.A. Levine: Phys. Rev. A **47,** 2039 (1993)
6.63 W. Spies, A. Müller, J. Linkemann, A. Frank, M. Wagner, C. Kozhuharov, B. Franzke, K. Beckert, F. Bosch, H. Eickhoff, M. Jung, O. Klepper, W. König, P.H. Mokler, R. Moshammer, F. Nolden, U. Schaaf, P. Spädtke, M. Steck, P. Zimmerer, N. Grün, W. Scheid, M.S. Pindzola, N.R. Badnell: Phys. Rev. Lett. **69,** 2768 (1992)
6.64 W. Spies, O. Uwira, A. Müller, J. Linkemann, L. Empacher, A. Frank, C. Kozhuharov, P.H. Mokler, F. Bosch, O. Klepper, B. Franzke, M. Steck: Nucl. Instrum. Methods. B **98,** 158 (1992)
6.65 D.A. Knapp, P. Beiersdorfer, M.H. Chen, J.H. Scofield, D. Schneider: Phys. Rev. Lett. **74,** 54 (1995)
6.66 O. Uwira, W. Spies, C. Kozhuharov, A. Müller, J. Linkemann, B. Franzke, K. Beckert, F. Bosch, H. Eickhoff, M. Jung, O. Klepper, P.H. Mokler, R. Moshammer, F. Nolden, U. Schaaf, P. Spädtke, M. Steck: 18th Int'l Conf. on the Physics of Electronic and Atomic Collisions (Århus, 1993) Abstr. Contributed Papers, p. 377
6.67 Y.I. Grineva, V.I. Karev, V.V. Korneev, V.V. Krutov, S.L. Mendelstam, L.A. Vainshtein, B.N. Vasilyev, I.A. Zhitnik: Sol. Phys. **29,** 441 (1973)
6.68 G.A. Doschek, R.W. Kreplin, U. Feldman: Astrophys. J. **233,** L157 (1979)
6.69 M. Bitter, K.W. Hill, N.R. Sauthoff, C. Efthimion, E. Merservey, W. Roney, S. von Goeler, R. Horton, M. Goldman, W. Stodiek: Phys. Rev. Lett. **43,** 129 (1979)
6.70 G.A. Doschek, U. Feldman, R.D. Cowan: Astrophys. J. **245,** 315 (1981)
6.71 J.F. Seely, U. Feldmann, U.I. Safronova: Astrophys. J. **304,** 838 (1986)
6.72 M. Bitter, H. Hsuan, J.E. Rice, K.W. Hill, M. Diesso, B. Grek, R. Hulse, D.W. Johnson, L.C. Johnson, S. von Goeler: Rev. Sci. Instrum. **59,** 2131 (1988)
6.73 U. Fano: Phys. Rev. **124,** 1866 (1961)
6.74 U. Fano, J.W. Cooper: Phys. Rev. **137,** A1364 (1965)
6.75 M.H. Chen: Phys. Rev. A **45,** 4604 (1992)
6.76 L.N. Labzowsky, A.V. Nefiodov: Phys. Rev. A **49,** 236 (1994)
6.77 A.V. Nefiodov, V.V. Karasiev, V.A. Yerokhin: Phys. Rev. A **50,** 4975 (1994)
6.78 S.L. Haan, V.L. Jacobs: Phys. Rev. A **40,** 80 (1989)
6.79 M.S. Pindzola, N.R. Badnell, D.C. Griffin: Phys. Rev. A **46,** 5725 (1992)
6.80 V.M. Shabaev: Phys. Rev. A **50,** 4521 (1994)
6.81 C. Brandau, A. Müller, C. Kozhuharov: Annual Rept. 1995, ed. by R. Kunz and H.H. Pitz, Institut für Strahlenphysik, Universität Stuttgart (1996) p. 66
6.82 B.H. Bransden: Rep. Prog. Phys. **35,** 949 (1972)
6.83 J.S. Briggs, K. Dettmann: Phys. Rev. Lett. **33,** 1123 (1974)
6.84 G. Raisbeck, F. Yiou: Phys. Rev. Lett. **4,** 1858 (1971)

6.85 H.W. Schnopper, H.D. Betz, J.P. Delvaille, K. Kalata, A.R. Sohvalt: Phys. Rev. Lett. **29,** 898 (1972)
6.86 P. Kienle, M. Kleber, B. Povh, R.M. Diamond, F.S. Stephens, E. Grosse, M.R. Maier, D. Proetel: Phys. Rev. Lett. **31,** 1099 (1973)
6.87 Y. Hahn: Phys. Scr. **43,** 386 (1991)
6.88 J. Eichler: Phys. Rep. **193,** 165 (1990)
6.89 J. Eichler, W.E. Meyerhof: *Relativistic Atomic Collisions* (Academic, San Diego 1995)
6.90 M. Kleber, D.H. Jakubassa: Nucl. Phys. A **252,** 152 (1975)
6.91 R. Hagedorn: *Relativistic Kinematics* (Cummings Publ. Comp., Reading, MA 1980)
6.92 E. Byckling, K. Kajantie: *Particle Kinematics* (Wiley, London 1973)
6.93 A. Ichihara, T. Shirai, J. Eichler: Phys. Rev. A **49,** 1875 (1994)
6.94 C.C.J. Roothaan: Rev. Mod. Phys. **32,** 179 (1960)
6.95 C.C.J. Roothaan, P. Bagus: *Methods in Computational Physics* Vol. 2 (Academic, New York 1963)
6.96 E. Clementi, C. Roetti: At. Data Nucl. Data Tables **14,** 177 (1974)
6.97 H.A. Bethe, E.E. Salpeter: *Quantum Mechanics of One- and Two-Electron Atoms* (Plenum, New York 1977)
6.98 F.F. Komarov, M.M. Temkin: J. Phys. B **9,** L255 (1976)
6.99 M. Stobbe: Ann. Physik (Leipzig) **7,** 661 (1930)
6.100 H.F. Beyer, K.D. Finlayson, D. Liesen, P. Indelicato, C.T. Chantler, R.D. Deslattes, J. Schweppe, F. Bosch, M. Jung, O. Kleppner, W. König, R. Moshammer, K. Beckert, H. Heuckhoff, B. Franzke, A. Gruber, F. Nolden, P. Spädtke, M. Steck: J. Phys. B **26,** 1557 (1993)
6.101 B.G. Williams: Phys. Scr. **15,** 92 (1977)
6.102 R.H. Pratt, A. Ron, H.K. Tseng: Rev. Mod. Phys. **45,** 273 (1973)
6.103 A. Ichihara, T. Shirai, J. Eichler: At. Data Nucl. Data Tables **55,** 63 (1993)
6.104 Th. Stöhlker, C. Kozhuharov, P.H. Mokler, A. Warczak, F. Bosch, H. Geissel, R. Moshammer, C. Scheidenberger, J. Eichler, A. Ichihara, T. Shirai, Z. Stachura, P. Rymuza: Phys. Rev. A **51,** 2098 (1995)
6.105 H. Tawara, P. Richard, K. Kawatsura: Phys. Rev. A **26,** 154 (1982)
6.106 C.R. Vane, S. Datz, P.F. Dittner, J. Giese, N.L. Jones, H.F. Krause, T.M. Roessel, P.S. Peterson: Phys. Rev. A **48,** 1847 (1994)
6.107 J.A. Tanis, M.W. Clark, K.H. Berkner, E.M. Bernstein, W.G. Graham, R.J. McDonald, R.H. McFarland, J.R. Mowat, D.W. Müller, A.S. Schlachter, J.W. Stearns, M.P. Stöckli: J. Phys. B **48,** 207 (1987)
6.108 Th. Stöhlker, C. Kozhuharov, A.E. Livingston, P.H. Mokler, Z. Stachura, A. Warczack: Z. Phys. D **23,** 121 (1992)
6.109 R. Anholt, S.A. Andriamonje, E. Morenzoni, Ch. Stoller, J.D. Molitoris, W.E. Meyerhof, H. Bowman, J.-S. Xu, Z.-Z. Xu, J.O. Rasmussen: Phys. Rev. Lett. **53,** 234 (1984)
6.110 A. Ron, I.B. Goldberg, J. Stein, S.T. Manson, R.H. Pratt, R.Y. Yin: Phys. Rev. A **50,** 1312 (1994)
6.111 L.C. Tribedi, V. Nanal, M.R. Press, M.B. Kurup, K.G. Prasad, P.N. Tandon: Phys. Rev. A **49,** 374 (1994)
6.112 B.R. Appleton, R.H. Ritchie, J.A. Biggerstaff, T.S. Noggle, S. Datz, C.D. Moak: Phys. Rev. B **19,** 4347 (1979)
6.113 S. Andriamonje, M. Chevallier, C. Cohen, J. Dural, M.J. Gaillard, R. Genre, M. Hage-Ali, R. Kirsch, A. L'Hoir, B. Mazuy, J. Mory, J. Moulin, J.C. Poizat, J. Remillieux, D. Schmaus, M. Toulemonde: Phys. Rev. Lett. **59,** 2271 (1987)
6.114 F. Sauter: Ann. Physik **9,** 217 (1931); *ibid.* **11,** 454 (1931)

6.115 E. Spindler: *Untersuchungen des strahlenden Elektroneneinfangs in Ion-Atomstößen*, Ph.D. Thesis, TU München (1979)
6.116 E. Spindler, H.-D. Betz, F. Bell: Phys. Rev. Lett. **42,** 832 (1979)
6.117 J. Eichler, A. Ichihara, T. Shirai: Phys. Rev. A **51,** 3027 (1995)
6.118 Th. Stöhlker, H. Geissel, H. Irnich, T. Kandler, C. Kozhuharov, P.H. Mokler, G. Münzenberg, F. Nickel, C. Scheidenberger, T. Suzuki, M. Kucharski, A. Warczak, P. Rymuza, Z. Statchura, A. Kriessbach, D. Dauvergne, B. Dunford, J. Eichler, A. Ichihara, T. Shirai: Phys. Rev. Lett. **73,** 3520 (1994)
6.119 J.A. Tanis, S.M. Shafroth, J.E. Willis, M. Clark, J. Swenson, E.N. Strait, J.R. Mowat: Phys. Rev. Lett. **47,** 828 (1981)
6.120 J.A. Tanis, E.M. Bernstein, W.G. Graham, M.W. Clark, S.M. Shafroth, B.M. Johnson, K.W. Jones, M. Meron: Phys. Rev. Lett. **49,** 1325 (1982)
6.121 W.G. Graham, W. Fritsch, Y. Hahn, J.A. Tanis (eds.): *Recombination of Atomic Ions* (Plenum, New York 1992)
6.122 J.A. Tanis: Nucl. Instrum. Methods. A **262,** 52 (1987)
6.123 Y. Hahn, K.L. LaGattuta: Phys. Rep. **166,** 196 (1988)
6.124 D.J. McLaughlin, R.A. Bellantone: Phys. Scr. **47,** 546 (1993)
6.125 D. Brandt: Phys. Rev. A **27,** 1314 (1983)
6.126 J.A. Tanis, E.M. Bernstein, M.W. Clark, W.G. Graham, M. Meron: Phys. Rev. A **31,** 4040 (1985)
6.127 M. Clark, D. Brandt, J.K. Swenson, S.M. Shafroth: Phys. Rev. Lett. **54,** 544 (1985)
6.128 W.G. Graham, E.M. Bernstein, M.W. Clark, J.A. Tanis, K.H. Berkner, P. Gohil, R.J. McDonald, A.S. Schlachter, J.W. Stearns, R.H. McFarland, T.J. Morgan, A. Müller: Phys. Rev. A **33,** 3591 (1986)
6.129 W.G. Graham, K.H. Berkner, E.M. Bernstein, M.W. Clark, B. Feinberg, M.A. McMahan, T.J. Morgan, W. Rathbun, A.S. Schlachter, J.A. Tanis: Phys. Rev. Lett. **65,** 2773 (1990)
6.130 J.A. Tanis, E.M. Bernstein, W.G. Graham, M.P. Stöckli, M. Clark, R.H. McFarland, T.J. Morgan, K.H. Berkner, A.S. Schlachter, J.W. Stearns: Phys. Rev. Lett. **53,** 2551 (1984)
6.131 A. Itoh, T.J.M. Zouros, D. Schneider, U. Stettner, W. Zeitz, N. Stolterfoht: J. Phys. B **18,** 4581 (1985)
6.132 M.S. Pindzola, N.R. Badnell: Phys. Rev. A **42,** 6526 (1992)
6.133 M.H. Chen: Phys. Rev. A **41,** 4102 (1990)
6.134 W.G. Graham: Nucl. Instrum. Methods. B **87,** 58 (1994)
6.135 T. Kandler, P.H. Mokler, Th. Stöhlker, H. Geisel, H. Irnich, C. Kozhuharov, A. Kriessbach, M. Kucharski, G. Münzenberg, F. Nickel, P. Rymuza, C. Scheidenberger, Z. Stachura, T. Suzuki, A. Warczak, D. Dauvergne, R.W. Dunford: Phys. Lett. A **204,** 274 (1995)
6.136 T. Kandler, P.H. Mokler, H. Geisel, H. Irnich, C. Kozhuharov, A. Kriessbach, M. Kucharski, G. Münzenberg, F. Nickel, P. Rymuza, C. Scheidenberger, Z. Stachura, Th. Stöhlker, T. Suzuki, A. Warczak, D. Dauvergne, R.W. Dunford: Nucl. Instrum. Methods. B **98,** 320 (1995)
6.137 I.I. Sobelman, L.A. Vainstein, E.A. Yukov: *Excitation of Atoms and Broadening of Spectral Lines*, 2nd edn., Springer Ser. Atoms Plasmas, Vol. 15 (Springer, Berlin , Heidelberg 1995)
6.138 W. Lotz: Z. Phys. **216,** 241 (1968)
6.139 H. van Regemorter: Astrophys. J. **136,** 906 (1962)
6.140 P. Mansbach, J. Keck: Phys. Rev. **181**, 275 (1969)
6.141 H.F. Beyer, D. Liesen, O. Guzman: Part. Accel. **24**, 163 (1989)
6.142 D.R. Bates, A.E. Kingston, R.W.P. McWhirter: Proc. Roy. Soc. A **267,** 297 (1962)

6.143 D.R. Bates, A. Dalgarno: In *Atomic and Molecular Processes*, ed. by D.R. Bates (Academic, New York 1962) p. 245
6.144 D.R. Bates: In *Case Studies in Atomic Physics IV*, ed. by E.W. McDaniel, M.R.C. McDowell (North Holland, Amsterdam 1975) p. 57
6.145 M.J. Seaton: Mon. Not. R. Astron. Soc. **119**, 81 (1959)
6.146 S. Byron, R.C. Stabler, P.J. Bortz: Phys. Rev. Lett. **8**, 376 (1962)
6.147 J. Stevefelt, J. Boulmer, J.-F. Delpech: Phys. Rev. A **12**, 1246 (1975)
6.148 H.R. Griem: *Plasma Spectroscopy* (McGraw-Hill, New York 1964)
6.149 J. Vogel, C. Toepffer: Nucl. Instrum. Methods. A **272**, 639 (1988)
6.150 M.E. Glinsky, T.M. O'Neil: Phys.Fluids B **3**, 1279 (1991)
6.151 L.I. Menshikov, P.O. Fedichev: Sov. Phys. - JETP **81**, 78 (1995)
6.152 A. Müller, A. Wolf: Hyp. Interact. to be published (1997)

Appendices

A.1 W.R. Johnson, G. Soff: At. Data Nucl. Data Tables **33**, 405 (1985)
A.2 V.A. Boiko, V.G. Pal'chikov, I.Yu. Skobelev, A.Ya. Faenov: *The Spectroscopic Constants of Atoms and Ions* (CRC, Boca Raton 1994)
V.G. Pal'chikov: 1996 unpublished
A.3 D.R. Plante, W.R. Johnson, J. Sapirstein: Phys. Rev. A **49**, 3519 (1994)
A.4 W.R. Johnson, D.R. Plante, J. Sapirstein: Adv. At. Mol. Opt. Phys. **35**, 255 (1995)
A.5 U.I. Safronova, M.S. Safronova, N.J. Snyderman, V.G. Pal'chikov: Phys. Scr. **50**, 29 (1994)
A.6 L.A. Vainshtein, U.I. Safronova: Atom. Data Nucl. Data Tables **21**, 49 (1978)
A.7 V.G. Pal'chikov, V.P. Shevelko: *Reference Data on Multicharged Ions*, Springer Ser. Atoms Plasmas, Vol. 16 (Springer, Berlin, Heidelberg 1995)
A.8 J. Nilsen, J.H. Scofield: Phys. Scr. **49**, 588 (1994)

Subject Index

Springer and the environment

At Springer we firmly believe that an international science publisher has a special obligation to the environment, and our corporate policies consistently reflect this conviction.

We also expect our business partners – paper mills, printers, packaging manufacturers, etc. – to commit themselves to using materials and production processes that do not harm the environment. The paper in this book is made from low- or no-chlorine pulp and is acid free, in conformance with international standards for paper permanency.

GPSR Compliance
The European Union's (EU) General Product Safety Regulation (GPSR) is a set of rules that requires consumer products to be safe and our obligations to ensure this.

If you have any concerns about our products, you can contact us on

ProductSafety@springernature.com

In case Publisher is established outside the EU, the EU authorized representative is:

Springer Nature Customer Service Center GmbH
Europaplatz 3
69115 Heidelberg, Germany

www.ingramcontent.com/pod-product-compliance
Ingram Content Group UK Ltd.
Pitfield, Milton Keynes, MK11 3LW, UK
UKHW021828190726
13853UKWH00003B/1256

* 9 7 8 3 6 6 2 0 3 4 9 6 5 *